Y0-CAJ-237

NCRP REPORT No. 168

Radiation Dose Management for Fluoroscopically-Guided Interventional Medical Procedures

Recommendations of the
NATIONAL COUNCIL ON RADIATION PROTECTION AND MEASUREMENTS

July 21, 2010

National Council on Radiation Protection and Measurements
7910 Woodmont Avenue, Suite 400 / Bethesda, MD 20814-3095

LEGAL NOTICE

This Report was prepared by the National Council on Radiation Protection and Measurements (NCRP). The Council strives to provide accurate, complete and useful information in its publications. However, neither NCRP, the members of NCRP, other persons contributing to or assisting in the preparation of this Report, nor any person acting on the behalf of any of these parties: (a) makes any warranty or representation, express or implied, with respect to the accuracy, completeness or usefulness of the information contained in this Report, or that the use of any information, method or process disclosed in this Report may not infringe on privately owned rights; or (b) assumes any liability with respect to the use of, or for damages resulting from the use of any information, method or process disclosed in this Report, *under the Civil Rights Act of 1964, Section 701 et seq. as amended 42 U.S.C. Section 2000e et seq. (Title VII) or any other statutory or common law theory governing liability.*

Disclaimer

Any mention of commercial products within NCRP publications is for information only; it does not imply recommendation or endorsement by NCRP.

Library of Congress Cataloging-in-Publication Data

Radiation dose management for fluoroscopically guided interventional medical procedures.
p. ; cm. -- (NCRP report ; no. 168)
Includes bibliographical references and index.
ISBN 978-0-9823843-6-7 (alk. paper)
1. Fluoroscopy--Equipment and supplies. 2. Radiation--Dosage. 3. Radiation--Safety measures. I. National Council on Radiation Protection and Measurements. II. Series: NCRP report ; no. 168.
[DNLM: 1. Fluoroscopy--instrumentation. 2. Equipment Safety. 3. Radiation Dosage. 4. Radiation Protection. 5. Radiography, Interventional--adverse effects. WN 220]
RC78.5.R33 2011
616.07'572--dc22

2011002846

[For detailed information on the availability of NCRP publications see page 314.]

Preface

This Report was developed under the auspices of Program Area Committee 2 of the National Council on Radiation Protection and Measurements, the committee that provides oversight for operational radiation safety.

Fluoroscopically-guided interventional medical procedures (*e.g.*, diagnostic angiography, angioplasty, stent placement) are now commonly performed in the United States, are often less invasive and less costly, and result in shorter hospital stays than surgical procedures. They are performed by radiologists and other medical specialists (*e.g.*, cardiologists, orthopedic surgeons) with the assistance of medical support staff.

This Report will be of benefit to these physicians and the medical support staff, particularly those who currently participate in fluoroscopically-guided interventional procedures but do not have sufficient training in the practical radiation protection aspects of the use of the equipment for the procedures (*e.g.*, knowledge of equipment operation, optimal imaging techniques, radiation dose management for patients and medical staff, benefit-risk tradeoffs, and the potential for early or late detrimental radiation effects). An extensive bibliography is provided to facilitate access to the primary literature.

This Report is also intended for policymakers who can place radiation-dose management requirements on those who conduct fluoroscopically-guided interventional procedures with regard to:

- optimizing imaging protocols;
- managing procedure time;
- utilizing available radiation protective equipment and dose-management features;
- tracking and trending radiation doses to patients and medical staff; and
- credentialing and privileging physicians to use the fluoroscopic equipment for these specialized procedures.

This Report was prepared by Scientific Committee 2-3 on Radiation Safety Issues for Image-Guided Interventional Medical Procedures. Serving on Scientific Committee 2-3 were:

Stephen Balter, *Chairman*
Columbia University Medical Center
New York, New York

Vice Chairmen

Beth A. Schueler
Mayo Clinic
Rochester, Minnesota

Donald L. Miller
Uniformed Services University
of the Health Sciences
Bethesda, Maryland

Members

Jeffrey A. Brinker
Johns Hopkins Hospital
Baltimore, Maryland

Charles E. Chambers
Pennsylvania State College
of Medicine
Hershey, Pennsylvania

Kenneth F. Layton
Baylor University Medical
Center
Dallas, Texas

M. Victoria Marx
University of Southern
California Medical Center
Los Angeles, California

Cynthia H. McCollough
Mayo Clinic
Rochester, Minnesota

Keith J. Strauss
Harvard Medical School
Boston, Massachusetts

Louis K. Wagner
University of Texas Medical
School
Houston, Texas

Consultants

John F. Angle
University of Virginia
Charlottesville, Virginia

Lionel Desponds
GE Healthcare
Buc, France

Andrew Einstein
Columbia University Medical
Center
New York, New York

Norman J. Kleiman
Columbia University
New York, New York

John W. Hopewell
Green Templeton College
University of Oxford
Oxford, United Kingdom

Matthew Williams
Columbia University
New York, New York

NCRP Secretariat
Marvin Rosenstein, *Staff Consultant*
Cindy L. O'Brien, *Managing Editor*
David A. Schauer, *Executive Director*

The Council expresses its appreciation to the Committee members for the time and effort devoted to the preparation of this Report. NCRP gratefully acknowledges the financial support provided by the National Cancer Institute (NCI) under Grant Number R24 CA074296-10. The contents of this Report are the sole responsibility of NCRP, and do not necessarily represent the official views of NCI, National Institutes of Health.

Thomas S. Tenforde
President

Contents

1. Executive Summary

1.1 General

The "fluoroscope" is defined as an instrument used chiefly in industry and in the practice of medicine for observing the internal structure of objects (such as the living body) by means of the shadow cast by the object examined upon a fluorescent screen when placed between the screen and a source of x rays.

The simple "fluoroscopic screen" referred to in this definition has been replaced in current medical practice by vacuum-tube image intensifiers and solid-state detectors. These devices are collectively called "image receptors."

Conventional radiography in medicine is preferred for the acquisition of static images (*e.g.*, chest radiography). Medical fluoroscopy is normally reserved for the observation of moving objects. Fluoroscopes have two main modes of operation:

- *fluoroscopy*: intended to observe moving objects for relatively long periods of time (seconds to minutes) without the intent of preserving the images; and
- *fluorography*: intended to record images of moving objects for a few seconds at a time (*e.g.*, cinefluorography of the heart).

The production of fluorographic images requires air-kerma rates (incident on the patient's skin) that are usually 10 to 100 times higher than for fluoroscopic images of the same patient and anatomical view. Traditionally, fluoroscopic, fluorographic, and radiographic images were acquired and managed using somewhat different technologies. At present, a state-of-the-art fluoroscopic system is able to produce all three classes of images using a single image-acquisition and image-processing chain. The production of the different classes of images is determined by the selection of technical factors in the imaging system. The boundary between fluoroscopy and fluorography has been further blurred in those systems that offer the possibility of retrospectively storing the last several seconds of fluoroscopy. This can reduce total patient dose by eliminating the need for fluorographic documentation in those situations where the previous fluoroscopic sequence meets clinical requirements.

These state-of-the-art fluoroscopic systems are now extensively used to conduct diagnostic or therapeutic interventional medical procedures performed *via* percutaneous or other access routes in order to:

- localize or characterize a lesion, diagnostic site, or treatment site;
- to monitor the procedure; or
- to control and document therapy.

The term *fluoroscopically-guided interventional* (FGI) procedure is used to describe this practice of medicine.

This Report is focused on the use of fluoroscopic systems as a tool for guiding diagnostic and therapeutic procedures because higher radiation doses (compared to conventional radiography and fluoroscopy) are received regularly from some types of FGI procedures and occasionally from many other types of FGI procedures. Other medical applications of fluoroscopy (*e.g.*, examination of the gastrointestinal system, guiding open surgical procedures) are outside the scope of this Report. Computed-tomography-guided interventional (CTGI) procedures are not discussed in detail due to continuing changes in the technology driven by the evolution of multi-slice computed tomography (CT) detectors. However, the principles presented in this Report are generally applicable to these domains. Most of the recommendations contained in this Report should be applied in all settings where fluoroscopic guidance is used.

Within the context of radiation dose management, the goal of this Report is to supply information that helps optimize patient outcomes without compromising worker safety. However, radiation is not the only risk to which patients and workers are exposed. In many cases, radiation is a minor component of overall risk. In these situations, too great a focus on radiation safety (*e.g.*, the use of unnecessarily thick lead aprons) may reduce the overall safety of patients or workers.

Some beneficial, clinically-justified FGI procedures, even when optimized for radiation protection, deliver substantial doses of radiation to patients. This puts the patient at risk for radiogenic stochastic effects and occasionally induces radiogenic deterministic effects. However, a complete risk analysis usually identifies many other procedural hazards and will often conclude that radiation is one of the lesser hazards from FGI procedures. While the decision to conduct an FGI procedure assumes that the use of ionizing radiation is warranted by the disease state for which the patient undergoes treatment, the benefits, risks, and alternative procedures that do not require the use of ionizing radiation should be considered.

FGI procedures should only be performed when there is the expectation of a benefit to the patient. At present, many medically necessary x-ray image-guided procedures (including FGI procedures using fixed or mobile equipment and CTGI procedures) can only be performed by physicians and support staff positioned adjacent to the patient. These individuals are unavoidably irradiated while performing their duties. Provided that appropriate radiation protection methods are applied, staff occupational exposure is justified if the procedure is itself justified, and the occupational exposure adheres to the as low as reasonably achievable (ALARA) principle. The ALARA principle should be applied without compromising either patient safety or the clinical procedure and without unacceptably increasing workers' nonradiation risks (*e.g.*, spinal injuries related to wearing inappropriately heavy lead aprons). In this context, worker irradiation is considered to be an unavoidable part of the social cost of providing FGI procedures.

Sections 2 through 6 of this Report are organized by the following topics:

- general information (Section 2) (includes scope, FGI procedures, efficacy and benefit-cost, justification and benefit-risk, physics and dosimetry, radiation biology);
- fluoroscopy equipment and facilities (Section 3);
- protection of the patient (Section 4);
- protection of staff (Section 5); and
- administrative and regulatory considerations (Section 6).

This material is intended to provide background information for readers, including facility administrative and management executives, professionals participating in or supporting FGI procedures, and regulatory officials. The contents of many of the sections are supplemented by materials in the appendices. Because of the diversity of FGI procedures it is impracticable to supply a complete handbook on this topic.

1.2 Recommendations

The National Council on Radiation Protection and Measurements (NCRP) recommendations in this Report are listed in Table 1.1 for ready reference. The recommendations are consecutively numbered in the order they appear in the text of this Report, and the subsection in which each statement appears and is discussed is noted in the right-hand column. The recommendations should not be read in isolation. *The reader should consult the indicated subsections for more complete explanations and further information.*

TABLE 1.1—*NCRP Report No. 168 recommendations.*

Number	Recommendation	Section
1	Radiation risk *should* be one of the many risks included in the benefit-risk analysis of FGI procedures.	2.4
2	The measured dose quantities air kerma-area product (P_{KA}) and air kerma at the reference point ($K_{a,r}$) *should* be used to compare similar FGI procedures. In this Report, P_{KA} and $K_{a,r}$ refer to the cumulative value for the FGI procedure.	2.5
3	Effective dose (E) *shall not* be used for quantitative estimates of stochastic radiation risk for individual patients or patient groups (the appropriate approach to obtain quantitative estimates is discussed in Section 2.6.3.3). Effective dose (E) *may* be used as a qualitative indicator of stochastic radiation risk for classifying different types of procedures into broad risk categories (as suggested in Table 2.4).	2.6.3.3
4	Peak tissue dose *shall* be used to evaluate the potential for deterministic effects in specific tissues. Examples include peak dose to the skin or to the lens of the eye.	2.6.4.2
5	An FGI procedure *should* be classified as a potentially-high radiation dose procedure if more than 5 % of cases of that procedure result in $K_{a,r}$ exceeding 3 Gy or P_{KA} exceeding 300 Gy cm^2.	3
6	Potentially-high radiation dose procedures *should* be performed using equipment designed for this intended use.	3.1.2
7	If fluoroscopes are intended to be routinely used for procedures that have the potential for high patient doses (*i.e.*, $K_{a,r} > 3$ Gy), the units either *should* be equipped or upgraded with add-on dose-monitoring equipment that monitors $K_{a,r}$ or the units *should* be replaced with a modern machine.	3.1.2

8	Equipment that is routinely used for pediatric procedures *should* be appropriately designed, equipped and configured for this purpose.	3.1.3
9	For newly designed and remodeled existing facilities, all spaces outside the procedure room (including control rooms) *should* be designed to limit E to not more than 1 mSv y^{-1}.	3.2
10	For newly designed and remodeled existing facilities, spaces within the FGI-procedure room intended exclusively for routine clinical monitoring of patients (or similar activities) *should* be shielded to limit E to not more than 1 mSv y^{-1}. Individuals working behind such barriers *shall* be monitored for radiation exposure.	3.2
11	Door interlocks that interrupt x-ray production *shall not* be permitted at any entrances to FGI-procedure rooms.	3.2
12	Procedure planning for FGI procedures on pregnant patients *shall* include feasible modifications to minimize dose to the embryo and fetus.	4.2.2
13	Fluoroscopy time *should not* be used as the only dose indicator during potentially-high radiation dose FGI procedures. All available dose indicators *shall* be used in such procedures.	4.2.4.1
14	Interventionalists *shall* be responsible for patient radiation levels during FGI procedures and *shall* ensure that radiation dose accumulation is continuously monitored during the procedure.	4.3.3.2
15	Patient dose data *shall* be recorded in the patient's medical record at the conclusion of each procedure. This *shall* include all of the following that are available from the system: peak skin dose ($D_{skin,max}$), $K_{a,r}$, P_{KA}, fluoroscopy time, and number of fluorographic images.	4.3.4.1

TABLE 1.1—*(continued)*

Number	Recommendation	Section
16	If a substantial radiation dose level (SRDL) (Table 4.7 and Section 4.3.4.2) is exceeded while performing an FGI procedure, the interventionalist *shall* place a note in the medical record, immediately after completing the procedure, that justifies the radiation dose level used.	4.3.4.3
17	If an SRDL is exceeded for an FGI procedure, the patient and any caregivers *should* be informed, prior to discharge, about possible deterministic effects and recommended follow-up. If fluoroscopy time exceeds the SRDL, but other measured dose metrics do not exceed the SRDL, patient information and follow-up *may not* be necessary.	4.3.4.4
18	Follow-up for possible deterministic effects *shall* remain the responsibility of the interventionalist for at least 1 y after an FGI procedure. Follow-up *may* be performed by another healthcare provider who remains in contact with the interventionalist. All relevant signs and symptoms (Table 2.5) *should* be regarded as radiogenic unless an alternative diagnosis is established.	4.3.4.4
19	Facilities *shall* have a process to review radiation doses for patients undergoing FGI procedures. Advisory data based on measured dosimetric quantities (in particular P_{KA} or $K_{a,r}$ to manage overall performance, and $K_{a,r}$ to manage deterministic effects) *should* be used for quality assurance purposes.	4.3.5.3
20	Each individual present in an FGI-procedure room while a procedure is in progress *shall* have appropriate radiation protection training.	5.2
21	Each individual present in an FGI-procedure room while a procedure is in progress *shall* be provided with and *shall* use appropriate radiation protective equipment.	5.5.5

22	Policies and procedures *should* be in place so that in the event of a time-critical urgent or emergent situation, as defined in this Report (Table 5.3), advanced provision exists for exceeding an annual occupational dose limit.	5.7.1
23	Determinations of occupational doses *shall* take into account the personal protective equipment used by each individual in the FGI-procedure environment in order to properly assess compliance with occupational dose limits.	5.7.3
24	Two personal dosimeters, one worn under the protective apron and a second worn at neck level above protective garments, are preferred and *should* be used in the FGI-procedure environment. A single personal dosimeter worn at neck level above protective garments *may* be used in the FGI-procedure environment. A single personal dosimeter worn under the protective apron *shall not* be used in the FGI-procedure environment.	5.7.3
25	Monitoring of equivalent dose to the lens of the eye *should* be performed with a personal dosimeter placed either at the collar level outside any radiation protective garment or near the eyes.	5.7.3
26	Investigations *should* occur if personal-dosimeter readings for an individual are substantially *above* or *below* the expected range for that individual's duties.	5.7.4
27	Every person who operates FGI equipment or supervises the use of FGI equipment *shall* have current training in the safe use of that specific equipment.	6
28	An FGI procedure *shall* be performed or supervised only by a physician or other medical professional with fluoroscopic and clinical privileges appropriate to the specific procedure.	6.3.1

TABLE 1.1—(*continued*)

Number	Recommendation	Section
29	Standards and guidelines provided by professional societies *shall* be considered when establishing radiation-related resources, and quality and performance requirements.	6.4.2
30	Interventionalists and qualified physicists *should* participate in the process for purchase and configuration of new fluoroscopes and fluoroscopy facilities.	6.4.2
31	A qualified physicist *shall* perform acceptance and commissioning tests before first clinical use of new, newly-installed, or newly-repaired fluoroscopy equipment, and *shall* perform subsequent periodic tests as part of a technical quality-control program.	6.4.2

These NCRP recommendations are expressed in terms of *shall* (or *shall not*), *should* (or *should not*), and *may,* each term italicized, where:

- *shall* (or *shall not*) indicates a recommendation from NCRP that is necessary to meet the currently accepted standards of radiation protection;
- *should* (or *should not*) indicates an advisory recommendation from NCRP that is to be applied when practicable or practical (*e.g.*, cost-effective); and
- *may* (or *may not*) indicates a reasonable practice that is permissible.

When the terms should and may appear in this Report in the context of their general usage, they are not italicized.

In this Report, the terms *interventionalist* and *qualified physicist* are defined as follows:

- *interventionalist*: an individual who has been granted clinical privileges to perform or supervise FGI procedures in a facility, and who is personally responsible for the use of radiation during a specific FGI procedure in that facility.
- *qualified physicist*: a medical physicist or medical health physicist who is competent to conduct the radiation protection functions for FGI-procedure equipment and facilities described in this Report. The qualified physicist is a person who is certified by the American Board of Radiology, American Board of Medical Physics, American Board of Health Physics, or Canadian College of Physicists in Medicine.

2. General Information

2.1 Scope and Applicability of this Report

This Report deals with interventional diagnostic and therapeutic procedures performed *via* percutaneous or other access routes, usually with local anesthesia and/or intravenous sedation, which use external ionizing radiation in the form of fluoroscopy to localize or characterize a lesion, diagnostic site, or treatment site; monitor the procedure; and/or control and document therapy (ICRP, 2000a). This Report refers to this practice of medicine as fluoroscopically-guided interventional (FGI) procedures.

The boundary between computed tomography (CT) and fluoroscopy has become blurred. CT performed with a continuous beam can be used for real-time guidance to localize a target or guide a needle. The C-arm gantry of a fluoroscopy suite can be continuously rotated during image acquisition, and the data thus obtained used to reconstruct CT images of the body, a process sometimes referred to as cone-beam CT, C-arm CT, or angiographic CT (Orth *et al.*, 2008; Wallace *et al.*, 2008). How and when each of these modalities is used are a function of equipment availability, operator training and preference, target location, the planned route to the target, and the purpose of the procedure.

Detailed discussions of procedures using CT have been excluded from this Report because this technology is evolving too rapidly for specific recommendations. However, most of the recommendations in this Report are also generally applicable to computed-tomography-guided interventional (CTGI) procedures.

Fluoroscopy is widely used in surgical settings in ways that can be comparable to its use in other FGI procedures. Patient and worker radiation safety concerns along with training, credentialing, and privileging of surgical staff should be similar to those for all other FGI procedures. The imaging systems typically used in surgical settings are mobile fluoroscopy C-arms. These range from simple analog devices to digital systems having many of the x-ray production and image-processing characteristics of a fixed fluoroscopy system. Hybrid rooms are of increasing importance. These facilities consist of a fixed fluoroscopy imaging system installed in a formal operating room. Conversely, nonsurgical FGI procedure rooms are often upgraded to operating room standards. Although not discussed in detail, most of the recommendations contained in this Report should be applied in all settings where fluoroscopic guidance is used.

Procedures performed with methods of imaging guidance that do not use ionizing radiation (*i.e.*, ultrasound, magnetic resonance imaging) are not considered in this Report.

2.2 Procedures

This section provides examples of the medical procedures that are the subject of this Report and describes the operators who perform them.

2.2.1 *Definition of a Fluoroscopically-Guided Interventional Procedure*

The International Commission on Radiological Protection (ICRP), in Publication 85, defines fluoroscopically-guided interventional procedures as "procedures comprising guided therapeutic and diagnostic interventions, by percutaneous or other access, usually performed under local anesthesia and/or sedation, with fluoroscopic imaging used to localize the lesion/treatment site, monitor the procedure, and control and document the therapy" (ICRP, 2000a). The World Health Organization (WHO, 2000) defines percutaneous procedures with a predominantly diagnostic objective as "invasive" procedures, and reserves the term "interventional" for procedures with a therapeutic intent. WHO (2000) also uses the term "interventional radiology" to describe all image-guided therapeutic interventions.

This Report uses the ICRP (2000a) definition, with the following comments on the WHO (2000) approach. The distinction between diagnostic and therapeutic procedures suggests that there is a difference between the two in terms of the magnitude of patient dose, and this is not always the case (Ruiz Cruces *et al.*, 1998; Tsalafoutas *et al.*, 2006). In addition, these procedures are performed by physicians in many medical specialties (Table 2.1). The term "interventional radiology" suggests that these procedures are performed only by radiologists. This may be misleading if it causes the reader to believe that all operators have equivalent training in radiation safety and radiation protection.

2.2.2 *Types of Procedures*

There are a large number of distinct FGI procedures. Annex A of ICRP (2000a) lists 36 different FGI procedures as examples.

TABLE 2.1—*Examples of FGI procedures.*

Organ System or Region (physician specialities)	Procedure
Central nervous system (radiology, neurosurgery, neurology)	Diagnostic angiography
	Embolization
	Thrombolysis
Chest (radiology, vascular surgery, internal medicine)	Biopsy
	Thoracentesis
	Chest-tube placement
	Pulmonary angiography
	Pulmonary embolization
	Thrombolysis
	Tumor ablation
Heart (cardiology, cardiac surgery)	Diagnostic angiography
	Angioplasty
	Stent placement
	Radiofrequency ablation
	Pacemaker placement
Gastrointestinal tract (radiology, gastroenterology)	Percutaneous gastrostomy
	Percutaneous jejunostomy
	Biopsy
	Stent placement
	Diagnostic angiography
	Embolization
Liver and biliary system (radiology, gastroenterology)	Biopsy
	Percutaneous biliary drainage
	Endoscopic retrograde cholangiopancreatography
	Percutaneous cholecystostomy
	Stone extraction
	Stent placement
	Transjugular intrahepetic portosytemic shunt
	Chemoembolization
	Tumor ablation

TABLE 2.1—*(continued)*

Organ System or Region (physician specialities)	Procedure
Kidney and urinary tract (radiology, urology)	Biopsy
	Nephrostomy
	Stent placement
	Stone extraction
	Tumor ablation
Reproductive tract (radiology, obstetrics/gynecology)	Diagnostic angiography
	Embolization
	Hysterosalpingography
Musculoskeletal system (radiology, orthopedics, neurosurgery, anesthesiology, neurology)	Biopsy
	Vertebroplasty
	Kyphoplasty
	Embolization
	Tumor ablation
	Nerve blocks
Vascular system (radiology, cardiology, vascular surgery, nephrology, neurosurgery)	Diagnostic angiography
	Diagnostic venography
	Angioplasty
	Stent placement
	Embolization
	Stent-graft placement
	Venous access
	Inferior vena cava filter placement
Other (radiology)	Fluid collection aspiration
	Abscess drainage

WHO (2000) describes more than 30 such procedures. Radiologists alone typically perform more than 40 different types of FGI procedures (Miller *et al.*, 2003a; Storm *et al.*, 2006). Radiologists also perform CTGI procedures. Other medical specialists perform numerous additional kinds of FGI procedures, but they typically do not perform CTGI procedures. Table 2.1 lists examples of some of these procedures, but the list is not intended to be exhaustive. Additional examples are provided in NCRP Report No. 133 (NCRP, 2000a).

2.2.3 *Temporal Trends*

The number of FGI procedures performed annually throughout the world has increased over the past 20 y. Percutaneous coronary intervention (PCI) (coronary angioplasty, with or without stent placement) is one example. In Germany, Japan and Spain the annual rate of increase in this procedure from 1994 to 1998 was between 10 and 20 % (ICRP, 2000a). In 1998, the number of these procedures performed was between 1.5 and 2 per 1,000 population in Germany, between 0.5 and 1 per 1,000 population in Japan, and ~0.5 per 1,000 population in Spain (ICRP, 2000a). In the United States, the rate of PCIs more than doubled from 1996 to 2000, from 0.66 to 1.63 per 1,000 population (CDC, 2004). In 2002, ~450,000 hospital stays in the United States included a PCI (CDC, 2004). Approximately 1,265,000 PCIs were performed on inpatients in the United States in 2005 (AHA, 2008).

Table 2.2 presents estimates of the annual numbers of selected procedures performed in the United States in 2008. Data reported as "cardiac cases" are from the IMV Medical Information Division (IMV, 2009a) with survey dates from February 2008 through March 2009. The survey included 373 hospital and nonhospital sites with cardiac catheterization laboratories, with results projected to the 2,020 hospital and nonhospital sites in the United States that provide these services. A "cardiac catheterization laboratory" is defined as a dedicated cardiac catheterization laboratory or an angiography suite where cardiac catheterization volume is at least 50 % of the total volume. A "case" is defined as a patient visit on any given day, regardless of the number of procedures performed during the visit.

Data reported as "interventional angiography procedures" are from IMV (2009b) with survey dates from October 2008 through May 2009. The survey included 628 hospital sites with 150 or more beds that provide angiography services including noncoronary interventional angiography services. The survey did not include smaller hospitals or nonhospital sites. The results were projected to the 1,720 hospital sites in the United States with 150 or more beds that provide these services. To qualify as an interventional angiography laboratory, the volume of cardiac catheterization or coronary angiography procedures had to be <50 % of the total volume of the angiography department.

In 2008, an estimated 3,750,000 cases were performed at the 2,020 cardiac catheterization sites, which included 4,225 fixed cardiac catheterization laboratories. Between 2002 and 2006, the annual growth rate in case volume averaged ~2 %. Between 2006 and 2008, however, case volume declined. There was a decrease of

TABLE 2.2—*FGI cases and procedures performed per year in the United States, 2008.*

Procedure	Estimated Number per Year (thousands)
Cardiac cases[a]	
Diagnostic only	1,441
Therapeutic only	373
Diagnostic and therapeutic combined	662
Electrophysiology	189
Cardiac device placement	361
Cardiac noncoronary	142
Pediatric	47
Other	45
Noncoronary vascular angiography procedures	
Noncoronary diagnostic procedures[b]	
Neurologic (excluding carotid)	70
Carotid and aortic arch	85
Neurologic (including carotid)	90
Pulmonary	35
Renal	135
Runoffs[c]	355
Other peripheral	195
Other	60
Noncoronary therapeutic procedures[b]	
Angioplasties	320
Stents	255
Inferior vena cava filters	185
Embolization	180
Thrombolytic therapy	125
Other	10

[a]Adapted from IMV (2009a).
[b]Adapted from IMV (2009b).
[c]Runoff = arteriogram of the abdominal aorta and arteries of the pelvis and legs.

~10 % from 2006 to 2007, and a further decrease of ~1 % from 2007 to 2008.

In 2008, an estimated 4,800,000 cases were performed at the 1,720 angiography sites in hospitals with 150 or more beds, which included 3,180 angiography systems. Between 2004 and 2008, the number of total procedures performed annually increased by ~20 %, with an average annual growth rate of 4.6 %. This is essentially the same as the estimated average annual growth rate of 4.3 % observed from 2000 to 2004 (IMV, 2009b). The interventional angiography data do not include procedures performed at smaller hospitals or free-standing sites outside the hospital, and also do not include nonvascular procedures (*e.g.*, nephrostomy, vertebroplasty, endoscopic retrograde cholangiopancreatography).

2.2.4 *Individuals Performing Interventional Procedures (interventionalists)*

The procedures listed in Table 2.1 are performed by physicians trained in a variety of medical specialties. As indicated in Table 2.1, some procedures involve multiple organ systems (*e.g.*, procedures in radiology), and some involve only one or two organ systems (*e.g.*, procedures in gastroenterology or urology). In some circumstances, some of these procedures may be performed by legally-authorized individuals who are not physicians. Each state independently defines those nonphysicians who are permitted to use fluoroscopy equipment. Individual facilities may elect to further limit fluoroscopic privileges of both physicians and nonphysicians to individuals with specific credentials. Clinical fluoroscopic privileges are discussed in Section 6.3.

Interventionalists vary in their level of training in radiation safety. In the United States, all radiologists receive training in radiation physics, radiation biology, and radiation safety as part of their radiology residency, and are required to pass an examination on these topics to receive certification by the American Board of Radiology. Interventional cardiology fellows receive training in radiation physics and safety, and the board examination in interventional cardiology includes questions in these subject areas. Physicians seeking board certification in pain management procedures are required to study radiation physics, radiation bioeffects, and safety, and are examined in these areas as part of the certification process. Physicians in other medical disciplines and at other levels of training receive variable amounts of education in radiation-related topics, and may or may not be examined in these areas as part of the board certification process. Section 6.2 discusses training in more detail.

2.3 Efficacy, Effectiveness, Benefit-Cost Analyses

Efficacy is defined as "the probability of benefit to individuals in a defined population from a medical technology applied for a given medical problem under ideal conditions of use," while effectiveness "reflects performance of a medical technology under ordinary, rather than ideal, conditions" (NCRP, 1995a). For FGI procedures, evaluations of benefit are derived from case series, retrospective studies, and some randomized controlled trials reported in the literature.

2.3.1 *Diagnostic Procedures*

The six-tiered model of efficacy (Fryback and Thornbury, 1991) presented in NCRP (1995a) is designed to evaluate diagnostic studies. The six levels are:

1. technical efficacy;
2. diagnostic accuracy efficacy;
3. diagnostic thinking efficacy;
4. therapeutic efficacy;
5. patient outcome efficacy; and
6. societal efficacy.

They can be applied to any diagnostic technology. Demonstration of efficacy at a higher level implies efficacy at a lower level, although the reverse is not true. The six-tiered model is generally applicable to diagnostic studies, such as percutaneous biopsies and angiograms, but not to therapeutic procedures.

Because of their invasiveness and risk, FGI diagnostic procedures are generally performed only when noninvasive procedures are inadequate to answer the clinical question. The clinician typically needs specific information for treatment planning. Examples of this information are the patient's specific cancer type, the organism responsible for the patient's infection, the severity of the disease process affecting the patient, and the specific location, length and severity of the patient's vascular disease. Any test that can provide answers to questions of this type is effective in that it significantly raises or lowers pretest diagnostic probabilities, changes the differential diagnosis, or suggests a new diagnosis. Such a test satisfies Level 3 of the six-tiered model (*i.e.*, diagnostic thinking efficacy). There is little question that knowledge of the patient's tumor type or infectious organism is essential to select the optimal surgical procedure, chemotherapeutic regimen or antibiotic drug, and that optimal therapy is necessary for optimal results. These tests therefore also satisfy Level 5 of the six-tiered model (*i.e.*, patient outcome efficacy).

Evaluations of Level 6 (*i.e.*, societal efficacy) require benefit-cost analyses and cost-effectiveness analyses. The only practical alternative to percutaneous biopsy or aspiration is a surgical procedure. Surgery is typically more expensive and higher risk than a percutaneous diagnostic procedure. Because of the wide variety of organs and structures that can be biopsied or aspirated, and the wide variety of indications for these procedures, a comprehensive review is impractical. Example of benefit-cost analyses for diagnostic procedures are given in Appendix A.1.

2.3.2 *Therapeutic Procedures*

Benefit-cost analyses of FGI therapeutic procedures differ from benefit-cost analyses of FGI diagnostic procedures. The six-tiered model of efficacy does not apply to FGI therapeutic procedures. Instead, benefit is measured in terms of the success rate for the intended clinical outcome of the procedure (*i.e.*, cure, palliation, relief of symptoms, or other intended benefit). Examples of this kind of benefit-cost analysis are given in Appendix A.2.

2.4 Justification, Appropriateness, and Benefit-Risk Analyses

ICRP (2007a) states: "The Commission recommends that, when activities involving an increased or decreased level of radiation exposure, or a risk of potential exposure, are being considered, the expected change in radiation detriment should be explicitly included in the decision-making process. The consequences to be considered are not confined to those associated with the radiation – they include other risks and the costs and benefits of the activity. Sometimes, the radiation detriment will be a small part of the total. Justification thus goes far beyond the scope of radiological protection. It is for these reasons that the Commission only recommends that justification require that the net benefit be positive."

A medical procedure should only be performed when it is appropriate for a particular patient. In this context, appropriateness (appropriate care) means that the expected health benefit (*e.g.*, increased life expectancy, relief of pain, reduction in anxiety, improved functional capacity) exceeds the expected negative consequences (*e.g.*, mortality, morbidity, anxiety of anticipating the procedure, pain produced by the procedure, misleading or false diagnoses, time lost from work) by a sufficiently wide margin that the procedure is worth doing (NHS, 1993; Sistrom, 2008). Professional societies such as the American College of Radiology (ACR) and the American College of Cardiology have used the concept of appropriateness to rank the suitability of FGI and other imaging procedures for the diagnosis and treatment of a variety of disease states (ACR, 2009; Hendel *et al.*, 2006).

This consideration of benefits and risks is called a benefit-risk analysis. Even if the benefit-cost analysis is favorable, FGI procedures should not be substituted for the alternative surgical procedures unless the benefit-risk ratio is equivalent to or better for the FGI procedure. The preprocedure analysis should include all of the benefits and risks evaluated for procedures that do not require ionizing radiation for imaging guidance. While radiation risk is one of the most difficult procedural risks to quantitate, it should also be included in the analysis (ICRP, 2000a).

Studies of radiation doses from FGI procedures typically focus on dose data (Miller *et al.*, 2003a; Tsalafoutas *et al.*, 2006). Most clinical studies of FGI procedures that evaluate benefit and risk do not consider radiation risk at all (Boden *et al.*, 2007; Goodney *et al.*, 2006; Pron *et al.*, 2003; Spies *et al.*, 2004). Radiation risk should be incorporated into the benefit-risk analysis when applicable.

Recommendation 1

Radiation risk *should* be one of the many risks included in the benefit-risk analysis of FGI procedures.

2.4.1 *Patient Benefit-Risk Analysis Models*

Different benefit-risk models are applicable to the three exposed groups of individuals considered in this Report. These groups are:

- interventionalists and staff performing FGI procedures;
- patients undergoing FGI diagnostic procedures; and
- patients undergoing FGI therapeutic procedures.

Benefit-risk analyses for medical procedures are traditionally viewed from the perspective of the patient, who has both the most to gain and the most to lose. In the context of this Report, two types of procedures can be modeled: FGI diagnostic procedures and FGI therapeutic procedures.

2.4.1.1 *Diagnostic Studies*

2.4.1.1.1 *Benefit.* The medical benefit to the patient from a diagnostic procedure can be very difficult to quantify (*e.g.*, the difficulty of placing a quantitative value on a diagnostic procedure that excludes significant disease or on a diagnostic procedure that establishes a diagnosis). Wall *et al.* (2006) observed that "it is very difficult to quantify the benefits of diagnostic x-ray examinations in any way that is comparable with the radiation risks, so an accurate quantitative weighing of benefits against risks is usually impossible." Perhaps in part because of this difficulty and in part because the radiation risks are not generally considered for an individual patient (with the exception of pediatric patients and pregnant women), benefit-risk calculations for FGI diagnostic procedures focus on success rates and procedure risk.

2.4.1.1.2 *Procedure risk.* For the patient, this is typically the principal concern before an interventional diagnostic study. Procedure-related complications may be a direct mechanical result of the invasiveness of the procedure (*e.g.*, an arterial injury in a patient undergoing an arteriogram) or may be biochemical (*e.g.*, renal failure from contrast material or a vasovagal reaction during a biopsy). In general, these risks have been well known for many years and have been quantified (Rose, 1983). They are reviewed with each patient as part of the informed consent process.

2.4.1.1.3 *Radiation risk.* Radiation risk may be divided into stochastic risk to the individual, stochastic risk to society, deterministic risk to the individual, and pregnancy risks (Section 2.6). Deterministic effects as a result of a diagnostic study (invasive or noninvasive) are exceedingly rare, but may occur when a number of these studies are performed on the same body part in close temporal succession (Imanishi *et al.*, 2005).

2.4.1.2 *Therapeutic Procedures*

2.4.1.2.1 *Benefit.* Unlike diagnostic procedures, all successful interventional therapeutic procedures provide clear, obvious and easily identifiable benefits to the patient. These benefits are readily quantified (Hehenkamp *et al.*, 2008; Keeling *et al.*, 2008). They include relief of symptoms, improvement in the quality of life, increased lifespan, and decreased morbidity and shorter recovery time as compared to other possible treatments.

When FGI procedures are compared to open surgical procedures, the benefits of the FGI procedure generally include a smaller incision, with less post-procedural pain, a shorter recovery time, a shorter hospitalization, and a more rapid return to work. As a result, there is a benefit in terms of quality of life. These benefits are present even without consideration of the relative success rates of FGI and open surgical procedures, which are often similar. In addition, because FGI procedures are less invasive than their open surgical counterpart, the complication rates for interventional procedures tend to be lower. Examples of benefit and risk for specific procedures are given in Appendix B.

2.4.1.2.2 *Procedure risk.* There may be significant procedure-related risk for patients undergoing interventional therapeutic procedures. The nature of the risk varies depending on the procedure and the patient. A patient undergoing carotid artery stent placement is at risk for stroke and death (Gurm *et al.*, 2008). A patient undergoing PCI is at risk for myocardial infarction and death (King *et al.*, 2008). A patient undergoing transjugular intrahepatic portosystemic shunt (TIPS) creation is at risk for major hemorrhage, encephalopathy, liver injury, and death (Harrod-Kim and Waldman, 2005). These procedure risks tend to be substantially greater than the radiation risk (Miller, 2008).

The degree of risk depends on the patient. The likelihood of bleeding is related to the patient's clotting function. Stroke risk and myocardial infarction risk are related to the severity of the patient's carotid artery and coronary artery disease. The likelihood of renal failure is related to the patient's underlying renal function and the presence or absence of diabetes mellitus.

Procedure risk is also related to the conduct of the procedure itself. Less experienced interventionalists tend to use more time and radiation to perform procedures than do more experienced interventionalists (Bor *et al.*, 2008; Verdun *et al.*, 2005). The risk of renal failure appears to increase with high doses of iodinated contrast material. More complex procedures require more contrast material and more catheter manipulation, and have a higher complication rate. Procedure risk can be estimated from studies on populations of patients, but should be individualized for each patient.

2.4.1.2.3 *Radiation risk.* As with interventional diagnostic studies, radiation risk may be divided into stochastic risk to the individual, stochastic risk to society, deterministic risk to the individual, and pregnancy risks. These risks are discussed in detail in Section 2.6 and are mentioned briefly here. Because patient radiation doses tend to be higher for some therapeutic procedures than for diagnostic procedures (Sections 4.1.1 and 4.1.2 present detailed dose data), the radiation risk is greater. Deterministic effects as a result of an interventional therapeutic procedure are uncommon, but do occur (Frazier *et al.*, 2007; Koenig *et al.*, 2001a; Monaco *et al.*, 2003; O'Dea *et al.*, 1999; Shope, 1996; Vlietstra *et al.*, 2004; Wong and Rehm, 2004).

Until relatively recently, radiation protection with regard to interventional therapeutic procedures had been primarily concerned with stochastic risks to the individual patient and to society. However, the potential for deterministic risk due to these procedures has been stressed since the early to mid-1990s (Rosenthal *et al.*, 1998; Shope, 1996; Trianni *et al.*, 2005). For interventional therapeutic procedures, stochastic risk is generally of less concern to the patient than deterministic risk. Stochastic effects generally do not manifest until years to decades after exposure. Adult patients who require interventional therapeutic procedures tend to be both older and more ill than the general population, and are therefore less likely to survive long enough for a stochastic effect to become clinically evident. Stochastic risk is of greater concern in the pediatric population, both because of the longer expected survival of children, and because of their greater sensitivity to the stochastic effects of radiation.

In contrast, deterministic effects will manifest, at least in their earliest stages, within days to weeks after the interventional therapeutic procedure. Moreover, if skin dose can be estimated within reasonable bounds, the patient can be advised about the potential for a deterministic effect and its likely severity.

For some patients and procedures, both stochastic and deterministic risks should be considered. This is especially true for larger adolescents who have the physical stature of an adult, but a higher risk for stochastic effects than an adult. Figure 2.1 illustrates the case of a 17 y old female (Vano *et al.*, 1998a). The patient underwent a cardiac ablation which lasted 5 h. She underwent a second 5 h ablation 13 months later. Fluoroscopy time from the second session was estimated to be around 100 min. An erythmatous plaque appeared in the right axilla. Red macula and blister lesions were observed one month later. The muscles in her right arm have been partially affected resulting in limited movement. In addition, substantial amounts of breast tissue were irradiated, placing the patient at increased risk for breast cancer.

FGI-equipment operators have been injured by their occupational exposure to radiation. Vano *et al.* (1998b) published the case of radiogenic eye changes resulting from performing FGI procedures using an over-table x-ray tube. Balter (2001a) reported hair loss on the legs of an interventionalist (below the margin of the lead-apron) resulting from years of occupational exposure. Photographs of these cases may be found in the cited publications.

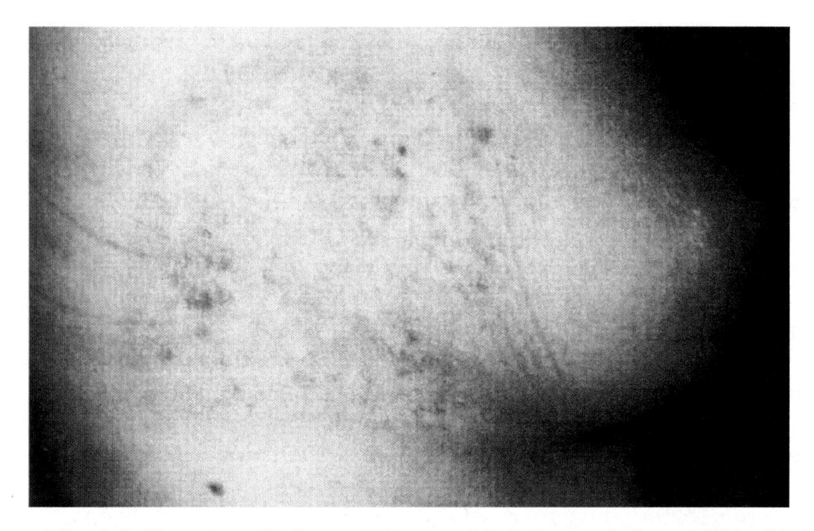

Fig. 2.1. Chronic radiodermatitis; atrophic indurated plaque. This 17 y old patient also has an increased risk of breast cancer. Photograph taken 2 y after two consecutive intra-cardiac ablations (Vano *et al.*, 1998a).

2.4.2 *Applying Benefit-Risk Analyses*

The benefit-risk estimate for an FGI procedure is a continuing process that begins when the procedure is first contemplated for a specific patient and that continues until the procedure is completed or stopped. This is not a formal process, but rather a conscious and subconscious weighing of benefits and risks. Ideally, it is guided by data in the published literature. Scoring systems have been developed to help determine the suitability of certain procedures for specific groups of patients, but these are sometimes imperfect, not available for every procedure, and may not apply to every patient (Faizer *et al.*, 2007; Harrod-Kim and Waldman, 2005). Tailoring the analysis to a specific patient often forces the interventionalist to rely on clinical experience as well.

The initial analysis determines whether the procedure will be recommended to the patient. It includes not only the benefits and risks of the proposed procedure, but also the benefits and risks inherent in *not* performing the procedure and the benefits and risks of alternative procedures. If the projected benefit is less than the estimated risk, the procedure is generally not performed. Exceptions may be made when no alternative exists, there is a substantial benefit if the intervention is successful, and there is a poor prognosis if no intervention is performed.

The benefit-risk analysis does not stop at this point. A change in the patient's condition prior to the procedure, or the appearance of new information, will cause the analysis to be repeated.

The benefit and risk are continuously evaluated during the procedure itself. An unfavorable ratio may cause the procedure to be modified or halted. Many factors are weighed, including contrast material dose, patient and lesion anatomy, technical issues related to guidewire, catheter and stent manipulation, the patient's tolerance and ability to cooperate, the adequacy of sedation or anesthesia, and changes in the patient's clinical status.

Radiation risk should be part of this ongoing intraprocedural analysis, but never the most important part. If the procedure is stopped prior to accomplishing the clinical objective, all of the radiation administered will have added to the radiation risk but provided no benefit. Radiation risk can be managed with the same approach clinicians currently use to manage the risk of renal failure from contrast material load. For radiation the corresponding actions would be:

- establish the level of risk before the procedure;
- monitor radiation dose during the procedure;
- limit radiation use if possible as the delivered dose increases; and
- consider the relative risk of a high-radiation dose as compared to the relative clinical risk if the procedure is modified or stopped prior to a successful conclusion.

The principle of optimization of radiation protection states that radiation exposure should be consistent with the ALARA principle, which takes into account economic and societal factors (ICRP, 2007a; 2007b; NCRP, 1993). The level of protection should be the best under the prevailing circumstances. For FGI procedures, this is applied in the design, appropriate selection and use of equipment, equipment replacement schedule, and in day-to-day working procedures, in order to maximize the net benefit of the procedure. In a practical sense, this is best described as management of the radiation dose to the patient so that it is commensurate with the medical purpose, avoiding radiation dose that is clinically unnecessary or unproductive. The dose efficiency of fluoroscopy equipment is continuously improving, which facilitates achievement of the ALARA principle.

The principle of dose limitation states that the total dose to any individual from all the regulated sources in planned situations other than medical exposure of patients should not exceed recommended limits (ICRP, 2007a; 2007b; NCRP, 1993). Provided that medical exposures of patients have been properly justified and that the associated doses are commensurate with the medical purpose, it is not appropriate to apply dose limits to medical exposure of patients, because such limits would often do more harm than good. Most likely, there are concurrent chronic, severe or even life-threatening medical conditions that are of greater concern than the radiation exposure. This is especially the case with FGI procedures.

2.4.3 *Risks to Interventionalists and Staff*

Benefit-risk analyses have not traditionally been performed for interventionalists and staff. From a radiation safety perspective, the only risk usually considered is radiation risk, and benefit is not considered at all. However, other risks, such as orthopedic injuries and infection, exist, and both the individual interventionalist and society as a whole benefit when these procedures are performed on patients who require them.

2.4.3.1 *Radiation Risk.* The interventionalist and all staff in the procedure room (*e.g.*, nurses, technologists) are at risk of the development of stochastic effects, principally an increased risk of cancer. This risk is relatively small, and is described in Section 2.6.3. Deterministic effects occurring in these individuals as a result of modern practice are rare, and usually due to lack of training, failure to use available protective equipment, or when unsuitable or poorly maintained equipment is used (Vano *et al.*, 1998c). All workers involved in FGI procedures have an obligation to properly manage their personal radiation dose (*i.e.*, apply the ALARA principle) while still providing the appropriate patient care.

The use of personal radiation protective equipment, spatial and temporal variability of the scattered radiation field, and the relatively low energy of radiation used in FGI procedures combine to produce highly nonuniform dose distributions in interventionalists and other staff in the FGI-procedure room. Estimates of effective dose (E) and equivalent dose to the skin, hands, feet, and lens of the eye received during FGI procedures should always take into account the personal protective equipment used in individual cases in order to properly assess compliance with occupational dose limits (Section 5.7.3). The dose limits recommended by NCRP (1993) for individuals in occupational situations are normally applicable. However, a discussion on when a need for an urgent or emergent FGI procedure may necessitate allowing the annual dose limits to be exceeded is presented in Section 5.7.1.

2.4.3.2 *Medical Risks.* Radiation is not the only risk during an FGI procedure. Interventionalists and staff are at risk of contracting infections from their patients. An obvious source is exposure to blood and body fluids through open wounds due to sharps (needles and scalpels), mucous membrane splashes, and skin exposure. In a 3 y study of 24,000 hospital workers, the overall annual incidence of occupational exposure was estimated at 3.5 incidents per 100 workers per year (Denis *et al.*, 2003). Of these, the most serious are sharps injuries. Despite substantial recent efforts to reduce the likelihood of sharps injuries, including worker education and device reengineering, these injuries remain common. An estimated 600,000 to 800,000 needlestick and other percutaneous injuries are reported annually in U.S. healthcare workers (Makary *et al.*, 2007). A detailed discussion of these risks is presented in Appendix C.

2.4.3.2.1 *Ergonomic risks to interventionalists.* Back pain is common in the United States. In a 2006 NIOSH survey, 22.7 % of individuals with a bachelors degree or higher experienced low back pain in the three months prior to the survey (Pleis and Lethbridge-Cejku, 2007). However, there appear to be specific ergonomic occupational hazards for interventionalists that increase the likelihood of developing back pain. Cervical and lumbar spinal disc disease is believed to be related to wearing lead aprons for radiation protection during FGI procedures. Anecdotal evidence has suggested a causal relationship for many years (Dehmer, 2006; Moore *et al.*, 1992; Ross *et al.*, 1997). A survey of 236 radiologists performed in 1992 was unable to substantiate this relationship. However, individuals with preexisting back pain were excluded from the analysis, and the survey did not include questions about cervical spine disease (Moore *et al.*, 1992).

Two subsequent surveys of interventional cardiologists concluded that there is evidence of a relationship between wearing lead aprons and spine problems. Ross *et al.* (1997) conducted a survey to compare the prevalence of back and neck pain among interventional cardiologists, orthopedic surgeons, and rheumatologists. The interventional cardiologists spent much of their workday standing and, on average, wore a lead apron for 8.4 h d^{-1}. Orthopedic surgeons spent much of their workday standing, but wore lead aprons a mean of 2 h d^{-1}. Rheumatologists did little standing and wore lead aprons a mean of 0.2 h d^{-1}. Axial skeletal complaints were reported significantly more frequently by interventional cardiologists than by orthopedic surgeons or rheumatologists. Interventional cardiologists also had a significantly greater frequency of missed days from work due to back or neck pain. Of the 385 interventional cardiologists who responded to the survey, 203 (53 %) reported having been treated for neck or back pain. Cervical spine disc disease was significantly more common among interventional cardiologists, and all survey respondents who reported cervical spine disc disease also reported lead apron use. Involvement of spinal discs at multiple levels was noted only in interventional cardiologists. The authors concluded that spinal disc disease was a "distinct occupational hazard" of interventionalists.

Goldstein *et al.* (2004) also investigated the prevalence of orthopedic problems among interventional cardiologists, using a survey of members of the Society for Cardiovascular Angiography and Interventions. Of the 423 respondents, 42 % reported spine problems (of these, 70 % were lumbosacral and 30 % were cervical), and 28 % reported hip, knee or ankle problems. The prevalence of spine problems was related to the interventionalists' annual procedural caseload and number of years in practice. Over one-third of respondents reported that spine problems had caused them to miss work.

Numerous apron designs have been developed and marketed as ergonomically superior, but no truly successful design exists. Substitution of other combinations of metals for lead has made aprons lighter than in years past, but they remain heavy, cumbersome and uncomfortable (Klein *et al.*, 2009). Requiring interventionalists to wear more radiation protective clothing than is essential may reduce radiation risk, but at the expense of increasing interventionalist discomfort and increasing ergonomic risks.

Radiation protective clothing is not the only potential source of ergonomic hazards in the fluoroscopy suite (Klein *et al.*, 2009). Current designs for fluoroscopy equipment and suites promote awkward positions and posture (Ross *et al.*, 1997). Goldstein *et al.* (2004) concluded that the workplace setting and physical working lifestyle of interventionalists "pose significant occupational hazards that exact a toll on physicians' health."

2.4.3.2.2 *Economic benefit to operator and staff.* As with all radiation workers, physicians, nurses, technologists, and all other healthcare workers are paid for their efforts. This is not because of the radiation risk they assume, but because of the healthcare that they provide.

2.4.4 *Benefit to Society*

The fundamental principle of healthcare is that sick persons should always receive care (Zuger and Miles, 1987). However, some traditional ethical models do not necessarily compel individual physicians to treat patients who pose a risk to them, regardless of the degree of risk (Zuger and Miles, 1987). Certainly, both in the past and more recently, some physicians have been unwilling to risk their own health to care for their patients (Ehrenstein *et al.*, 2006; Wynia and Gostin, 2004). As a result of the appearance of the human immunodeficiency virus (HIV) epidemic in the United States in the last quarter of the twentieth century, and the lack of unanimity among the medical profession at the time, legal entitlements to medical care were created by the Americans with Disabilities Act (White, 1999). Severe acute respiratory syndrome (SARS) and the reappearance of biological and chemical terrorism in recent years (*e.g.*, anthrax in the United States and sarin in Japan) have also led to a rethinking of medical ethics and the physician's role in epidemics (Huber and Wynia, 2004: Ruderman *et al.*, 2006; Wynia and Gostin, 2004). In the 2003 SARS outbreak in Toronto, 51 % (73/144) of the SARS patients were healthcare workers (Booth *et al.*, 2003). In the United States, medical professional societies have now clearly stated that physicians are expected to provide medical care even if it puts their own health at risk (AMA, 2004; Wynia and Gostin, 2004). While these statements have been developed in the context of scenarios of epidemics of infectious disease, they would seem to apply equally to the medical risk and stochastic radiation risk associated with performing FGI procedures. Within broad limits, the benefit to society is considered to outweigh the risk to the individual practitioner. As Huber and Wynia (2004) succinctly state, "Like other public service professionals including the fire and police forces, some risk is simply part of the job description for medical professionals."

2.4.5 *Discussion*

A wide variety of FGI procedures exist that are performed by physicians in a number of medical specialties with varying amounts of education and training in radiation safety. There is continual growth in the number of procedures performed annually, due to their effectiveness, favorable benefit-risk ratio, and cost-effectiveness.

Benefit-risk estimates for patients should include all benefits and all risks, but typically omit radiation risk. The radiation risk from all medical procedures and particularly from FGI procedures should be incorporated into benefit-risk models. For patients, the medical risk of FGI procedures is almost always less than that of any alternative surgical procedure (Miller, 2008). Either the medical benefit of FGI procedures outweighs all risks, both medical and radiation related, or the procedure should not be performed. Radiation risk can and should be managed in the same way as other risks from medical procedures are managed.

When interventionalist and staff risk from FGI procedures is analyzed, it is common to consider only radiation risk. This is not the only risk to which healthcare workers are subject. Radiation risk should be compared with other risks to medical personnel, principally the risk of transmission of infectious disease from patient to healthcare provider and the ergonomic risk inherent in the use of radiation protective devices, particularly lead aprons. The ethics of decreasing nonradiation risks by reducing radiation protection must be evaluated on an individual basis. Managing ergonomic risk may require redesigning the FGI-procedure workspace and installing appropriate features to minimize staff irradiation without increasing other risks. Optimally this may permit the clinical work to be done with less ergonomic risk and no loss of radiation protection. As a last resort, modifying personal radiation protective devices can be considered, even at the potential cost of somewhat reduced protection against radiation risk.

Both radiation risk and medical risk to the interventionalist and staff are low in magnitude. Both risks should be managed to the best extent possible. While both risks should be managed appropriately, the benefits to the patient and to society, which are often ignored in a radiation protection context, should also be considered. It is now well-established, both in ethics and in law, that even if there is a real risk of transmission of infectious disease to healthcare workers, medical care should still be provided to patients. Risk to medical personnel, even the risk of serious illness or death, is considered acceptable by society if attempts by healthcare workers to avoid these risks would result in the inability of patients to receive needed medical care.

2.5 Physics and Dosimetry

The basis for radiation protection is the scientific observation that ionizing radiation produces biological effects that can lead to health effects (*e.g.*, cancer or skin changes). The literature reporting the health effects of radiation is usually written in terms of the consequences of delivering a "dose of radiation" to a person (patient or worker). The word "dose" is used in many different contexts. Disparate definitions are found in the dictionary, where dose is defined as: (1) the measured quantity of a therapeutic agent to be taken at one time and (2) the quantity of radiation administered or absorbed.

Definition (1) is applicable to the concept of a patient swallowing two aspirin tablets. Even in this context, dose is further confused by the common practices of administering the same amount of drug to patients of different weights ("take two aspirin") or adjusting drug dose on the basis of patient weight, depending on the drug being prescribed. The biological distribution and pharmacokinetics of prescription drugs are carefully studied during the drug development process. However, actual drug concentrations in body tissues (principally the blood) of individual patients are measured and explicitly considered during routine clinical therapy for relatively few classes of drugs.

The radiological definition (2) describes something quite different. In radiological terms, the dose of radiation is the amount of energy locally delivered to a very small mass of material at a specified point (absorbed dose). Absorbed dose is expressed in joule per kilogram (gray). Knowing the absorbed dose at a point seldom provides information regarding the absorbed dose at any other point or regarding the total amount of radiation energy delivered to an object or individual.

When an individual is exposed to a radiation field, energy is transferred from the radiation field to the individual. For x-ray fields used in FGI procedures, the resulting dose distributions in the individual (either patient or staff member) are always highly nonuniform in both space and time. The causes of these nonuniformities include initially nonuniform radiation fields, differential shielding of different parts of the individual's body from the radiation, and scattered radiation produced within the irradiated individual.

To the extent possible, this Report uses the definitions and notations given in Report 74 of the International Commission on Radiation Units and Measurements (ICRU, 2005). This Report extends the formalism given in ICRU (2005) to include specific quantities for FGI procedures. Additional quantities include peak skin dose ($D_{\text{skin,max}}$), air kerma at the reference point ($K_{\text{a,r}}$),[1] and air kerma-area product (P_{KA}).[2] Appendix D.1 briefly reviews established radiation quantities needed to describe the FGI-procedure environment. For convenience, Table D.2 in Appendix D.1 provides conversion factors from some common presentations in other literature and regulatory documents to those used in this Report.

The x-ray spectra used for FGI procedures are often modified by the addition of up to 1 mm of copper filtration to the beam. This can have a profound effect on both skin dose and image contrast. Dose measurements should account for this. Further information is given in Appendix D.2.

Currently, skin dose estimates are not generally available in real time during the procedure. Automated dose documentation (Section 4.3.4.1) has the potential to support real-time skin-dose

[1]Referred to as reference air kerma by IEC (2000; 2004; 2010).

[2]Sometimes abbreviated KAP, and similar to dose-area product.

maps. At the present time, several research projects are directed toward this aim (Chugh *et al.*, 2004; den Boer *et al.*, 2001; Morrell and Rogers, 2006). Most institutions in the United States can use $K_{a,r}$ for dose monitoring. The quantity P_{KA} may also be used if $K_{a,r}$ is not available. Fluoroscopy time correlates poorly with other dose metrics (Fletcher *et al.*, 2002; Miller *et al.*, 2003b; 2004). In general, fluoroscopy time should be used with caution to monitor patient irradiation during interventional procedures (Stecker *et al.*, 2009). Figure 2.2 illustrates the relationship between $K_{a,r}$ and fluoroscopy time for both cardiac and noncardiac procedures (Balter, 2008). Because fluoroscopy time is such a poor indicator of radiation dose, its use is generally discouraged.

Most fluoroscopes used for FGI procedures are capable of recording the cumulative values of P_{KA} and $K_{a,r}$ for a procedure (FDA, 2009a; IEC, 2000; 2010).[3] These quantities provide information on the amount of radiation received by a patient during a procedure and can serve as indicators of stochastic risk and deterministic risk, respectively. The quantity P_{KA} is best used to compare FGI procedures performed in the same anatomical region (*i.e.*, that irradiate the same group of organs and tissues) using similar radiographic techniques and beam orientations.

Recommendation 2

The measured dose quantities air kerma-area product (P_{KA}) and air kerma at the reference point ($K_{a,r}$) *should* be used to compare similar FGI procedures. In this Report, P_{KA} and $K_{a,r}$ refer to the cumulative value for the FGI procedure.

All statements of patient dose contain some degree of uncertainty. Even the most sophisticated dose-measurement instrumentation has unavoidable uncertainties related to variations in instrument response with changes in beam energy, dose rate, and collimator size.

Converting measurements from such instruments to skin dose requires additional assumptions about the patient's size and position relative to the reference point for $K_{a,r}$. Additional information on the physics and dosimetry associated with FGI procedures is provided in Appendix D. Also, both $K_{a,r}$ and P_{KA} ignore the effect of backscatter from the patient. Backscatter can increase skin dose by about one-third depending on the beam area and energy. Therefore,

[3]In this Report, P_{KA} and $K_{a,r}$ refer to the cumulative value for the FGI procedure.

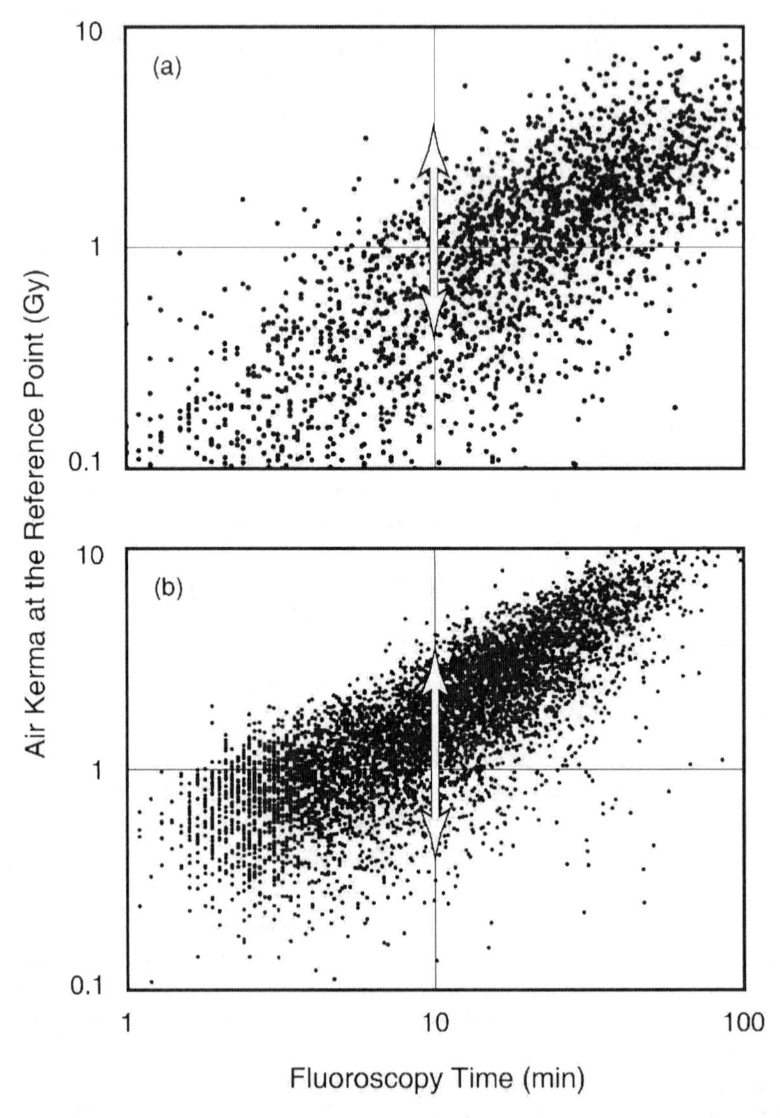

Fig. 2.2. Relationship between $K_{a,r}$ and fluoroscopy time for FGI procedures (Balter, 2008). Scatter plots of $K_{a,r}$ versus fluoroscopy time in minutes (F_{min}): (a) 2,100 noncardiac interventions; (b) 1,700 coronary-artery procedures. The linear regression equations ($K_{a,r}$ as a function of F_{min}) and the regression coefficients (R^2) are: for (a) $K_{a,r} = 0.41 + 0.037 F_{min}$, R^2 = 0.50; for (b) $K_{a,r} = 0.53 + 0.12\ F_{min}$, $R^2 = 0.68$. Data points below 1 min or above 100 min are not displayed but were used in the regressions. The white arrows demonstrate that $K_{a,r}$ varies by more than a factor of 10 at most fluoroscopy times.

the actual skin dose due to irradiation of a particular site on the skin may differ from the contribution that irradiation makes to $K_{a,r}$ by a factor of two or more (Hirshfeld *et al.*, 2004). However, a particular site on the skin may not always be directly in the irradiation field. Users of dose data should be aware of these considerations.

2.6 Radiation Health Effects

The health effects resulting from radiation exposure are divided into the two broad classifications, stochastic and deterministic. The health effects discussed in this Report are cancer induction (stochastic) and tissue reactions (deterministic), which include reactions affecting hair follicles, skin, subcutaneous tissues, and the lens of the eye. However, the classification of some effects (such as cataract) as deterministic or stochastic is uncertain (Section 2.6.4.2 and Appendix E.2).

Stochastic effects (*e.g.*, cancer induction) are believed to be attributable to the consequences initiated by the misrepair of damage to the deoxyribonucleic acid (DNA) of a single cell that results in a genetic transformation (ICRP, 2007a). The probability of such an event depends on the type of cell. The expression of changes in progeny cells can be affected by the characteristics of the individual (*e.g.*, age, gender, health status, genetics). Depending on the type of cell, once such cells begin to divide it can take years to decades to accumulate enough daughter cells to produce a clinically-observable cancer (Hall and Giaccia, 2006).

Deterministic effects localized to the individual tissues of interest for this Report (*i.e.*, hair follicles, skin, subcutaneous tissues, and the lens of the eye) are largely caused by the reproductive sterilization of stem cells or damage to supportive tissues (ICRP, 2007a). This is radiation dose related. Damage is expressed when these cells attempt division or differentiation, with concomitant cell death. Unless sufficient numbers of cells are affected, nearby uninjured cells will be able to make up the loss, without a clinically-visible injury. The "threshold dose" for such localized deterministic effects is the level at which cell loss overwhelms repopulation or repair. The timing of expression of such damage is therefore related to the cell proliferation kinetics of the specific tissues, providing an explanation for the variability in the time of appearance of the effects.

2.6.1 *Demographic Factors*

Demographic factors include patient gender, age and weight. Very young patients have a greater stochastic risk for a given radiation dose than do adults because of their longer life expectancy and greater susceptibility to stochastic radiation effects (Section 2.6.3).

For an infant, overall stochastic risk is approximately five times higher than for a 60 y old. The risk of a fatal breast cancer in a newborn female is eight times higher than for a 40 y old; this risk drops by an additional factor of 18 between 40 and 80 y of age (NA/NRC, 2006). There is an increased risk of a radiogenic breast cancer in the heavily irradiated young woman shown in Figure 2.1.

Obese patients are at greater risk of skin injury because of poor radiation penetration in obese patients and the accompanying closer proximity of the x-ray source to the patient (Bryk *et al.*, 2006). The result is that the entrance-surface air kerma ($K_{a,e}$) (or entrance-surface absorbed dose) for obese patients can be as much as 10 times higher than in some nonobese patients (Wagner *et al.*, 2000). It is not surprising that many of the documented effects associated with FGI procedures are seen in patients with thicker bodies and are the result of high radiation doses (Koenig *et al.*, 2001a). Medical co-morbidities, such as diabetes mellitus, may increase skin sensitivity to radiation effects, requiring additional attention to radiation dose management (Bryk *et al.*, 2006).

Ethnic differences in skin coloration are also associated with differences in radiation sensitivity; individuals with light-colored hair and skin are most sensitive (Geleijns and Wondergem, 2005).

2.6.2 *Medical History Factors*

Important parts of the patient's medical history include coexisting diseases and genetic factors, medication use, radiation history, and pregnancy. Other patient-related factors that increase susceptibility to radiation effects include smoking, poor nutritional status, and compromised skin integrity (Hymes *et al.*, 2006).

Defects in DNA repair genes may predispose some individuals to increased radiation sensitivity. This includes individuals with the ataxia telangiectasia mutated (ATM) gene, an autosomal recessive gene which is responsible for ataxia telangiectasia. It has been suggested that many patients with serious and unanticipated radiation effects may be heterozygous for the ATM gene, or harbor some other ATM abnormality (Hymes *et al.*, 2006). Heterozygosity for ATM occurs in ~1 % of the population. Irradiation of patients with heritable nevoid basal-cell carcinoma (Gorlin syndrome) may result in widespread cutaneous tumors. Other disorders with a genetic component that affects DNA breakage or repair have been found to increase radiation sensitivity. These include Fanconi anemia, Bloom syndrome, and xeroderma pigmentosum. Familial polyposis, Gardner syndrome, heritable malignant melanoma and dysplastic nevus syndrome also increase radiation sensitivity (Hymes *et al.*, 2006).

Preexisting autoimmune and connective tissue disorders predispose patients to the development of severe radiation effects in an unpredictable fashion. The etiology is not known. These disorders include scleroderma, systemic lupus erythematosus, and possibly rheumatoid arthritis (Hymes *et al.*, 2006). Hyperthyroidism and diabetes mellitus are also associated with increased radiation sensitivity (Koenig *et al.*, 2001b). Diabetes is believed to predispose to radiation effects secondary to small vessel vascular disease and consequent decreased healing capacity (Bryk *et al.*, 2006; Herold *et al.*, 1999).

A number of drugs are known to increase radiation sensitivity. These include actinomycin D, bleomycin, doxorubicin, 5-fluorouracil, and methotrexate (Koenig *et al.*, 2001b). When given in conjunction with radiation therapy, docetaxel, paclitaxel, and possibly tamoxifen can result in cutaneous toxicity (Hymes *et al.*, 2006).

2.6.3 *Cancer Risks*

A major portion of radiobiological research has been devoted to understanding the relationship between the irradiation of groups of individuals (populations) and their subsequent risk of stochastic effects (ICRP, 1957; 1977; 1991a; 2007a; NA/NRC, 1990; 2006; NCRP, 1987; 1989a; 1993; 1997; 2005; UNSCEAR, 1986; 2000a; 2000b; 2001). This body of work documents the observation of increased probabilities of cancer in humans and heritable effects in plants and animals attributable to ionizing radiation. However, there have been no unequivocal reported observations of radiation-induced heritable effects in humans.

This subsection addresses the implications that the radiogenic cancer risk experienced by worker and patient populations in the FGI-procedure environment has on managing dose during these procedures. The age distribution and general health status of both groups differ from each other and from those populations used to derive radiogenic risk factors.

Radiation-induced cancer induction is a stochastic effect. The BEIR VII report (NA/NRC, 2006) provides estimates of lifetime attributable risk resulting from the irradiation of specific organs as a function of gender and age-at-irradiation. Lifetime attributable risk is an approximation of the risk of exposure-induced death (or exposure-induced incidence) which estimates the probability that during a lifetime an individual will die (or develop cancer) due to the exposure. Lifetime attributable risk includes deaths (or incident cases) of cancer that would have occurred without exposure but occurred at a younger age because of the exposure. As an example to illustrate the impact of gender and age, Figure 2.3 is a plot of the lifetime attributable risk for cancer mortality for all cancers as a function of gender and age-at-irradiation after a uniform whole-body dose of 1 Gy (*i.e.*, all organs and tissues receive the same mean absorbed dose) (NA/NRC, 2006). For comparison, the nominal 5 % per sievert risk factor for cancer mortality that is often applied to the general population[4] is indicated by the dotted line. A similar plot can be obtained for cancer incidence (NA/NRC, 2006). However, FGI procedures do not result in uniform whole-body irradiation and the appropriate starting point would be the lifetime attributable risk as a function of gender and age-at-irradiation for each of the specific organs or tissues of interest, such as found in Table 12.D.1 (for incidence) and Table 12.D.2 (for mortality) in NA/NRC (2006). Radiogenic risk and its uncertainty are discussed in other reports by NCRP (1997; 2005). Additional factors that

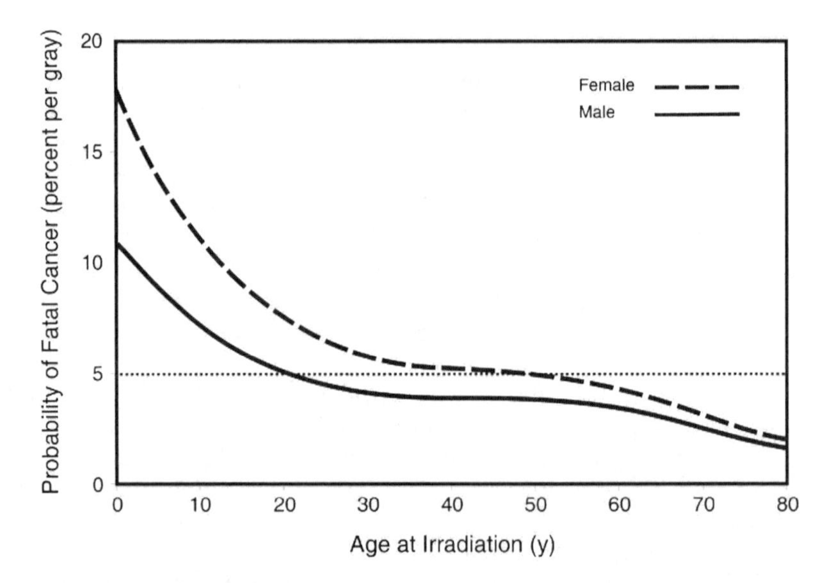

Fig. 2.3. Lifetime attributable risk for cancer mortality as a function of gender and age-at-irradiation after a uniform whole-body dose of 1 Gy (derived from Table 12.D.2 in NA/NRC, 2006). For comparison, the nominal 5 % per sievert risk factor for cancer mortality that is often applied to the general population is indicated by the dotted line. Lifetime attributable risk as a function of gender and age-at-irradiation for specific organs and tissues are found in Table 12.D.1 (for incidence) and Table 12.D.2 (for mortality) in NA/NRC (2006).

[4]The general population is a reference population of equal numbers of both males and females and a wide range of ages (ICRP, 1991a).

impact the risk estimates for the specific considerations of FGI procedures are discussed in Section 2.6.3.3.

The accumulated radiation doses of workers and patients from FGI procedures seldom reach levels high enough to detect an excess in the occurrence of such events with statistical validity. Estimates of individual risk for radiation workers have been well discussed elsewhere (Berrington *et al.*, 2001; Sigurdson *et al.*, 2003; Yoshinaga *et al.*, 2004). Estimates of individual patient risk are also confounded by differences between patient and reference populations.

2.6.3.1 *Worker Cancer Risk.* Workers conducting FGI procedures are likely to be similar in age distribution and health status to other radiation workers. The existing system of dose limits for radiation protection of workers is generally appropriate for these workers. However, the nature of the radiation fields and the use of personal radiation protective equipment during FGI procedures result in highly nonuniform dose distributions in the organs and tissues of workers. Many of the standard methods used for evaluating the measured operational dose quantities provided by personal dosimeters do not take into account the personal protective equipment used and therefore greatly overestimate the values of E (and therefore the stochastic risk) for the workers in the FGI-procedure environment. Differences in monitoring methods and algorithms can result in a fivefold difference in estimates of E (Chida *et al.*, 2009a). This situation has been previously discussed in depth by NCRP (1995b). This Report recommends that E be determined appropriately (Sections 2.4.3.1 and 5.7.3).

2.6.3.2 *Patient Cancer Risk.* The cancer risk associated with irradiation of most organs is known to be a function of both the age and gender of the irradiated individual (NA/NRC, 2006) (Section 2.6.3). Figure 2.3 demonstrates the differences in cancer mortality risk between any particular age group of patients (by gender) and the general population. Most patient populations undergoing specific types of procedures (*e.g.*, FGI procedures) have a different age distribution and may be less healthy than the general population. Risk estimation tools designed for a general population have uncertainties that are beyond the scope of this Report. Their application to FGI patients is likely to have increased uncertainty.

Cancer induction is most often one of the lesser risks for a patient undergoing an FGI procedure. Major FGI procedures (those typically associated with potentially-high radiation dose levels) have known nontrivial frequencies of nonradiation short-term risks of death and serious complications. Alternatives to FGI procedures such as medical management or surgery also entail short-term risks of death and serious complications. Properly selected and performed FGI procedures result in less morbidity and a lower serious-complication rate than the alternatives (Miller, 2008). This is an especially important consideration in palliative procedures such as a TIPS creation.

Individuals, usually patients, may receive additional irradiation in conjunction with medical research. In some cases, all of the radiation delivered to a patient in the course of the research is attributable to the research protocol. In other cases, the research component is an extension of a fully-justified clinical procedure. In some cases, the radiation may be due entirely to a clinically-justified medical procedure that would be performed even if no research were conducted (*e.g.*, validating a new stent design by randomizing between the new and approved stent). With regard to evaluating radiation risk for a research protocol, only the additional risk from radiation attributable to the research protocol (not the clinical procedures that are already justified) should be considered (DHHS, 2009). The radiation risk coefficient (risk per unit dose) is the same for a given patient population whether the FGI procedure is for healthcare or for research.

2.6.3.3 *Stochastic Risk-Related Aspects of Dosimetry.* The quantity E [as well as its predecessor effective dose equivalent (H_E)] was developed as a practical quantity for use in the general system of radiation protection, particularly with regard to applying the principles of optimization of radiation protection and dose limitation for stochastic effects. The quantity E is defined by ICRP (2007a) using the method shown in Figure 2.4. This method starts by calculating the mean absorbed dose in each of the relevant organs and tissues (D_T) for the ICRP reference male and the ICRP reference female (ICRP, 2009). After applying radiation weighting factors (w_R) to each D_T value, the equivalent doses for each organ or tissue (H_T) are gender averaged. The tissue weighting factors (w_T) are then applied to obtain E.

In the formulation of E, judgments were made on the relative severity of various factors that influence radiation risk with regard to an adult worker population and the general population (ICRP, 1991a; 2007a). The risk of stochastic effects is dependent on age, gender, and other characteristics of the exposed population (*e.g.*, life expectancy, health status). In particular, the age and gender distributions and health status for workers and the general population (for which E was derived) can be quite different from that of the overall size, age and gender distribution, and health status for

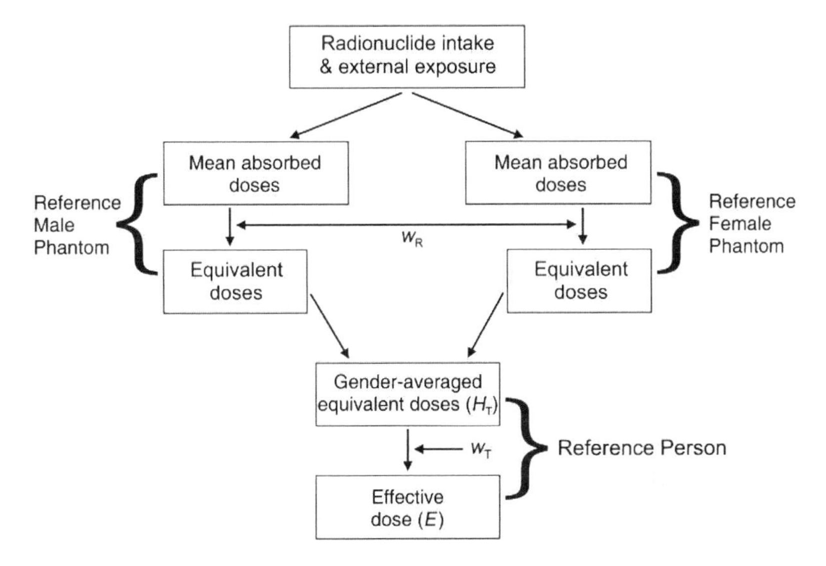

Fig. 2.4. Schematic illustrating the method of computing E (ICRP, 2007a).

the population undergoing medical procedures using ionizing radiation, and will also differ from one type of medical procedure to another, depending on the characteristics of the individuals with the medical condition being evaluated.

An "effective dose" for an FGI procedure is often evaluated using P_{KA} and a procedure-specific conversion coefficient. Published conversion coefficients are obtained using Monte-Carlo simulations or thermoluminescent-dosimeter measurements in phantoms. The variation of conversion coefficient with imaging procedure techniques and beam qualities can be investigated using Monte-Carlo simulations with procedural data obtained from a series of real procedures (*e.g.*, Smans *et al.*, 2008). Conversion coefficients for the same procedure can vary by factor of three. Table 2.3 reproduces a portion of the findings reported in Smans *et al.* (2008). A method for estimating "effective dose" to pediatric patients undergoing FGI procedures using anthropomorphic phantoms has also been described (Miksys *et al.*, 2010).

The "effective dose" calculated in this manner for an individual patient and procedure is often regarded as the ICRP quantity E (McCollough *et al.*, 2010). However, the ICRP quantity E applies to a "representative member of the population" without further consideration of age or gender or any other confounding factors. ICRP never intended to use its formulation of E to make quantitative estimates of stochastic risk for either individual patients or for

TABLE 2.3—*Monte-Carlo calculated ratio ("effective dose" / P_{KA}) ($mSv\ Gy^{-1}\ cm^{-2}$) for representative FGI procedures (Smans et al., 2008).*

Monte-Carlo Calculated Ratio ("effective dose" /P_{KA})[a] ($mSv\ Gy^{-1}\ cm^{-2}$)			
Hospital	Abdomen	Pelvis	Upper Leg
Center 3	0.212	0.130	0.030
Center 5	0.070	0.030	0.005
Center 6	0.063	0.055	0.007
Center 7	0.127	0.075	0.020
Center 8	0.113	0.066	0.016
Center 9	0.093	0.054	0.013
Center 10	0.132	0.070	0.015
Center 11	0.137	0.073	0.047
Center 12	0.270	0.215	0.047

[a]Tissue weighting factors from ICRP Publication 103 (ICRP, 2007a) were used to compute "effective dose".

patient populations (ICRP, 2007a; 2007b). Quantitative assessments of stochastic radiation risk for FGI procedures should be made using appropriate risk coefficients (*i.e.*, NA/NRC, 2006) for the individual tissues at risk, and for the age and gender distribution of the patient population undergoing the specific type of clinical procedure. The methods for doing such quantitative assessments are outside the scope of this Report. Detailed discussions of the methods are presented in NA/NRC (2006).

The health status of patients undergoing FGI procedures (*i.e.*, the reduced lifespan of patients with serious health conditions) is also a consideration. Other considerations that may impact the risk assessment include the possibility of increased biological effectiveness due to the lower-energy photons used in FGI procedures (relative to higher-energy photons) (Kocher *et al.*, 2005; Land *et al.*, 2003; NA/NRC, 2006), and the possibility of removing the dose and dose-rate effectiveness factor due to the higher dose rate associated with FGI procedures (relative to the low dose rate at which the dose and dose-rate effectiveness factor is applied) (NA/NRC, 2006). Methods for doing quantitative assessments of these considerations are also outside the scope of this Report.

However, an indication of the stochastic radiation risk attributable to a particular type of procedure is helpful in planning and selecting appropriate procedures. Also, an indication of the stochastic risk attributable to research procedures (or research components of clinical procedures) is usually required in the risk evaluation of the proposed research. The classification scheme shown in Table 2.4 may be used in both these situations, and is derived from proposals made in ICRP (1991b) and Martin (2007). With this scheme, estimates of E may be used as a qualitative indicator of stochastic radiation risk for classifying different types of procedures into broad risk categories.

Recommendation 3

Effective dose (E) *shall not* be used for quantitative estimates of stochastic radiation risk for individual patients or patient groups (the appropriate approach to obtain quantitative estimates is discussed in Section 2.6.3.3).

Effective dose (E) *may* be used as a qualitative indicator of stochastic radiation risk for classifying different types of procedures into broad risk categories (as suggested in Table 2.4).

The classification scheme shown in Table 2.4 does *not* apply to deterministic effects, and the classification scheme does not attempt to quantify specific probabilities for stochastic effects because of the unresolved considerations discussed in Section 2.6.3.3. The column labeled "NCRP (this Report)" suggests qualitative descriptors of stochastic risk presented in a manner similar to those presented in ICRP (1991b) and Martin (2007). However, the overall benefits and risks of an FGI procedure used in research when $E > 100$ mSv should be carefully evaluated. In particular, the relative impact on radiation risk of the age distribution for a given patient population is fairly well understood and should be taken into account in applying the classification scheme (Section 2.6.3.3 and footnotes to Table 2.4). The last column of Table 2.4 suggests the minimum level of benefit desired for approval of use of an FGI procedure in proposed research that entails the qualitative level of risk noted in the "NCRP (this Report)" column.

2.6.4 *Deterministic Effects*

Deterministic effects occur when individual tissues or organs are irradiated to a dose level that overwhelms repopulation and repair processes or causes sufficient damage to supporting tissues (Section 2.6). Evaluating the possibility and/or severity of a deterministic effect requires knowledge of the locally delivered dose. Dosimetric quantities used to assess stochastic risk do not provide the information needed for these evaluations.

TABLE 2.4—*Suggested classification scheme for use of E as a qualitative indicator of stochastic risk for FGI procedures (adapted from ICRP, 1991b; Martin, 2007).*[a,b]

Range of E (mSv)	Radiation Risk Descriptor			Expected *Minimum* Individual or Societal Benefit
	ICRP (1991b)	Martin (2007)	NCRP (this Report)	
<0.1	Trivial	Negligible	Negligible	Describable
0.1 – 1	Minor	Minimal	Minimal	Minor
1 – 10	Intermediate	Very low	Minor	Moderate
10 – 100	Moderate	Low	Low	Substantial
>100	—	—	Acceptable (in context of the expected benefit)	Justifiable expectation of very substantial individual benefit

[a]ICRP (1991b) states that for children the radiation detriment per unit E is two to three times larger than for adults, for people aged 50 y or over the detriment is one-fifth to one-tenth that for younger adults, and for those suffering from serious and possibly terminal disease the likely risk will be even lower.

[b]To observe how lifetime attributable risk for radiation-related cancer varies with age and gender, refer to Figure 2.3 in this Report (for mortality, all cancer) and Tables 12.D.1 (for incidence, specific organs and tissues) and 12.D.2 (for mortality, specific organs and tissues) in NA/NRC (2006).

2.6.4.1 *Tissue Reactions: Skin and Hair.* Radiation-induced skin injuries are a well known complication of radiation therapy and have been recognized as a potential complication of FGI procedures (Hymes *et al.*, 2006; ICRP, 2000a; Shope, 1996; Sovik *et al.*, 1996). Tissue reactions (deterministic effects) occur in a localized area and can be predicted by the dose delivered in that area. This dose may have been delivered by one procedure or by a series of procedures. The most severe reaction occurs at the location in the skin or other organ that has received the highest dose (peak tissue dose).

Most reported skin injuries have been the result of FGI procedures in cardiology, but an appreciable fraction have been caused by FGI procedures in radiology or neuroradiology (Aerts *et al.*, 2003; Koenig *et al.*, 2001a). Some FGI procedures in cardiology, especially PCI for the recanalization of chronic total vascular occlusions and radiofrequency (RF) ablation for treatment of cardiac dysrhythmias, often require particularly high radiation doses and are associated with a higher frequency of skin injuries (Koenig *et al.*, 2001a; Suzuki *et al.*, 2008). Skin injuries can also be caused by the cumulative dose from multiple diagnostic procedures, each of which may impart a radiation dose insufficient to cause injury by itself (Imanishi *et al.*, 2005).

Radiation reactions in patients undergoing potentially-high radiation dose procedures often occur inadvertently (*i.e.*, due to lack of proper radiation dose management) (*e.g.*, Shope, 1996). On some occasions, the radiation reactions are a justifiable outcome of a deliberate clinical benefit-risk decision. Numerous publications have described techniques for patient dose management in the fluoroscopy environment and methods for minimizing $D_{\mathrm{skin,max}}$ (Archer and Wagner, 2000; Balter and Moses, 2007; Hirshfeld *et al.*, 2004; Johnson *et al.*,1992; Mahesh, 2001; Miller *et al.*, 2002; 2010a; Norbash *et al.*, 1996).

Radiation reactions can be severe and clinically devastating. Although commonly referred to as skin injuries, severe radiation injuries may extend into the subcutaneous fat and muscle (Monaco *et al.*, 2003). Patients may face years of pain, multiple surgical procedures, and permanent disfigurement (Frazier *et al.*, 2007; ICRP, 2000a; Koenig *et al.*, 2001a). Injuries are typically on the patient's back when radiation is delivered in the frontal plane, because the patient is usually placed supine on the procedure table and the x-ray tube is positioned underneath the table in order to minimize scattered radiation to the chest and head of the interventionalist (Koenig *et al.*, 2001a; Schueler *et al.*, 2006). If the frontal beam is angled steeply, as is sometimes done in certain cardiac procedures, injuries to the skin of the shoulder can occur. When the lateral plane is used, injuries to the arm or breast may be seen (ICRP, 2000a; Koenig *et al.*, 2001a).

Damage may be expressed in the epidermis, dermis and subcutaneous tissues. When this damage becomes clinically evident is dependent on the level of radiation dose, and on individual sensitivity. Lesions can be loosely classified as being prompt, early, mid-term, or late in terms of their time of expression. A summary of skin and hair effects, as a function of dose and time, is given in Table 2.5 (Balter *et al.*, 2010). Due to dosimetric uncertainty and variability in individual sensitivity, boundaries between the rows and columns in Table 2.5 are not rigid. There is overlap between events in any one-time dose zone and adjacent zones. Detailed discussions of tissue reactions are presented in Appendix F. Images of tissue reactions of different severities are presented in Appendix G.

The characteristics and timing of the signs and symptoms of radiation-induced injuries are also influenced by a variety of aggravating and mitigating factors. Specifically, anything that damages irradiated skin (sunburn, abrasion and biopsy) is likely to aggravate the tissue response and may increase the probability of infection. Because of clinical variability, it should be assumed that any skin changes observed following an FGI procedure are radiogenic in origin unless an alternative diagnosis is established.

If surgical intervention is required to repair acute ulceration, secondary ulceration, or dermal necrosis, the total thickness of irradiated skin and some subcutaneous tissue is usually involved. Provided that good healing is achieved, no subsequent radiation-induced reactions would be expected to occur. If these lesions are small and only involve a small portion of the exposed site, unassisted healing will result in the formation of extensive scar tissue. The healed wound will not attain full strength and will be at risk for secondary breakdown. This can be precipitated by mild trauma, among other things. Atrophic dermal tissue is also more likely to show the effects of trauma, with delayed necrosis. All of these events tend to be random in nature, but instructions given to radiation-therapy patients should be applied equally to patients receiving substantial radiation doses from FGI procedures (BCCA, 2006). Detailed information on the management of radiation injuries is beyond the scope of this Report. Selected general information may be found in the following sections.

2.6.4.2 *Tissue Reactions: Lens of the Eye.* One of the important tissue reactions associated with ionizing radiation exposure is reduced transparency of the lens of the eye, a pathology called "radiation cataract." Current ocular guidelines are predicated on the view that the development of a radiation-induced cataract is a deterministic effect with a relatively-high threshold dose (2 to 5 Gy) (NCRP, 1993; 2000b). Appendix E reviews the anatomy and radiation pathology of the lens of the eye along with findings from human epidemiological studies following acute or chronic low-dose radiation exposure. Table E.1 in Appendix E gives a protocol for a slit-lamp examination to detect radiation-induced lens changes.

Many early studies of radiation cataracts had short follow-up periods, failed to take into account the increasing latency period as dose decreases, did not have sufficient sensitivity to detect early lens of the eye changes, and had relatively few subjects with doses below a few gray (Cogan and Dreisler, 1953; Cogan *et al.*, 1953; Leinfelder and Kerr, 1936; Merriam and Focht, 1962). More recent human epidemiological data suggest that opacification of the lens of the eye may occur following exposure to significantly-lower doses of ionizing radiation. These data are derived from studies of individuals undergoing CT scans (Klein *et al.*, 1993) or radiation therapy (Hall *et al.*, 1999), atomic-bomb survivors (Minamoto *et al.*, 2004; Nakashima *et al.*, 2006; Neriishi *et al.*, 2007), infants treated with ^{226}Ra for skin haemangioma (Wilde and Sjostrand, 1997), residents of contaminated buildings (Chen *et al.*, 2001, Hsieh *et al.*, 2010), Chernobyl nuclear reactor accident liquidators (Worgul *et al.*, 2007), and radiologic technologists (Chodick *et al.*, 2008).

Worgul *et al.* (2007) found detectable changes in the lens of the eye consistent with ionizing radiation exposure in a large cohort of Chernobyl nuclear reactor accident cleanup workers. The findings suggested a dose-effect threshold significantly below 1 Gy, a result inconsistent with the current ICRP threshold of 5 Gy for "detectable opacities" from protracted exposures. In its current recommendations, ICRP (2007a) states that "new data on the radiosensitivity of the eye with regard to visual impairment are expected — because of the uncertainty concerning this risk, there should be particular emphasis on optimization in situations of exposure of the eyes" (ICRP, 2007a). In another study, Chodick *et al.* (2008) studied the risk of cataract among radiologic technologists in a prospective cohort of 35,705 individuals aged 24 to 44 y. They found that workers in the highest (mean 60 mGy) versus lowest (mean 5 mGy) categories had an adjusted hazard ratio (an estimate of relative risk) of cataract of 1.18 (95 % confidence interval, 1.06 to 1.47). These findings also challenge the ICRP assumption that the lowest cumulative ionizing radiation cataractogenic dose to the lens of the eye is ~2 Gy.

New data from exposed human populations and animal models suggest that posterior lens opacities occur at doses that are lower

TABLE 2.5—*Tissue reactions from a single-delivery radiation dose to the skin of the neck, torso, pelvis, buttocks or arms.*[a,b]

Band	Single-Site Acute Skin-Dose Range (Gy)[c,d,e]	NCI (2006) Skin Reaction Grade	Approximate Time of Onset of Effects[e,f]			
			Prompt <2 weeks	Early 2 – 8 weeks	Mid term 6 – 52 weeks	Long term >40 weeks
A1	0 – 2	Not applicable	No observable effects expected at any time			
A2	2 – 5	1	Transient erythema	Epilation	Recovery from hair loss	None expected
B	5 – 10	1 – 2	Transient erythema	Erythema, epilation	• Recovery • At higher doses: prolonged erythema, permanent partial epilation	• Recovery • At higher doses: dermal atrophy induration
C	10 – 15	2 – 3	Transient erythema	• Erythema, epilation • Possible dry or moist desquamation • Recovery from desquamation	• Prolonged erythema • Permanent epilation	• Telangiectasia[g] • Dermal atrophy induration • Skin likely to be weak

D	>15	3 – 4	• Transient erythema • After very-high doses: edema and acute ulceration, long-term surgical intervention likely to be required	• Erythema, epilation • Moist desquamation	• Dermal atrophy • Secondary ulceration due to failure of moist desquamation to heal, surgical intervention likely to be required • At higher doses: dermal necrosis, surgical intervention likely to be required	• Telangiectasia[g] • Dermal atrophy induration • Possible late skin breakdown • Wound might be persistent and progress into a deeper lesion • Surgical intervention likely to be required

[a]This table is applicable to the normal range of patient radiosensitivities in the absence of mitigating or aggravating physical or clinical factors.

[b]This table does not apply to the skin of the scalp.

[c]Skin dose refers to actual skin dose (including backscatter). This quantity is *not* air kerma at the reference point ($K_{a,r}$).

[d]Skin dosimetry based on $K_{a,r}$ or P_{KA} is unlikely to be more accurate than ±50 %.

[e]The dose range and approximate time period are not rigid boundaries. Also, signs and symptoms can be expected to appear earlier as the skin dose increases.

[f]Abrasion or infection of the irradiated area is likely to exacerbate radiation effects.

[g]Refers to radiation-induced telangiectasia. Telangiectasia associated with an area of initial moist desquamation or the healing of ulceration may be present earlier.

than those generally expected. These data are statistically consistent with the absence of a dose threshold (Appendix E). Many of these studies included subjects with posterior subcapsular opacities that had not yet progressed to significant visual disability. While the initial stages of ionizing radiation-induced lens opacification do not usually result in changes in visual acuity, the severity of these changes generally increases with dose and over time until vision is impaired and cataract surgery is required (Lett *et al.*, 1991; Merriam and Worgul, 1983; NCRP, 2000b; Neriishi *et al.*, 2007). In general, the latency of these changes is inversely related to dose. Furthermore, unlike cortical or nuclear lens changes, subcapsular cataracts appear to result in decrements in contrast sensitivity as well as visual acuity (Stifter *et al.*, 2005; 2006). Recent experimental work suggests that individual genetic differences in radiosensitivity may help explain the wide variation in reported time of cataract onset, degree of lens opacification, and subsequent progression to visual disability among various exposed individuals (Hall *et al.*, 2005; Kleiman, 2007).

In light of these findings, current NCRP and ICRP occupational guidelines for dose limits to the lens of the eye, 150 mGy y^{-1}, may need to be reevaluated. Further data may be developed from future follow-up of atomic-bomb survivors, populations exposed to the Chernobyl nuclear reactor accident, or interventional medical personnel cohorts.

Practitioners who perform FGI procedures may be exposed to a relatively-high ocular dose of x rays over the course of their careers (Kim *et al.*, 2008; Vano *et al.*, 2006). A well-designed epidemiological study in selected individuals with well-documented exposures over a long career is likely to be important in determining cataract risk in this population. In this context, it is interesting to note that radiation cataracts have been reported in workers in interventional radiology and cardiology laboratories where inadequate eye protection was provided (Vano *et al.*, 1998b). Reconstructed yearly dose estimates of lens of the eye exposure ranged from 450 to 900 mSv (equivalent dose). Worgul *et al.* (2004), in a pilot study of interventional radiologists, reported that the prevalence and severity of posterior subcapsular cataracts was associated with physician age and years of practice.

Similarly, Kleiman *et al.* (2009), in another pilot study, reported that posterior subcapsular changes in the lens of the eye consistent with x-ray exposure were noted in almost 50 % of the interventional cardiologists examined. Interventional cardiology nurses and technicians, who generally work at a greater distance from the x-ray source, appeared to have substantially less risk of developing these changes in the lens of the eye. This finding is supported by associated estimates of cumulative ocular exposure, where a sixfold higher estimated equivalent dose to the lens of the eye was noted in interventional cardiologists (6 Sv), as compared to nurses and technicians (1.5 Sv) (Vano *et al.*, 2010). In that same study, as compared to a cohort of 93 nonmedical professionals who had no exposure to x rays in the head or neck region, the prevalence of posterior lens opacification consistent with ionizing radiation exposure in a similarly aged cohort of 58 interventional cardiologists was significantly greater ($p < 0.005$).

Recommendation 4

Peak tissue dose *shall* be used to evaluate the potential for deterministic effects in specific tissues. Examples include peak dose to the skin or to the lens of the eye.

2.6.5 *In Utero Irradiation*

There are radiation-associated risks to the embryo and fetus during pregnancy related to the stage of pregnancy and the absorbed dose to the embryo and fetus. The assessment of these risks by ICRP (2007a; 2007b) is presented below:

- *Lethal effects*: There is sensitivity to the lethal effects of irradiation in the preimplantation period of embryonic development (up to 14 d after conception). At doses below 100 mGy, these lethal effects will be very infrequent. There is no reason to believe that significant risks to health will be expressed after birth.
- *Malformations*: During the period of major organogenesis, conventionally taken to be from the third to the eighth week after conception, malformations may be caused, particularly in the organs under development at the time of exposure. These effects have a threshold of ~100 mGy.
- *Central nervous system effects*: From 8 to 25 weeks after conception, the central nervous system is particularly sensitive to radiation. A reduction in intelligence quotient cannot be identified clinically at absorbed doses to the fetus below 100 mGy. During the same time period, absorbed doses to the fetus on the order of 1 Gy result in a high probability of severe mental retardation. The sensitivity is highest from 8 to 15 weeks after conception, and lower from 16 to 25 weeks

of gestational age. No mental retardation cases were detected before 8 weeks or after 25 weeks.

- *Leukemia and childhood cancer*: Radiation has been shown to increase the probability of leukemia and many types of cancer in both adults and children. Throughout most of pregnancy, the embryo and fetus are assumed to be at approximately the same risk for potential carcinogenic effects as children (*i.e.*, about three times that of the population as a whole).

Consideration of the effects listed above is important when pregnant patients undergo FGI procedures. For each patient and procedure, the benefit to the healthcare of the patient should outweigh the potential for detrimental health effects to the embryo and fetus.

2.6.6 *Heritable Effects*

The latest assessment by ICRP (2007a) for the risk of heritable effects reaches the following conclusions:

- There continues to be no direct evidence that exposure of parents to radiation leads to excess heritable disease in their offspring.
- There is compelling evidence that radiation causes heritable effects in experimental animals.
- The risk of radiation-associated heritable diseases tended to be overestimated in the past.
- The current methodology used by ICRP (2007a) is similar to that used by NA/NRC (2006) and UNSCEAR (2001) and is based on a modified version of the previously used concept of the doubling dose for disease-associated mutations, but through the second generation only.
- The present estimate of genetic risks (through the second generation) for exposure of the whole population is ~0.2 % per gray [absorbed dose in the reproductive organs (*i.e.*, testes and ovaries)].

3. Fluoroscopy Equipment and Facilities

Each specific type of imaging equipment is designed for optimal performance of a specific subset of all possible medical-imaging procedures. These procedures are called the "intended uses" of the equipment (IEC, 2004). The list of intended uses is initially proposed by the manufacturer of the equipment and may be later modified based on clinical experience. Section 4 discusses FGI procedures in some detail.

For the purposes of defining the necessary equipment features for FGI procedures, it is useful to separate these procedures into those that are potentially high-dose and those that are not. Patient doses from FGI procedures are reviewed in Section 4.1. As discussed in Section 4.3.3.2, during a procedure the operator should be notified of patient dose when it reaches the notification levels shown in Table 4.7. The first such notification level is at a $K_{a,r}$ of 3 Gy or a P_{KA} of 300 Gy cm^2. This Report considers that any FGI procedure where more than 5 % of the cases of that procedure exceed one of these values for the first notification should be considered a potentially-high dose procedure. Examples of potentially-high dose procedures are given in Table 4.6.

Recommendation 5

An FGI procedure *should* be classified as a potentially-high radiation dose procedure if more than 5 % of cases of that procedure result in $K_{a,r}$ exceeding 3 Gy or P_{KA} exceeding 300 Gy cm^2.

This Report proposes that all equipment used for FGI procedures be assigned to one of two classes: (1) suitable for procedures that are potentially-high radiation dose or (2) not suitable for these procedures.

This can be done by considering both the characteristics of the equipment and the intended use for that fluoroscope in a particular setting. For fluoroscopic equipment with intended uses for FGI procedures, manufacturers may design their equipment to conform to IEC Standard 60601-2-43 and may then declare their equipment as compliant with this standard (IEC, 2000; 2010). Equipment conforming to this standard is preferable for performing potentially-high radiation dose procedures. The dose component of the decision can be made using the information provided in this Report, the published literature, or local experience. However, the assignment of a procedure to the potentially-high radiation dose category or the use of equipment intended for potentially-high radiation dose procedures does not justify the actual use of more radiation than is necessary for each individual patient.

Some procedures can be performed anywhere with simple imaging equipment and require minimal radiation doses (*e.g.*, peripherally-inserted central catheter). Procedures at the other extreme might have to be done in an appropriately equipped operating-room or an interventional room, using highly-specialized equipment. Some, but not all, procedures in the potentially-high radiation dose category require high radiation doses (*e.g.*, major vascular reconstruction). Most FGI procedures fall somewhere within this range (*e.g.*, neurointerventions and complex angioplasty). They usually require specialized spaces and equipment. Imaging-intensive procedures are likely to require high radiation doses. Potentially-high radiation dose procedures should be done using equipment specifically designed for such intended uses.

The purpose of a properly-designed state-of-the-art FGI-procedure suite is to provide an environment that allows effective and efficient patient care during FGI procedures. The imaging equipment should be designed, configured, and operated to provide high-quality images of seriously-ill patients with complex problems. The facility should provide a clean and sterile environment for infection control and satisfy the environmental demands of the imaging equipment. Appropriate ancillary equipment (*e.g.*, anesthesia machines, physiological monitoring equipment, pumps, and injectors) is required. Multiple staff members are needed to operate and manage the variety of equipment. The suite should provide all staff members adequate work space for their associated equipment and allow effective teamwork to ensure quality patient care and minimize hazards to the patient and staff.

Economics is always a matter of concern in the healthcare environment. However, safety and efficacy goals are best realized using facilities and equipment that are fully appropriate for the performance of specific procedures. Resources should always be available to optimize safety. In addition, budgetary planning should provide the necessary resources for the initial configuration of a procedure room and for appropriate upgrades to accommodate changes in procedure types and advances in technology. Alternatively, one medical center may develop a referral relationship with another fully equipped center for the treatment of patients requiring potentially-high radiation dose procedures.

3.1 Equipment Configurations

3.1.1 *General Considerations*

It is difficult to optimize a general purpose fluoroscope for each of a wide range of procedures. Even though equipment looks the same superficially, there are often major differences in the hardware components, software, and configuration settings needed for different applications. The use of equipment with inappropriate components or configurations for the planned procedure can be hazardous to the patient or operator. The equipment supplier's application specialists, the facility's qualified physicists, and the interventionalists should participate in matching equipment to procedures. For example, cardiac systems usually have a relatively small image receptor and are equipped with fluoroscopy and cinefluorography capabilities. Interventional radiology systems, on the other hand, have larger image receptors, to accommodate larger body areas during a procedure, and are typically equipped with fluoroscopy and digital-subtraction angiography (DSA) capabilities. Using a cardiac system for abdominal procedures may require additional runs to obtain anatomical coverage as compared to using an interventional radiology system. This can result in additional irradiation of patients and staff. On the other hand, the large image receptor typically found on an interventional radiology system may mechanically limit achievable views of a patient's heart. This might degrade the clinical performance of a cardiac procedure.

Most FGI and mobile C-arm systems are configured so that the x-ray tube is normally closer to the floor than the image receptor (*i.e.*, under-table systems). This is done so that the most intense scattered radiation is directed away from the operator's head (Section 5.4 and Figure 5.1). With an under-table configuration, the breast tissue of a supine patient is seldom irradiated by the entrance beam. Fluoroscopes are also available with the x-ray tube normally over the patient (*i.e.*, over-table systems). Additional caution for both patient and staff is warranted when these systems are used for FGI procedures.

As noted, this Report divides procedures into two categories based on the expected $K_{a,r}$ or P_{KA}. Fluoroscopy time should only be used if no other metric is available. This classification facilitates the overall tabulation of desirable equipment characteristics shown in Table 3.1. The recommended features of interventional

TABLE 3.1—*Examples of desirable equipment features.*[a,b]

Equipment Feature	Requirement for Potentially-High Radiation Dose Procedures[c]	Acceptable Characteristic for Other Procedures
Mechanical geometry	Isocentric	Procedure dependent
Fluoroscopic modes	Variable-rate pulsed	Continuous[d] or variable-rate pulsed
Added filtration	Automatic spectral filters	Fixed spectral filter suggested but not mandatory
Collimation adjustment	Virtual: Adjustable without radiation	Adjustable with live radiation
Anatomical programming	Configures image production and processing controls	Suggested, but not mandatory
Digital fluoroscopy	Required	Highly recommended
Digital acquisition of images	Required	Required
Stored fluoroscopy	Dynamic loops	Last image hold
Radiation monitors	$K_{a,r}$ meter and P_{KA} meter	$K_{a,r}$ meter or P_{KA} meter
Configurable radiation dose-level alerts (not currently available on FGI-procedure equipment)	Required	Procedure dependent

Table-mounted shielding	Required	Procedure dependent
Ceiling-suspended shielding	Required	Procedure dependent
Compliance	IEC 60601-2-43 (IEC, 2000; 2010) FDA (2009a)	FDA (2009a)

[a]This table supplies examples of key features; it is not intended to be a complete list of all desirable features.
[b]Additional equipment information is found in Appendix H.
[c]An FGI procedure *should* be classified as a potentially-high radiation dose procedure if more than 5 % of cases of that procedure result in $K_{a,r} > 3$ Gy or $P_{KA} > 300$ Gy cm^2
[d]See Section 3.1.1, second bullet (fluoroscopic modes).

equipment are listed for each radiation-dose category. Since patient procedural and radiation risks are not trivial, properly equipped fluoroscopy systems are required to permit proper dose management. A fluoroscope without the vast majority of recommended features listed in Table 3.1 for the anticipated risk level should not be used for that examination. Additional equipment requirements for different types of procedures are shown in Appendix H.

The following comments apply to the different features contained within Table 3.1. More information concerning each parameter may be found in Appendix H.

- *Mechanical geometry*: Gantries designed for FGI procedures rotate the x-ray tube and image receptor about a fixed point in space called the isocenter, the point in space where the patient anatomy of interest is placed during FGI procedures. Mobile C-arms and other types of fluoroscopy equipment may not have a true isocenter.
- *Fluoroscopic modes*: During fluoroscopy, radiation may be emitted continuously or in repetitive pulses. A choice of different pulse rates may be available to the operator. Variable-rate pulsed fluoroscopy is desirable for all FGI procedures. However, continuous fluoroscopy is acceptable for low-dose FGI procedures performed using existing equipment, but new equipment for FGI procedures other than peripheral procedures should provide variable-rate pulsed fluoroscopy.
- *Added filtration*: Automatic spectral filters provide different thicknesses of beam filters with an atomic number greater than that for aluminum. The filter thickness typically changes automatically as a function of the patient's thickness and density. Mobile C-arm equipment typically does not offer automatic spectral filters. Some fixed interventional equipment requires the operator to choose the filter thickness.
- *Collimation adjustment*: Collimators should be adjusted to always provide a radiation-free margin at the image receptor. Virtual collimation provides a graphical display of the position of the collimator blades on the clinical image while the collimator blades are adjusted. This feature eliminates patient irradiation during adjustment of the collimator.
- *Anatomical programming*: Dose rates should be set to the minimum practicable value. Management of radiation dose and image quality during an FGI procedure requires the control of a large number of image production and processing parameters as discussed in Section 3.1.1. Anatomical programming allows the control of these variables by selecting the region and size of the patient's body to be imaged.

- *Digital acquisition*: Image processing of nondigital images is severely limited compared to digital images. Film or analog video-based images are not adequate for FGI procedures. Analog fluoroscopy is marginally adequate for simple procedures.
- *Stored fluoroscopy*: Last-image-hold continuously displays the last fluoroscopic image frame after the foot pedal is released and also allows storage of this single image. The dynamic loop feature allows the operator to store up to 300 images of the most recent fluoroscopic-imaging sequence. This stored sequence may be replayed as a continuous loop.
- *Radiation monitors*: Real-time display of $K_{a,r}$ and P_{KA} to the interventionalist allows for evaluation of patient radiation risk during the procedure. IEC (2000) allows a deviation in displayed $K_{a,r}$ and P_{KA} values of as much as 50 %. FDA (2009a) limits uncertainty in $K_{a,r}$ to 35 %. The revised IEC (2010) standard limits both $K_{a,r}$ and P_{KA} values to a maximum deviation of 35 %. For example, the displayed value can deviate from the $K_{a,r}$ that is accurately physically measured at the reference point by ±35 %.
- *Configurable radiation dose-level alerts (not currently available on FGI-procedure equipment)*: The notification values and the substantial radiation dose levels (SRDLs) discussed in Section 4.3.3.2 and presented in Table 4.7 are suggested default values. A facility should have the ability to configure such alerts to conform to local requirements.
- *Table-mounted shielding*: These shields typically consist of lead drapes mounted to the rail of the patient table. The drapes extend to the floor. The typical lead-equivalent thickness is 0.5 mm. Mobile C-arm fluoroscopes typically do not provide radiation protective shields of any kind.
- *Ceiling-suspended shielding*: The typical lead-equivalent thickness is 0.5 mm. These shields typically are mounted on articulating arms attached to the ceiling of the procedure room.
- *Audible alerts*: As specified by the second edition of IEC Standard 60601-2-43 (IEC, 2010), an audible alert is to be activated in the procedure room whenever x rays are produced. Different sounds for different modes of operation are desirable.
- *Compliance*: FGI procedures should be performed on imaging equipment that is compliant with the listed standards.

Machine parameters that are set by the image production controls (*e.g.*, pulse rate, pulse width, x-ray tube current, x-ray tube voltage, filtration, and focal-spot size), as well as the type of image receptor and the patient's size, determine the incident air kerma for the patient, radiation dose at the image receptor, and image quality. Proper adjustment of the machine parameters contributes to optimization of x-ray production. This improves image information content and reduces radiation dose to the patient.

3.1.2 *Specific Equipment Components*

Dose management during FGI procedures is possible through the appropriate use of the basic features of fluoroscopy equipment and the intelligent use of dose-reducing technology (Mavrikou *et al.*, 2008; Wagner, 2007). This technology should be present and available in the equipment. For some procedures, additional, highly-specialized equipment may be helpful for dose management. For example, there is evidence that the use of robotic, magnetically-steered catheters can reduce radiation dose for certain cardiac electrophysiology (EP) procedures (Steven *et al.*, 2008).

Fluoroscopy equipment that incorporates modern dose-reduction and dose-measurement technology should be used for potentially-high radiation dose procedures (Miller *et al.*, 2002; NCI, 2005). This equipment should be compliant with IEC Standard 60601-2-43 (IEC, 2000; 2010), which provides requirements for basic safety and essential performance for equipment used for FGI procedures.

Some FGI procedures, such as venous access procedures, are almost always performed with minimum amounts of radiation. There is usually little radiation risk associated with these procedures (Storm *et al.*, 2006). These procedures yield patient doses high enough to be of concern only in rare outlier cases. For these kinds of procedures, extensive dose-reduction technology is not essential and fluoroscopy equipment that is not specifically intended for use in FGI procedures may be used safely. It is prudent to use equipment with dose-monitoring and dose-reduction technology for all FGI procedures. Noncompliant equipment should be replaced as soon as practicable. The long-term goal is to use the lowest practicable dose for all procedures.

Recommendation 6

Potentially-high radiation dose procedures *should* be performed using equipment designed for this intended use.

Certain procedures usually reserved for the treatment of serious or life-threatening conditions may result in an absorbed dose to the skin of the patient that places the patient at risk for deterministic skin effects. Monitoring $K_{a,r}$ takes priority over fluoroscopy time and P_{KA} for these procedures. $K_{a,r}$ monitoring capability is required on all fluoroscopes manufactured after mid-2006 that are sold in the United States. Fluoroscopic units purchased before mid-2006 may not have this capability. This dose surrogate monitors the air kerma at a point in space located at a defined, fixed distance from the focal spot, the reference point for $K_{a,r}$. When economically practicable, fluoroscopes routinely used for potentially-high radiation dose procedures *should* either be upgraded with add-on dose-monitoring equipment that monitors $K_{a,r}$ or replaced with a modern fluoroscope. Ideally, an automated, real-time map of the patient's estimated skin dose would be available during the procedure, along with an indication of the $D_{skin,max}$ value and its location on the patient's skin. Various methods to produce skin dose maps have been described in the literature (Chugh *et al.*, 2004; den Boer *et al.*, 2001; Morrell and Rogers, 2006). Development and implementation of these or alternative methods by fluoroscopy equipment manufacturers or third-parties for routine clinical use is encouraged.

Recommendation 7

If fluoroscopes are intended to be routinely used for procedures that have the potential for high patient doses (*i.e.*, $K_{a,r} > 3$ Gy), the units either *should* be equipped or upgraded with add-on dose-monitoring equipment that monitors $K_{a,r}$ or the units *should* be replaced with a modern machine.

The power supplies in older fluoroscopes usually made noise when x rays were produced. Often the noise varied depending on technical conditions such as pulse and dose rates. Modern systems are often installed in a manner that eliminates these sounds. Staff members working in FGI procedures should know when radiation is being produced in order to optimize their personal radiation protection. Lack of awareness can lead to inadvertent worker irradiation. Different sounds for fluoroscopy and for image acquisition should alert the interventionalist and staff that radiation is being generated and which mode is in use.

Staff members in the room who are unfamiliar with the operation and purpose of the fluoroscopy foot switch (*i.e.*, initiation and termination of x-ray production based on the position of the foot pedal) could inadvertently cause the machine to produce radiation by such actions as accidently stepping or standing on the foot switch, or rolling a cart or piece of equipment on the foot switch. For this reason, an additional safety switch should be provided on equipment control panels, located ideally both in the control booth and at table-side. When activated, the safety switch prevents the production of x rays. An indication should be provided on the control panel that clearly indicates when the safety switch is activated. The safety switch should not provide any other function. The safety switch is included in the specifications provided by IEC 60601-2-43 (IEC, 2000). This function can easily be added to older facilities by using the external door-interlock circuit with a wall-mounted switch.

Any staff whose duties require their presence in the procedure room should be trained on the use of this safety feature and be encouraged to use it. This includes such individuals as operators, nurses, anesthesiologists, technologists, housekeepers, and stock clerks.

3.1.3 *Specific Equipment Components for Pediatric Procedures*

The unique imaging challenges presented by the pediatric patient, discussed in more detail in Appendix I, require equipment and techniques that differ from adult patients. For example, higher frame rates are often needed to accommodate the more rapid heart rates of small children. Equipment with appropriate hardware features for pediatric applications differs from equipment used for adults. The appropriate hardware features and configuration of the equipment are a function of patient size. The smallest focal-spot size provided on typical dual focal-spot x-ray tubes is larger than necessary for small pediatric patients. A triple focal-spot x-ray tube with nominal sizes of 0.3, 0.6, and 1 mm provides more flexibility for pediatric applications. Many pediatric interventional fluoroscopes consist of two complete imaging planes that allow the acquisition of images in two planes simultaneously with one injection of contrast material. This is necessary due to the toxicity of iodine in very small pediatric patients (Appendix I). Finally, the image-receptor size may be larger than suggested by the corresponding adult procedure due to the differing pathophysiology of some pediatric disease states. For example, some pediatric heart diseases require imaging of the lungs as well as the heart. Other information is provided in Appendix I.

If specialized pediatric equipment is not available, adult equipment can usually be configured to permit a pediatric appropriate procedure (Sidhu *et al.*, 2009). Section 4.3.2 discusses examples of the parameters that should be configured to appropriately image pediatric patients. Appendix H discusses these parameters in further detail. Appendix H.9 discusses the software-controlled management of these parameters and provides additional information on appropriate configuration of interventional fluoroscopes for pediatric examinations.

Recommendation 8

Equipment that is routinely used for pediatric procedures *should* be appropriately designed, equipped and configured for this purpose.

The configuration and operation of the equipment when performing pediatric procedures is critically important. The goal of the Alliance for Radiation Safety in Pediatric Imaging (ARSPI, 2008), which now includes numerous U.S. and international professional societies, is to change practice by raising the awareness of opportunities to lower radiation dose during pediatric imaging. ARSPI attempts to provide straightforward information to every member of the medical care team.

The international Image Gently® education and awareness campaign of ARSPI (2008) presents information pertaining to the various imaging modalities using ionizing radiation including FGI procedures. The FGI-procedure campaign is appropriately named Step Lightly (ARSPI, 2009; Sidhu *et al.*, 2009). Information provided by ARSPI (2009) includes:

- a PowerPoint® training presentation on application of the ALARA principle during interventional imaging;
- a procedure check list of dose-reduction steps for review during each case;
- dose-reduction and quality-maintenance steps to take in the department;
- targeted guidelines to radiologists, technologists, and qualified physicists;
- publication lists; and
- brochures for parents.

3.2 Room Design and Structural Radiation Protection

Facility design is an essential component of radiation safety and management. When planning an interventional facility, careful attention to worker radiation exposure from the beginning of the planning process can help to reduce occupational exposure (NCRP, 2004). Ample procedure room size is advantageous because it allows ancillary staff space to safely maneuver and operate during the procedure and also affords them the opportunity to maximize their distance from the patient consistent with their duties. A large control room with a generously-sized window can serve as an observation area. This permits nonessential staff to remain in a protected area. This design feature is particularly useful in a teaching facility. An intercom between the control and procedure rooms can also allow some staff to remain outside of the procedure room until they are needed. Designs that permit easy and comfortable access to mobile and fixed shielding increase the likelihood that this shielding will be used.

When selecting fluoroscopy equipment, over-table x-ray tube configurations should be avoided because of the high scattered radiation levels directed at the head and neck of the interventionalist and the potential for excessive hand exposure. All imaging equipment should incorporate the patient dose-reduction features discussed in Section 3.1.2.

Lights or signs at the procedure room entrance that indicate that the x-ray equipment is in use or that the x-ray beam has been activated will alert staff that radiation is present and that precautions should be taken if room entry is required. Caution lights indicating the (potential) production of x rays are to be installed outside each entry into a procedure room. It is important to note that caution lights should be regularly tested and maintained in good working order if they are to be an effective part of a facility's regular routine.

All staff present in the procedure room need a clear indication when x rays are being produced so they may adequately protect themselves. Since some fluoroscopy equipment does not emit noise during the production of x rays and is not designed to produce an audible tone as described in Section 3.1.2, staff in the procedure room cannot rely on an audible indication of the production of x rays. The presence of static or dynamic patient images on the display monitors cannot be used as a visual indication of the production of x rays since patient images remain after x-ray production ceases. In many cases staff cannot see the image monitors. In some installations, the room lights can be automatically dimmed during the production of x rays, and may be switchable within a given room to address the preferences of different operators. For all these reasons, caution lights (amber) of universal design are required in the procedure room, in multiple locations if necessary, so that they are visible to all staff working anywhere in the room. These lights are continuously illuminated while x rays are produced.

The walls, floor, ceiling, doors and windows surrounding an FGI-procedure room should provide sufficient attenuation to reduce the radiation in outside spaces to the levels recommended by NCRP (2004) for controlled and uncontrolled areas. The NCRP (2004) recommendations apply to newly designed and remodeled existing facilities. Facilities designed before the publication of NCRP (2004) and meeting the requirements of NCRP Report No. 49 (NCRP, 1976) need not be reevaluated (NCRP, 1993). In controlled areas, where access is restricted to radiation workers, NCRP (1993) recommended an annual limit for E of 50 mSv y^{-1} with the cumulative limit not to exceed the product of 10 mSv and the radiation worker's age in years. That notwithstanding, NCRP (1993) recommended that for design of new facilities, the limit should be a fraction of the 10 mSv y^{-1} implied by the cumulative limit. NCRP (2004) recommends a fraction of one-half of that value, or 5 mSv y^{-1}. In uncontrolled areas, which may be occupied by the public, patients, visitors, or other workers, NCRP (2004) recommends a limit on E of 1 mSv y^{-1}. Typically, a control room for an x-ray-imaging facility is considered to be a controlled area. However, for an FGI-procedure facility, the control room is frequently used for procedure observation by staff who may not be radiation workers. Therefore, this Report recommends that control rooms for these facilities be considered uncontrolled areas. They should be designed with structural shielding to a limit of 1 mSv y^{-1}, if practicable. The design of such spaces should assume full occupancy by each individual who might be present in the space. This will eliminate the need for radiation monitoring for individuals who are not otherwise classified as radiation workers by the facility. Using the structural shielding method described in NCRP (2004), a busy cardiac angiography suite usually requires not more than a 1.5 mm lead-equivalent barrier to reduce scattered radiation levels below the shielding design goal of 1 mGy y^{-1} (air kerma) at a distance of 3 m from the center of the imaging field-of-view.

Recommendation 9

For newly designed and remodeled existing facilities, all spaces outside the procedure room (including control rooms) *should* be designed to limit E to not more than 1 mSv y^{-1}.

Whenever possible, personnel performing routine clinical monitoring of patients during an FGI procedure should be located in the control room. However, clinical requirements may necessitate that staff responsible for anesthesia, clinical monitoring of patients, or other duties be physically located in the FGI-procedure room during the procedure. In many such situations, it is practicable to provide either fixed or mobile protective barriers without interfering with patient care. These barriers should be designed to limit E for any individual to 1 mSv y^{-1} without the use of protective garments. The design of these barriers should be based on the maximum expected occupancy by any single individual in all locations behind protective barriers during the work week. This recommendation is slightly different than the usual occupancy factor discussed in NCRP Report No. 147 (NCRP, 2004). The occupancy factor for in-room barriers should be determined by the working pattern of the individual who spends the greatest fraction of time behind the barrier. Individuals whose job assignments are limited to working behind in-room barriers should be monitored both to ensure the adequacy of the barrier and to detect exposure when in the FGI-procedure room and not behind the designated in-room barrier.

Clinical situations exist where full protective barriers are not clinically practicable. Protective garments are required in such situations. However, any practical partial shielding should also be used.

Recommendation 10

For newly designed and remodeled existing facilities, spaces within the FGI-procedure room intended exclusively for routine clinical monitoring of patients (or similar activities) *should* be shielded to limit E to not more than 1 mSv y^{-1}.

Individuals working behind such barriers *shall* be monitored for radiation exposure.

Procedure room doors rarely also serve as mandatory control booth shielding. In some facilities, it is common to add procedure room door interlocks that prevent x-ray production when the door is opened. While this practice eliminates dose to personnel who inadvertently open the procedure room door while exposure is activated, the interruption of imaging during a critical portion of the procedure could be detrimental to the patient. Termination of an angiogram imaging sequence prior to completion may require the imaging sequence be repeated, with additional radiation exposure and contrast material. Interruption of fluoroscopic guidance during interventional-device positioning could result in a need for additional fluoroscopy time or incorrect placement. Incorrect placement may have severe consequences for the patient. Improper stent placement can lead to occlusion or narrowing of arteries, with stroke, myocardial infarction or kidney failure, depending on the artery being treated. Improper placement of occlusion devices of embolic agents can lead to blockage of other arteries than those intended, with similar severe consequences.

The dose rate at the procedure door during fluoroscopy will generally be <0.1 μGy s^{-1}. For this low exposure level, the potential patient harm that could occur does not justify the installation of door interlocks. However, these doors should be kept closed during procedures unless emergency access to the room is needed. Warning lights indicating the potential production of x rays are to be installed at entries to the procedure room.

Recommendation 11

Door interlocks that interrupt x-ray production *shall not* be permitted at any entrances to FGI-procedure rooms.

Figure 3.1 is a typical layout of an FGI-procedure room with its associated control area. The thickness of lead shielding required in each wall of the procedure room is a function of how much radiation is created within the procedure room and the degree of occupancy by staff or members of the general public in the surrounding areas adjacent to the procedure room. In Figure 3.1, since the barrier on the right is an exterior wall, it may require reduced shielding if only momentary pedestrian traffic occurs in the adjacent space. If this exterior space is fully occupied (*e.g.*, an adjacent office), shielding requirements would not be reduced. The barrier along the corridor requires reduced levels of shielding since only momentary occupancy in this area occurs by a single individual. The area labeled administrative offices is treated as though it were occupied all the time by individuals who are not radiation workers (*i.e.*, an uncontrolled area). The control room can often be designed to limit full occupancy to <1 mSv y^{-1} (public dose limit) at reasonable cost. This simplifies the management of the many nonradiation workers who may occupy this space for variable amounts of time.

The areas at the location of the doors leading to the staff corridor and leading to the control room from the FGI-procedure room illustrate the presence of single-scattered radiation from the patient (darker-shaded beams from isocenter) into these areas when the doors are open. While the dose rate during fluoroscopy in these areas with the doors open (generally <0.1 μGy s^{-1}) does not justify door interlocks, the door leading to the staff corridor must remain closed whenever possible. Figure 3.1 is a typical design. In the control room, the lightly-shaded rectangular area labeled step-down

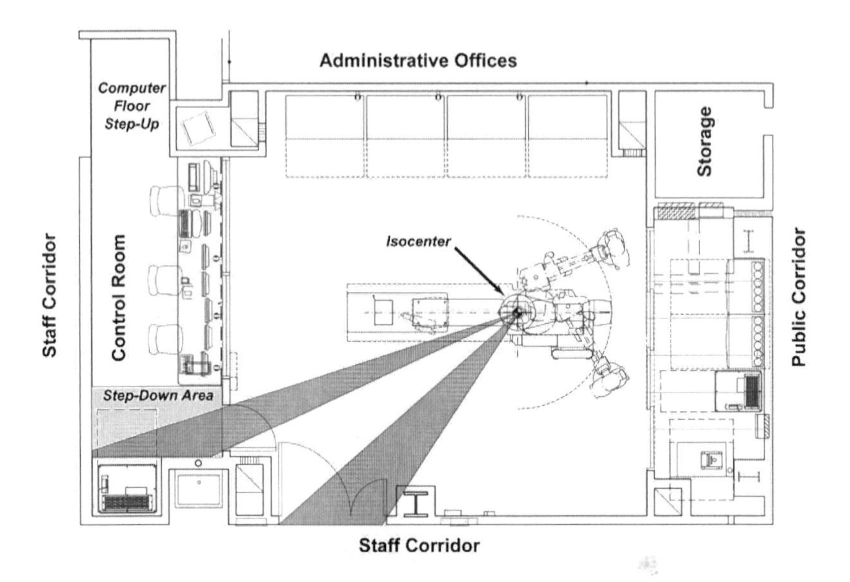

Fig. 3.1. Radiation protection aspects of a typical FGI-procedure room.

area is not part of the computer floor. Portions of this area receive single-scattered x rays from the patient (darker-shaded beam from isocenter). Typically the entire step-down area is declared an area in which staff should wear protective garments.

4. Protection of the Patient

Fluoroscopy is used to obtain clinical information about the patient and to provide necessary guidance for interventions. Because x rays are used, the interventionalist has the responsibility of balancing patient radiation dose against the expected clinical benefits of the procedure. This responsibility can only be fulfilled optimally if the interventionalist understands how the equipment's operation affects clinical image quality and emitted radiation intensity, the meaning of available real-time dosimetric displays, and the radiobiological consequences of irradiation.

This body of knowledge empowers the interventionalist to predict the likely health effects of the irradiation and to modify the procedure where safe and feasible. While other staff members present during a procedure have important roles in patient care and safety, the interventionalist is ultimately responsible for the patient's well being.

4.1 Patient Dose Review

There are substantial differences in the statistical nature of the reported data describing FGI diagnostic and therapeutic procedures, and conventional radiographic diagnostic-imaging procedures. The typical Gaussian statistical descriptors used to derive diagnostic reference levels (DRLs) for conventional radiographic procedures are seldom applicable to FGI procedures (Miller *et al.*, 2009). Many FGI diagnostic procedures and most FGI therapeutic procedures are better represented by a lognormal distribution where the data are positively skewed (Table 4.1) (Dauer *et al.*, 2009). The shape of such a distribution is remarkably constant regardless of the type of procedure or the dose metric used to describe it (Miller *et al.*, 2003a; 2003b; Storm *et al.*, 2006).

Patient dose depends on numerous factors, including the interventionalist's knowledge, skill and experience; the type of procedure; the location of the lesion; the complexity of the procedure; and the indication for the procedure (Bor *et al.*, 2008; Miller *et al.*, 2003a). The complexity of a procedure is affected by factors related to the patient's anatomy and to the location, anatomy and severity of the lesion being treated. The effect of procedure complexity on dose is well-established (Peterzol *et al.*, 2005; Vehmas, 1997). Increased

TABLE 4.1—*Distribution of $D_{skin,max}$ for all FGI procedures performed in 2006 for a patient population with cancer (Dauer et al., 2009).*

$D_{skin,max}$ (Gy)	Frequency	Percent	Cumulative Percent
<0.01	1,381	34.7	34.7
0.01 – 0.1	1,520	38.2	73.0
0.1 – 0.2	368	9.3	82.2
0.2 – 0.3	215	5.4	87.6
0.3 – 0.4	109	2.7	90.4
0.4 – 0.5	57	1.4	91.8
0.5 – 0.6	54	1.4	93.2
0.6 – 0.7	34	0.9	94.0
0.7 – 0.8	23	0.6	94.6
0.8 – 0.9	13	0.3	94.9
0.9 – 1.0	19	0.5	95.4
1 – 2	104	2.6	98.0
2 – 3	44	1.1	99.1
3 – 4	15	0.4	99.5
4 – 5	10	0.3	99.7
5 – 6	4	0.1	99.8
6 – 7	1	0.0	99.9
7 – 8	4	0.1	100.0
8 – 9	1	0.0	100.0

complexity results in increased patient dose (Balter *et al.*, 2008). As the complexity of these procedures has increased, radiation doses to patients and healthcare personnel have also increased.

These factors and others should be considered when analyzing procedures. One should not pass judgment on the appropriateness of the radiation dose associated with an FGI procedure without a careful review of all aspects of the procedure. A large patient who presents with multiple problems, or even one problem that is difficult to correct, may require a high radiation dose even if a highly-skilled interventionalist performs the procedure using optimized equipment. On the other hand, a much lower patient dose may be higher than necessary if a normal size patient undergoes a routine intervention performed by an individual with marginal skills or performed with inadequate or poorly maintained equipment.

There is substantial available literature reporting patient radiation doses from FGI procedures. Relatively large data sets are available for diagnostic coronary angiograms and for a range of interventional cardiology procedures (den Boer *et al.*, 2001; Leung and Martin, 1996; McFadden *et al.*, 2002; Padovani and Quai, 2005; Park *et al.*, 1996; Rosenthal *et al.*, 1998; Stisova, 2004). Most of the remaining literature has been published about procedures performed by radiologists (Dauer *et al.*, 2009; Miller *et al.*, 2003a; 2003b; Tsalafoutas *et al.*, 2006; Vano *et al.*, 2008a; 2009). This literature is characterized by a fairly large number of studies comprising relatively small series of patients, because many of these procedures are performed relatively infrequently, even at major medical centers (Andrews and Brown, 2000; Bergeron *et al.*, 1994; Gkanatsios *et al.*, 2002; Livingstone and Mammen, 2005; Marshall *et al.*, 1995; McParland, 1998a; Nikolic *et al.*, 2000; Ruiz-Cruces *et al.*, 1997; 1998; Theodorakou and Horrocks, 2003; Williams, 1997; Zweers *et al.*, 1998). Relatively little data exist for the same kinds of procedures performed by other physicians (*e.g.*, surgeons, gastroenterologists, urologists, orthopedic surgeons) (Buls *et al.*, 2002; Lipsitz *et al.*, 2000; Perisinakis *et al.*, 2004; Tsalafoutas *et al.*, 2008).

4.1.1 *Cardiac Procedures*

There are wide variations in patient dose for cardiac interventions. Most of the FGI-procedure dose literature has focused on cardiac procedures. The highest dose procedures are percutaneous transluminal coronary angioplasty (PTCA), a form of PCI, and RF cardiac ablation (performed for treatment of cardiac dysrhythmias). In a 2005 review, P_{KA} values ranged from 14 to 116 Gy cm^2 for PTCA in several series comprising 1,208 patients and from 95 to 257 Gy cm^2 for RF ablation procedures in several series comprising more than 960 patients (Padovani and Quai, 2005). Chida *et al.* (2006) reported a mean P_{KA} of 149 Gy cm^2 for PTCA (172 patients) and 110 Gy cm^2 for RF ablation (28 patients). Note that the mean P_{KA} for PTCA in Chida *et al.* (2006) is outside the range in Padovani and Quai (2005).

In Padovani and Quai (2005), a $D_{skin,max}$ of 1.8 Gy was reported for PTCA and mean values of $D_{skin,max}$ for RF ablation of 1.5 to 1.8 Gy were reported. Trianni *et al.* (2005) demonstrated a $D_{skin,max}$ of 3.4 Gy for PTCA and lower values of $D_{skin,max}$ for RF ablation. For extremely complex PTCA procedures in patients with chronic occlusions of the coronary arteries, Suzuki *et al.* (2006) observed a median $D_{skin,max}$ of 4.6 Gy; one patient received a $D_{skin,max}$ of 9.7 Gy.

In a study of 322 patients undergoing either diagnostic coronary angiography (134 patients) or PTCA (188 patients), den Boer *et al.* (2001) observed that 13 % (42/322) received a $D_{skin,max} > 2$ Gy, and 1 % (4/322) received a $D_{skin,max} > 4$ Gy. Rosenthal *et al.* (1998) observed a $D_{skin,max} > 2$ Gy in 22 % of 859 RF cardiac ablation procedures; the mean estimated $D_{skin,max}$ was 1.3 Gy. Six of the 624 adult patients (1 %) in this series received a $D_{skin,max} > 7$ Gy. In a series of 500 patients undergoing RF cardiac ablation, Park *et al.* (1996) found that 28 patients (5.6 %) received a $D_{skin,max} > 2$ Gy. In McFadden *et al.* (2002), of 50 patients undergoing RF cardiac ablation, 6 (12 %) received a $D_{skin,max} > 2$ Gy.

The International Atomic Energy Agency (IAEA) has organized several multi-national evaluations of radiation doses associated with FGI procedures. The dosimetric and demographic data collected in one of these studies are reproduced in Table 4.2 (Balter *et al.*, 2008; IAEA, 2009). The institutions participating in this study collected somewhat different data sets. The total number of procedures in each category (coronary angiography, Table 4.2a; coronary angioplasty, Table 4.2b; PCI, Table 4.2c) is shown in the header of each subtable. All institutions contributed cases in all three subcategories. The number of individual cases used to derive each distribution is indicated in the body of each subtable.

Note that the demographic information (for age, height, weight) demonstrates an essentially normal distribution. However, all of the radiation metrics and the data for contrast material use do not. The radiation metrics and contrast material volume more closely approximate a lognormal distribution. Table 4.1 illustrates this pattern, using a data set obtained outside the IAEA (2009) study.

4.1.2 *Other Interventional Procedures*

Published data have been extensively tabulated and summarized in recent publications (Blaszak *et al.*, 2009; Bleeser *et al.*, 2008; Brambilla *et al.*, 2004; Dauer *et al.*, 2009; Miller *et al.*, 2003a; 2003b; Tsalafoutas *et al.*, 2006; Vano *et al.*, 2008a; 2009; Verdun *et al.*, 2005). The wide variety of FGI procedures makes it difficult to provide generalized dose data. In one publication, 21 separate procedures were studied, as well as subtypes of these procedures categorized by lesion etiology and location, for a total of 35 procedure categories; this was not considered a comprehensive list (Miller *et al.*, 2003a). These subtypes sometimes demonstrated substantial differences in dose. For example, the mean P_{KA} value for nephrostomy was 26 Gy cm^2 when performed for relief of urinary obstruction, but 45 Gy cm^2 when performed for treatment of

TABLE 4.2—*Summary statistics for the IAEA (2009) multi-national evaluations of radiation doses associated with FGI procedures performed on adult patients.*

TABLE 4.2a—*Summary statistics for coronary angiography (diagnostic procedure only) (N = 2,265).*[a]

	N[b]	10 %[c]	25 %[c]	50 %[c]	75 %[c]	95 %[c]
Age (y)	1,133	47	56	65	73	80
Height (cm)	1,135	155	162	168	175	183
Weight (kg)	1,136	58	68	77	88	108
Fluoroscopy time (min)	1,826	2.1	3.1	5.2	8.7	19.1
P_{KA} (Gy cm^2)	2,265	13.8	20.8	31.8	49.4	96.6
$K_{a,r}$ (Gy)	830	0.3	0.5	0.7	1.0	1.7
Contrast material (mL)	449	60	90	115	155	260

TABLE 4.2b—*Summary statistics for coronary angioplasty (interventional procedure only) (N = 1,027).*

	N[b]	10 %[c]	25 %[c]	50 %[c]	75 %[c]	95 %[c]
Age (y)	851	49	56	64	71	80
Height (cm)	848	155	163	169	175	183
Weight (kg)	848	61	68	76	85	103
Fluoroscopy time (min)	916	4.5	6.8	11.7	19.8	40.6
P_{KA} (Gy cm^2)	1,027	18.8	29.9	53.3	98.4	201.1
$K_{a,r}$ (Gy)	200	0.6	1.1	1.9	3.0	6.4
Contrast material (mL)	170	100	155	250	390	695

TABLE 4.2c—*Summary statistics for PCI (combined coronary angiography and PTCA) (N = 817).*

	N^b	10 %[c]	25 %[c]	50 %[c]	75 %[c]	95 %[c]
Age (y)	502	49	56	65	73	80
Height (cm)	504	155	163	170	178	183
Weight (kg)	504	60	70	82	91	114
Fluoroscopy time (min)	728	7.5	10.3	14.7	23.6	42.1
P_{KA} (Gy cm^2)	817	36.6	59.1	92.9	138.3	260.9
$K_{a,r}$ (Gy)	556	0.9	1.3	1.9	2.7	4.9
Contrast material (mL)	397	170	219	290	380	567

[a]Total number of procedures in category.
[b]Number of individual cases used to derive each distribution.
[c]Percent of procedures equal to or less than the column entry.

kidney stone disease (Miller *et al.*, 2003a). Even for a single type of procedure at a single medical center there can be an extraordinarily large dose range; the ratio of maximum to minimum dose (P_{KA} or $K_{a,r}$) often exceeds 100, and may exceed 1,000 (Dauer *et al.*, 2009; Tsalafoutas *et al.*, 2006). This variability is due primarily to patient factors and procedure complexity.

Some FGI procedures, such as venous access procedures, are essentially always low dose (Storm *et al.*, 2006); that is, the 95th percentile $K_{a,r} < 3$ Gy. Others, particularly neuroembolization procedures, generally require high radiation doses, both as defined in ICRP Publication 85 (ICRP, 2000a) and in this Report. Some procedures yield patient doses high enough to be of concern only in rare outlier cases, and some procedures typically result in patient doses high enough to be of concern. TIPS creation, all embolization procedures, and angioplasty of arteries in the abdomen and pelvis fit within the latter group (Miller *et al.*, 2003a).

Some specific examples illustrate the opposite ends of the dose spectrum. Cerebral embolization, an FGI procedure, is typically performed for the treatment of life-threatening diseases [intracranial aneurysms, arteriovenous malformations (AVMs), or tumors]. Without question, this is a potentially-high radiation dose procedure. In a series of 356 patients, the mean P_{KA} value was 320 Gy cm^2 and the mean $K_{a,r}$ value was 3.8 Gy (Miller *et al.*, 2003b). The mean $D_{skin,max}$ was 2 Gy, 17 % of patients had a $D_{skin,max} > 3$ Gy, and 4 % of patients had a $D_{skin,max} > 5$ Gy. The highest $D_{skin,max}$ observed was 6.7 Gy (Miller *et al.*, 2003b). On the other hand, placement of a chest port for venous access is a relatively low-dose FGI procedure. In the Storm *et al.* (2006) series of 303 chest port placements, the median P_{KA} value was 3.7 Gy cm^2. In the same series, median $D_{skin,max}$ was 0.02 Gy, and the highest $D_{skin,max}$ observed was 0.76 Gy. Further, most patients who undergo this procedure have a life expectancy lower than that of the general population, because they are being treated for cancer.

Doses vary widely, both among procedures and for a given procedure (Table 4.3). Patient radiation doses for FGI procedures demonstrate lognormal distribution similar to the example in Table 4.1.

4.1.3 *Pediatric Interventional Procedures*

While the general comments in Section 4.1 concerning factors that affect patient radiation dose during adult FGI procedures also generally apply to similar pediatric FGI procedures, the importance of the parameters that affect patient dose differs to some degree. The small size of the pediatric body may either increase or decrease the radiation dose of a given procedure. The small size reduces the patient dose rate on properly configured equipment, but the increased complexity in imaging and gaining access into small regions of anatomy of the newborn may result in longer fluoroscopy times or the recording of more images due to greater frame rates. The variety of diseases presented in pediatric patients with cardiac anomalies as opposed to adult patients with routine coronary artery disease can significantly increase the complexity of the pediatric procedure. For example, routine coronary artery interventions in adults typically require <2 h to perform. The time needed to complete typical cardiac interventions in children averages 3 to 5 h.

Table 4.4 lists patient age, height, mass, fluoroscopy time, P_{KA}, and $K_{a,r}$ (gray) for three pediatric cardiac procedures (*i.e.*, interventional catheterization, EP study, and diagnostic catheterization). The fluoroscopy time, P_{KA}, and $K_{a,r}$ are listed for each imaging plane for procedures performed with biplane imaging equipment. Tables 4.4a, 4.4e, and 4.4f) presents data for a pediatric patient mass of 30 to 40 kg. For the most common procedure, interventional catheterization, the data are presented for four ranges of patient mass: 2 to 4, 4 to 10, 30 to 40, and 70 to 85 kg (Tables 4.4a, 4.4b, 4.4c, and 4.4d, respectively), in order to indicate the effect of patient size on patient dose. Since all these procedures are performed with biplane capability, data are presented for both imaging planes. All the data in Table 4.4 were collected from calibrated air-kerma and P_{KA} displays of the interventional equipment within the Catheterization Laboratory at Children's Hospital Boston from 2005 to 2009. The estimated values of P_{KA} and $K_{a,r}$ for adult-sized pediatric patients with congenital heart disease (Table 4.4d) are more than an order of magnitude higher (at the 50th percentile) than those for the smallest pediatric patients (Table 4.4a), while the fluoroscopy times are similar. All the other comparisons given in this paragraph are also at the 50th percentile. The frontal plane during an interventional catheterization (Table 4.4c) delivers similar P_{KA} and $K_{a,r}$ values as the lateral plane of an EP study (Table 4.4e). The frontal plane in an EP study (Table 4.4e) delivers a $K_{a,r}$ value that is 30 % of the $K_{a,r}$ value of the lateral plane. The lateral plane during an interventional catheterization (Table 4.4c) delivers about twice the P_{KA} and $K_{a,r}$ of the frontal plane. Another comparison suggests that the estimated $K_{a,r}$ during a diagnostic catheterization (Table 4.4f) is approximately half the $K_{a,r}$ during an interventional catheterization (Table 4.4c).

Table 4.5 lists patient mass and $K_{a,r}$ values (milligray) for pediatric noncardiac interventional procedures performed in different regions of the patient's body (*e.g.*, abdomen, thorax and head). The pediatric noncardiac procedures performed on each region of

TABLE 4.3—*Patient radiation dose distribution data for selected noncoronary procedures expressed as $K_{a,r}$ (gray) (adapted from Miller et al., 2009; Storm et al., 2006).*

Procedure	Number of Patients	Percentile[a]				
		10	25	50	75	95
TIPS creation	134	0.5	1.0	1.5	2.6	5.6
Biliary drainage	123	0.1	0.3	0.5	1.2	3.2
Nephrostomy for obstruction	76	0.0	0.1	0.1	0.4	0.8
Nephrostomy for stone access	62	0.1	0.2	0.3	0.6	2.4
Pulmonary angiogram	104	0.1	0.2	0.3	0.5	0.7
Inferior vena cava filter placement	274	0.0	0.1	0.1	0.2	0.4
Renal / visceral angioplasty without stent	53	0.4	0.5	0.7	1.6	3.8
Renal / visceral angioplasty with stent	103	0.6	0.9	1.2	2.0	4.0
Iliac angioplasty without stent	24	0.4	0.6	0.9	1.2	1.4
Iliac angioplasty with stent	93	0.5	0.7	1.1	1.6	3.4
Bronchial artery embolization	27	0.4	0.6	0.9	1.5	2.4
Hepatic chemoembolization	125	0.4	0.7	1.2	1.8	3.6

Uterine fibroid embolization	90	0.9	1.4	2.0	3.3	5.7
Other tumor embolization	88	0.5	0.8	1.1	2.1	4.2
Gastrointestinal hemorrhage localization / treatment	94	0.7	1.2	1.9	3.2	5.6
Embolization-head-AVM	134	1.6	2.4	3.6	5.4	9.2
Embolization-head-aneurysm	148	1.9	2.7	3.5	4.4	6.8
Embolization-head-tumor	51	2.0	2.5	3.4	5.0	7.5
Vertebroplasty	98	0.3	0.6	1.0	1.7	3.0
Pelvic artery embolization trauma / tumor	35	0.8	1.1	1.5	2.2	3.7
Embolization-spine-AVM / tumor	21	3.0	4.0	5.6	7.3	9.3
Peripherally-inserted central-catheter placement	480			0.02		
Chest-port placement	295			0.05		
Tunneled dialysis-catheter placement	66			0.04		

[a]Percentile for patients with $K_{a,r}$ values (gray) equal to or less than the column entry.

TABLE 4.4—*Summary statistics for the evaluations of radiation doses associated with FGI procedures performed on pediatric patients at Children's Hospital Boston (2005 to 2009).*

TABLE 4.4a—*Summary statistics for pediatric interventional catheterization (N = 358).*[a]
2 to 4 kg Patients: Catheterization Laboratory.

	N[b]	Mean	Standard Deviation	10 %[c]	25 %[c]	50 %[c]	75 %[c]	90 %[c]
Age (y)	358	0.11	0.16	0	0	0	0.2	0.4
Height (cm)	256	49.3	5	44	47	50	51	54
Weight (kg)	358	3.2	0.5	2.4	2.8	3.3	3.5	3.7
Frontal plane								
Fluoroscopy time (min)	197	24.6	17.9	7.1	12.4	19.7	31.8	50.6
P_{KA} (Gy cm^2)	358	7.8	6.8	1.6	2.8	5.7	10.8	16.4
$K_{a,r}$ (Gy)	358	0.11	0.1	0.024	0.04	0.08	0.15	0.23
Lateral plane								
Fluoroscopy time (min)	194	12.9	11.7	2.5	4.8	8.6	17.2	27.4
P_{KA} (Gy cm^2)	355	7.5	6.4	1.4	3.0	5.6	10	15.5
$K_{a,r}$ (Gy)	355	0.16	0.14	0.032	0.063	0.12	0.21	0.36

TABLE 4.4b—*Summary statistics for pediatric interventional catheterization (N = 900). 4 to 10 kg Patients: Catheterization Laboratory.*

	N^b	Mean	Standard Deviation	10 %[c]	25 %[c]	50 %[c]	75 %[c]	90 %[c]
Age (y)	900	0.83	0.7	0.2	0.4	0.6	1.1	1.7
Height (cm)	842	66	9	54	59	65	73	79
Weight (kg)	900	6.8	1.8	4.4	5.2	6.6	8.4	9.3
Frontal plane								
Fluoroscopy time (min)	485	29.9	20.9	8.4	14.2	25	41	62
P_{KA} (Gy cm^2)	900	16.5	15.4	3.0	5.8	11.9	22	36
$K_{a,r}$ (Gy)	900	0.21	0.2	0.038	0.071	0.15	0.29	0.46
Lateral plane								
Fluoroscopy time (min)	465	12.8	11	2.6	4.6	9.4	18	28
P_{KA} (Gy cm^2)	881	13.7	12.6	2.6	5	10	17.6	30.3
$K_{a,r}$ (Gy)	881	0.27	0.26	0.05	0.093	0.19	0.36	0.62

TABLE 4.4c—*Summary statistics for pediatric interventional catheterization (N = 195). 30 to 40 kg Patients: Catheterization Laboratory.*

	N[b]	Mean	Standard Deviation	10 %[c]	25 %[c]	50 %[c]	75 %[c]	90 %[c]
Age (y)	195	12	3.2	8.6	10	11.6	13.2	15.5
Height (cm)	191	138	22	129	135	141	147	152
Weight (kg)	195	34.8	3.1	30.6	32.2	35	37.4	39.2
Frontal plane								
Fluoroscopy time (min)	126	22.6	21.2	3.7	8.6	18.2	27	46
P_{KA} (Gy cm^2)	195	41	42	5	12.1	28	55	93
$K_{a,r}$ (Gy)	195	0.42	0.50	0.036	0.11	0.25	0.51	0.98
Lateral plane								
Fluoroscopy time (min)	110	11.6	13.2	0.8	3.3	7.7	17.6	25.7
P_{KA} (Gy cm^2)	180	50	51.6	7.1	18	34	62	107
$K_{a,r}$ (Gy)	180	0.75	0.81	0.1	0.22	0.50	1.04	1.58

TABLE 4.4d—*Summary statistics for pediatric interventional catheterization (N = 208). 70 to 85 kg Patients: Catheterization Laboratory.*

	N[b]	Mean	Standard Deviation	10 %[c]	25 %[c]	50 %[c]	75 %[c]	90 %[c]
Age (y)	208	30.3	15.6	15.6	17.9	24.2	40.2	56.8
Height (cm)	208	170	9.6	159	163	170	177	183
Weight (kg)	208	76.7	4.3	71.4	73	76.2	80.4	83.4
Frontal plane								
Fluoroscopy time (min)	127	25.6	19.3	6.3	12.8	22	33	50
P_{KA} (Gy cm^2)	208	109	111	17	40	79	140	229
$K_{a,r}$ (Gy)	208	0.92	0.92	0.16	0.32	0.62	1.2	2.2
Lateral plane								
Fluoroscopy time (min)	113	12.7	11.1	1.3	4.3	10.6	17.1	28
P_{KA} (Gy cm^2)	191	137	121	28	48	105	187	321
$K_{a,r}$ (Gy)	191	1.9	1.7	0.35	0.65	1.37	2.58	4.0

TABLE 4.4e—*Summary statistics for pediatric interventional EP (N = 104). 30 to 40 kg Patients: Electrophysiology Suite.*

	N[b]	Mean	Standard Deviation	10 %[c]	25 %[c]	50 %[c]	75 %[c]	90 %[c]
Age (y)	104	10.9	4.2	7.8	8.9	10.5	12	13.5
Height (cm)	104	136	25	130	135	141	145	150
Weight (kg)	104	35.1	3.2	30.5	32	35.3	38	39.5
Frontal plane								
Fluoroscopy time (min)	42	15.9	12.7	5.5	7.9	12	18.1	38
P_{KA} (Gy cm^2)	104	10.5	11.8	2.3	3.7	7.1	12.5	22.3
$K_{a,r}$ (Gy)	104	0.106	0.104	0.023	0.04	0.075	0.13	0.22
Lateral plane								
Fluoroscopy time (min)	40	11.9	9.2	2.6	6.1	8.9	14.8	25.5
P_{KA} (Gy cm^2)	99	22.4	20.2	4.4	8.9	16.2	29.4	49.1
$K_{a,r}$ (Gy)	99	0.38	0.34	0.065	0.15	0.26	0.49	0.92

TABLE 4.4f—*Summary statistics for pediatric diagnostic catheterization (N = 29). 30 to 40 kg Patients: Catheterization Laboratory.*

	N[b]	Mean	Standard Deviation	10 %[c]	25 %[c]	50 %[c]	75 %[c]	90 %[c]
Age (y)	29	14.5	4	9.7	10.9	14.4	16.7	20.4
Height (cm)	29	144	8.6	134	139	143	148	159
Weight (kg)	29	36.2	2.7	32.6	33.9	37	38.2	39.6
Frontal plane								
Fluoroscopy time (min)	29	12.2	6.6	4.9	6.5	11.8	16.9	19.1
P_{KA} (Gy cm^2)	29	17.7	15.8	3.2	6.2	11.7	25.2	33.9
$K_{a,r}$ (Gy)	29	0.15	0.14	0.022	0.038	0.080	0.21	0.34
Lateral plane								
Fluoroscopy time (min)	7	2.9	1.3	1	2.1	3.1	3.4	4.5
P_{KA} (Gy cm^2)	7	17.5	14.4	2.5	8.2	14.9	25.5	40.9
$K_{a,r}$ (Gy)	7	0.26	0.21	0.037	0.13	0.23	0.32	0.61

[a]Total number of procedures in category.
[b]Number of individual cases used to derive each distribution.
[c]Percent of procedures equal to or less than the column entry.

TABLE 4.5—*Patient dose data for selected noncoronary pediatric procedures expressed as $K_{a,r}$ (milligray)*[a] *at the Radiology Department of Children's Hospital Boston (2005 to 2009).*

Procedure	Mass (kg)	Number of Patients	Percentiles[b]				
			10	25	50	75	95
Abdominal	3 – 9	180	0.2	0.4	2	6	30
	10 – 18	252	0.2	0.3	0.6	3	20
	20 – 30	183	0.2	0.4	1	6	30
	50 – 70	253	0.8	2	9	22	110
Thorax	3 – 9	189	0.2	0.4	0.9	2	6
	10 – 18	148	0.3	0.4	1	2	38
	20 – 30	75	0.3	0.4	1	3	25
	50 – 70	137	0.3	0.5	1	5	35
Skull							
Frontal	3 – 9	29	0.4	1	23	170	370
Lateral	3 – 9	24	0.2	3	34	134	440

Frontal	10 – 18	70	1	14	21	75	450
Lateral	10 – 18	55	2	10	23	137	354
Frontal	20 – 30	77	5	15	29	69	343
Lateral	20 – 30	68	4	11	24	47	288
Frontal	50 – 70	81	14	28	63	385	905
Lateral	50 – 70	74	11	20	44	183	483

[a]1,000 mGy = 1 Gy.
[b]Percentile for patients with $K_{a,r}$ values (milligray) equal to or less than the column entry.

the body consist of a wider variety of imaging tasks than those for the pediatric cardiac procedures and that wider variety affects the range of delivered radiation dose. The data are presented at four patient masses: 3 to 9, 10 to 18, 20 to 30, and 50 to 70 kg. Since head procedures are performed with biplane capability, data are presented for both imaging planes. All the data in Table 4.5 were collected from calibrated air-kerma displays of the interventional equipment within the Radiology Department at Children's Hospital Boston from 2005 to 2009.

The estimated $K_{a,r}$ values of the interventional procedures performed in the Radiology Department (Table 4.5) in general are orders of magnitude less than the $K_{a,r}$ values during a pediatric cardiac catheterization (Table 4.4a, 4.4b, 4.4c, and 4.4d). The higher $K_{a,r}$ values listed for the head procedures (95th percentile) include cerebral embolizations, and involve estimated $K_{a,r}$ values approaching 1,000 mGy (1 Gy). The highest recorded values exceed 3,000 mGy (3 Gy).

4.2 Patient Radiation Dose-Management Program

Many factors affect patient radiation risk. These may be conveniently divided into demographic, medical history, radiation history, pregnancy, and type of procedure. The first two of these were discussed in Sections 2.6.1 and 2.6.2. The remaining three factors are discussed here.

4.2.1 *Radiation History*

Previous radiation doses to the same area of skin that will be irradiated for the planned interventional procedure can increase the risk of deterministic skin effects, depending on the radiation dose from previous procedures and the time interval between previous procedures and the planned procedure (Balter *et al.*, 2010). Similarly, if the planned interventional procedure uses the same radiation field as the skin entrance portal to be used for future radiation therapy, there may be an increased risk of deterministic skin effects from the radiation therapy. Therefore, substantial skin doses from FGI procedures should be considered in the radiation-therapy treatment planning process.

4.2.2 *Pregnancy*

Pregnancy requires special factors due to the risk for radiation-induced cancer in the unborn child and deterministic effects in the embryo and fetus at absorbed doses to the embryo and fetus above 50 mGy.

The pregnancy status of a prospective FGI-procedure patient should be determined. If the patient is pregnant, in consultation with a qualified physicist, the dose to the embryo and fetus should be assessed prospectively to determine the potential risk and an appropriate benefit-risk evaluation made. It may be possible to plan the procedure so that little or no direct exposure to the embryo and fetus occurs. When this is clinically practicable, the unborn child will be exposed to mostly scattered radiation which often results in very low levels of dose. The dose to the embryo and fetus can be minimized by using low-dose fluoroscopy mode, a nominal number of fluorographic images, beam collimation and other dose-reduction methods detailed in Section 4.2.4.2.

Recommendation 12

Procedure planning for FGI procedures on pregnant patients *shall* include feasible modifications to minimize dose to the embryo and fetus.

As discussed in Section 2.6.5, irradiation of the embryo and fetus may also produce deterministic effects. Absorbed doses to the embryo and fetus in excess of 50 mGy might lead to undetectable changes in developmental status, and doses in excess of 100 mGy can result in subtle to obvious changes in development, depending on dose. In most complex interventions in the abdomen or pelvis a potential for direct exposure to the embryo and fetus exists (ACR, 2008a).

4.2.3 *Type of Procedure*

As discussed in Section 3, some types of interventional procedures are known to be in the potentially-high radiation dose category. These procedures have resulted in skin doses that produced deterministic effects in average patients (Section 2.6.4). Some representative procedures are listed in Table 4.6, but this table is not intended to be a complete or comprehensive list. Many of the procedures listed in Tables 2.1 and 2.2 may also be potentially-high radiation dose procedures on occasion.

Potentially-high radiation dose procedures are more likely to result in skin injury if the radiation dose is high and there is also relatively little motion of the entrance radiation field on the patient's skin during the procedure. Procedures such as TIPS creation, chemoembolization, PCI, and complex placement of cardiac EP devices often require that the radiation beam remain at the same entrance-skin site for much of the procedure.

TABLE 4.6—*Examples of interventional procedures with the potential for high skin doses.*

- TIPS creation
- embolization (any location, any lesion)
- stroke therapy
- biliary drainage
- angioplasty with or without stent placement
- stent-graft placement
- chemoembolization
- angiography and intervention for gastrointestinal hemorrhage
- carotid stent placement
- RF cardiac ablation
- complex placement of cardiac EP devices
- PCI (single or multiple vessel)

Other potentially-high radiation dose procedures utilize either single-plane fluoroscopy and imaging with multiple beam entrance sites or biplane fluoroscopy and imaging, but are still associated with skin injuries because overall radiation doses tend to be high. Examples of these kinds of procedures include embolization of abnormalities in the brain and interventional treatment of strokes.

It is the interventionalist's responsibility to be aware of beam motion during every potentially-high radiation dose procedure. Almost any FGI procedure, when performed on certain patients, can be technically difficult or unusually prolonged and may result in skin doses sufficient to produce deterministic effects. This can be due to lesion anatomy or characteristics or to patient anatomy or characteristics. Regardless of the cause, technically difficult and prolonged procedures often result in higher than expected radiation doses.

Different types of interventional procedures place different portions of the skin in the path of the primary beam. Sensitivity to radiation effects depends to some extent on the location of the irradiated skin (Hymes *et al.*, 2006). The scalp, for example, is relatively resistant to the development of skin damage, but scalp hair is relatively sensitive to epilation as compared to hair elsewhere on the body (Geleijns and Wondergem, 2005).

4.2.4 *General Principles of Radiation Dose Management*

Fluoroscopy equipment is described in Section 3.1, which includes a discussion of the various controls and features that affect radiation production, image generation and manipulation, and radiation dose. Radiation quantities are discussed in detail in Appendix D and briefly reviewed in this section.

4.2.4.1 *Patient Dose Monitoring.* All fluoroscopy equipment provides one or more of the following dose surrogates for dose monitoring. Although none of these metrics measures patient risk directly, they are intended to give the interventionalist and staff enough of a real-time sense of radiation use to allow an ongoing benefit-risk evaluation. The same metrics are often used as dose data for a facility's quality management program:

- fluoroscopy time;
- number of fluorographic (cinefluorography or serial radiographic) frames;
- air kerma-area product (P_{KA}); and
- air kerma at the reference point ($K_{a,r}$).

The clinical utility of each of these metrics varies for different classes of procedures and for different patient populations.

Fluoroscopy time and P_{KA} are relevant for procedures where stochastic risks are of concern, such as patients whose life expectancy after the procedure is likely a decade or more. For all procedures, fluoroscopy time and P_{KA} are useful as quality assurance tools for assessing interventionalist performance. Fluoroscopy time monitors the efficiency of an interventionalist in completing a procedure. P_{KA} is useful for monitoring how carefully an interventionalist controls radiation use.

Fluoroscopy time should not be used as the only dose surrogate for procedures that might require intermediate or high radiation usage. Equipment with better dose-monitoring capability should be used for such procedures. Fluoroscopy time does not supply any information about dose rate at the patient's skin, field size or location, or the dose contribution of fluorographic modes (cinefluorography, DSA). Risk estimates can be somewhat improved if the number of fluorographic frames produced during the procedure is known. However, information about dose rate and field size is still unavailable.

Recommendation 13

Fluoroscopy time *should not* be used as the only dose indicator during potentially-high radiation dose FGI procedures. All available dose indicators *shall* be used in such procedures.

P_{KA} is a measure of the total x-ray energy absorbed by the patient. It is therefore an indicator of stochastic risk from FGI procedures, and is widely used in Europe for comparing doses among procedures and institutions.

P_{KA} is not a useful indicator for deterministic effects. It correlates poorly with $D_{skin,max}$ for individual patient procedures (Fletcher *et al.*, 2002; Miller *et al.*, 2003b). The value for P_{KA} when a small area receives a high radiation dose may be the same as the value for P_{KA} when a large area receives a low radiation dose.

$K_{a,r}$ is the cumulative air kerma at a point in space located at a fixed distance from the focal spot. As such, it is related to the incident air kerma ($K_{a,i}$) for the patient. However, the readout cannot be assumed to represent $K_{a,i}$ because the relationship depends on the clinical procedure and how the physician maneuvers the x-ray equipment during the case. Nevertheless, $K_{a,r}$ is a useful practicable metric for clinical patient management.

4.2.4.2 *Optimizing Patient Dose.* Dose optimization is possible through the appropriate use of the basic features of fluoroscopy equipment and the intelligent use of dose-reducing technology (Mavrikou *et al.*, 2008). Many features can be adjusted during the procedure to reduce radiation use or to improve image quality, depending on the demands of the situation (Wagner, 2007). Interventionalist skill and experience are also important factors in optimizing dose (Bor *et al.*, 2008; Verdun *et al.*, 2005).

Dose optimization requires attention to several basic principles, all of which have been discussed in detail in several excellent reviews (Koenig *et al.*, 2001a; Mahesh, 2001; Miller *et al.*, 2010a; Wagner *et al.*, 2000) and are summarized here. These include minimizing fluoroscopy time, minimizing the number of images obtained, and minimizing dose through control of technical factors (NCI, 2005).

Minimizing fluoroscopy time is the direct responsibility of the interventionalist. Fluoroscopy time can be minimized with the judicious use of intermittent fluoroscopy, last image hold, and, where available, "radiation-free" collimator settings. Fluoroscopy should only be used to observe motion or to guide devices within the body. Last-image-hold images and fluoroscopy loops should be used for intraprocedural review purposes, as they do not subject the patient to additional radiation during the review.

Fluoroscopy should never be used unless the interventionalist is looking at the fluoroscopy monitor. By definition, fluoroscopic images are not intended to be saved. Some newer systems offer retrospective storage of the last few seconds of fluoroscopy. This may eliminate the need for a subsequent fluorography run or permit looped replays. Generally, if fluoroscopic images are not observed when they are being produced, the associated radiation affects both patient and staff without any clinical benefit. Use of fluoroscopy when the operator is not looking at the fluoroscopy monitor is a known issue (Figure 4.1). No definitive data on this behavior are available. However, based on anecdotal reports, the frequency of such behavior ranges from negligible to >10 % of the total irradiation time.

In Figure 4.1, the patient's head (out of view) is located on the operator's left-hand side in each photograph. The camera is at the same height as the fluoroscopy monitor and positioned just to the head side of the monitor. The two photographs were taken a few minutes apart. Fluoroscopy is indicated by the illuminated x-ray warning light in the ceiling (above the head of the operator). The operator and assistant are watching the fluoroscopy monitor in the left-hand photograph. Even though radiation is being produced when the right-hand photograph was taken, the operator is looking at the display associated with an intravascular ultrasound device, not at the fluoroscopy monitor. The patient and staff are receiving radiation, but there is no clinical benefit from the radiation.

Minimizing the number of images obtained during a procedure requires awareness and planning. Modern fluoroscopy equipment with DSA capability can easily be set to acquire images at two or more images per second, and to perform the entire angiographic series at that rate. This is neither necessary nor desirable. Imaging

Fig. 4.1. Example of fluoroscopy with and without the interventionalist looking at the fluoroscopy monitor. The patient's head is on operator's left (reader's right) (see Section 4.2.4.2 for a description).

sequences with variable frame rates minimize the number of images obtained while assuring that no important information is lost. Fluoroscopy equipment with DSA capability can and should be preprogrammed with the same imaging sequences used previously with radiographic-film changers (film hard copy) (Miller *et al.*, 2002). If the only purpose of an image is to document what is seen on the last-image-hold, there is no need to perform an additional imaging run. Instead, the last-image-hold fluoroscopic image should be stored if it demonstrates the finding.

Dose can also be optimized through control of technical factors. Some of these are under the direct control of the interventionalist, and can be optimized with any fluoroscopy equipment. Optimizing technical factors includes maximizing source-to-skin distance (SSD), minimizing the air gap between the patient and the image receptor, grid removal (where appropriate), and limiting the use of electronic magnification. Incremental changes in operational parameters are multiplicative and markedly affect total dose delivered to a patient's skin. For long procedures, differences in doses of 8 Gy or more are possible for some combinations of operational techniques (Nickoloff *et al.*, 2007; Wagner *et al.*, 2000).

The assistance of a qualified physicist may be required to establish other technical factors, including beam filtration, fluoroscopic kilovolt peak, and fluoroscopic and digital imaging dose settings. However, these technical factors should not be changed to reduce dose to the point that it impairs clinical performance. Images that are inadequate for diagnosis and for guiding interventions introduce the risk of catastrophic and avoidable complications.

If dose-saving pulsed fluoroscopy is available, the interventionalist should employ it whenever possible. More than all other operational factors, the use of pulsed fluoroscopy has the greatest potential for reducing patient radiation dose (Wagner *et al.*, 2000). It is important to note that pulsed fluoroscopy can be accomplished with different methods, some of which do not reduce the dose rate. Some pulsed fluoroscopy modes actually yield a higher dose rate than conventional fluoroscopy (Miller *et al.*, 2002). It may be necessary to check with the manufacturer of the fluoroscopy equipment or to have a qualified physicist measure the dose rate for each pulsed fluoroscopy mode.

Outdated equipment should be replaced with new fluoroscopy equipment that incorporates current dose-reduction and dose-measurement technology (Miller *et al.*, 2002; NCI, 2005). This equipment should be compliant with IEC Standard 60601-2-43 (IEC, 2010), which provides requirements for basic safety and essential performance for fluoroscopy equipment used for FGI procedures.

Patient protection should not be sacrificed in the interest of economy. Purchasers and users of fluoroscopy equipment should ensure that all individuals who make the purchasing decision for new fluoroscopy equipment are aware of the importance of dose-optimization and dose-measurement technology. Some dose-optimization technology may be an extra-cost option. The reduction in patient radiation dose can be considerable, and the additional cost of dose-reduction technology is justified.

4.2.4.3 *Managing Peak Skin Dose*. Measures that reduce total radiation dose will also reduce $D_{\mathrm{skin,max}}$. Two simple, basic techniques are also available which are intended specifically to reduce $D_{\mathrm{skin,max}}$ (Miller *et al.*, 2002). The first technique is to change the position of the radiation field on the patient's skin using gantry angulation, table movement, or both. The second technique is to reduce the size of the radiation field using collimation. The purpose of these techniques is to reduce the maximal dose to any location on the skin. Spreading the skin dose over a larger area accomplishes two things. First, it reduces $D_{\mathrm{skin,max}}$. Second, it reduces the size of the skin area subjected to $D_{\mathrm{skin,max}}$.

Collimation of the irradiated field is as important as dose spreading. Even with the use of dose-spreading techniques, different irradiated fields can overlap on the skin surface. The overlap area receives a higher dose. Optimal collimation may prevent overlap, especially with biplane fluoroscopy equipment, and markedly improves the effectiveness of dose-spreading techniques (Miller *et al.*, 2002).

For cerebral procedures, special consideration should be given to minimizing dose to the patient's eyes. In the frontal projection, the x-ray tube should be positioned under the table so that the eyes are not in the entrance field. In the lateral projection, the beam size and location should be adjusted so that the lenses are outside of the imaging field.

It is not always possible to keep $D_{\mathrm{skin,max}}$ below the threshold for skin effects (Table 2.5, Section 2.6.4.1). Patient factors, anatomic variations, disease complexity, and the type of procedure may combine so that a prolonged procedure with a high radiation dose is unavoidable. This is not necessarily a contraindication to performing or continuing a procedure. It also does not necessarily indicate poor technique on the part of the interventionalist. As with all of medicine, it is necessary to consider all of the benefits and risks of the FGI procedure as well as all of the benefits and risks of alternative therapies, if any are available.

4.3 Sample Patient Radiation Dose-Management Process

Radiation dose management requires a comprehensive program including preprocedural planning, fluoroscopic set-up and configuration, intraprocedural management, and post-procedural care.

4.3.1 *Preprocedure Planning*

Radiation risk to the patient should be considered along with other procedural risks. Relevant demographic (Section 2.6.1), medical (Section 2.6.2), and procedural (Section 2.4.1) risk factors should be considered as part of the preprocedure patient evaluation. Pregnancy considerations are discussed in Sections 2.6.5 and 4.2.2. The patient's radiation exposure history, including a review of anatomical areas that have received or are planned to receive radiation therapy, should also be considered when planning the clinical approach to the current procedure (Section 4.2.1). This implies that the facility's medical information system provides timely access to relevant information.

If there is a history of previous radiation to the same area of skin that will be irradiated for the planned interventional procedure, the skin at the planned radiation entrance site should be examined for possible radiation changes. If these are present, modification of the procedure may be desirable, if this can be done without undue risk to the patient.

If radiation risk factors are discovered during the preprocedure evaluation, or if a potentially-high radiation dose procedure is to be performed, relevant radiation risks should also be discussed with the patient as part of the informed consent process (NCI, 2005). Informed consent is more than just a signed document; it is an active process between the physician and patient. Language in the informed consent form should not be the only means of communicating radiation risk to the patient. Sample language for a radiation risk discussion is given in Appendix J. Documentation of the contents of the radiation risk discussion, and that the patient understood these risks, should be placed in the patient's medical record, along with documentation of the other risks discussed (Stecker *et al.*, 2009). This implies that the individual seeking consent is knowledgeable about radiation risks and has the ability to communicate this information to the patient (Balter *et al.*, 2009).

4.3.2 *Fluoroscopic Set-Up and Configuration*

The fluoroscopy system should be checked for proper patient identification, configuration, and adequate image storage space prior to starting a procedure. Where possible, this should be done before placing the patient on the table. The fluoroscopy system should be initially configured to provide the lowest dose rate to the patient consistent with the image-quality requirements of the procedure. It is appropriate to verify these settings as part of the preprocedure time-out. The Joint Commission specifies the activities that are required as part of a mandatory "time-out" before surgical procedures (TJC, 2010). The time-out is applicable to FGI procedures (Angle *et al.*, 2008; Knight *et al.*, 2006).

The fluoroscopy equipment usually provides different operational configuration settings. These settings provide good image quality at properly managed patient doses that are a function of procedure type and the patient sizes examined within the clinical practice (especially pediatric patients). As part of a safety culture, during the preprocedure time-out process, the operator should verify that the appropriate size-based configuration has been selected for the patient. For the example of pediatric imaging, important configuration settings as a function of patient size include:

- reduced exposure pulse durations;
- lowest pulse rates suitable for the patient's heart rate;
- reduced tube currents that maintain x-ray tube voltages above an appropriate minimum;
- reduced focal-spot sizes;
- selection of appropriate beam filter thicknesses (when not automatically selected by the fluoroscopy equipment);
- removal of the anti-scatter grid when imaging small children (*i.e.*, <20 kg or <15 cm path length for x rays) or small body parts such as extremities (*i.e.*, <15 cm path length for x rays) of larger children or adults;
- when possible, larger fields of view (decreased magnification) with beam area limited appropriately using manual collimation;
- appropriate choice of operator selectable patient radiation exposure levels; and
- image-processing parameters.

A more detailed discussion of interventional fluoroscopes is presented in Appendix H.

4.3.3 *Intraprocedural Management*

Conceptually, the clinical management of radiation dose is similar to the clinical management of iodinated contrast media (Balter and Moses, 2007). The interventionalist should control the use and monitor the radiation dose continually during the procedure. As more radiation is used, attempts should be made to minimize further administration of radiation consistent with clinical requirements. As with iodinated contrast material, some patients will be more susceptible to harmful effects at a given radiation dose, as a result of the various sensitizing factors noted in Section 2.4.2. For these patients, special care should be taken to control radiation dose. If this is not possible, the benefit-risk ratio for the procedure should be reevaluated with consideration of this additional risk.

When an FGI procedure is performed on a pregnant patient, one goal is to manage the dose to the embryo and fetus to a practicable minimum for the procedure. Some procedures can be performed with little or no direct exposure to the embryo and fetus. If radiation exposure to the embryo and fetus is limited to scattered radiation, the result is often a very low dose (ACR, 2008a).

4.3.3.1 *Radiation Dose Management.* It is the responsibility of the interventionalist to properly manage radiation dose. A number of radiation dose-management techniques can be found in national guidelines and other literature (ACR, 2008b; Hirshfeld *et al.*, 2004; NCI, 2005; Wagner, 2007):

- radiation dose to the patient should be limited to that required for the procedure being performed;
- appropriate collimation should be used for the imaging task to reduce the size of the irradiated area when possible;
- patient should be positioned as close as reasonably possible to the image receptor;
- distance between the patient and the x-ray tube should be maximized to the extent practicable;
- each arm of the patient should be kept outside the radiation field unless an arm is intentionally imaged as part of the procedure;
- electronic magnification modes and high-dose-rate modes should be used only when necessary;
- lowest dose rate that is clinically acceptable should be used at all times;
- when electronic magnification is necessary, the lowest acceptable magnification factor should be used;
- fluoroscopy should be used sparingly and only when real-time imaging guidance is needed;
- last-image-hold feature or loop replay should be used;
- image acquisition should be activated only when higher-quality image review is essential, and it should be limited to

the frame rate and run duration necessary to accomplish the immediate task; and

- in some cases, retrospectively-stored fluoroscopy may reduce the need for image acquisition.

All personnel participating in the procedure share a responsibility for achieving radiation dose management and safety goals. Personnel should be able to recognize and correct unsafe practices or bring them to the attention of other personnel who can correct the situation (ACR, 2008b).

4.3.3.2 *Monitoring Patient Radiation Dose*. Radiation data should be available to the interventionalist during the course of the interventional procedure. Fluoroscopy equipment that cannot estimate and display $K_{a,r}$ or P_{KA} in real time should not be used for potentially-high radiation dose procedures. Radiation dose is monitored throughout the procedure (Stecker *et al.*, 2009). The purpose of dose monitoring is to ensure that the interventionalist is aware of how much radiation has been administered. It is the responsibility of the interventionalist to be informed about dose levels and to include radiation dose in the continuous benefit-risk balance used to determine the value of continuing a procedure (Stecker *et al.*, 2009).

Recommendation 14

Interventionalists *shall* be responsible for patient radiation levels during FGI procedures and *shall* ensure that radiation dose accumulation is continuously monitored during the procedure.

It is common for the interventionalist to concentrate on the clinical requirements of the interventional procedure and lose awareness of the patient's radiation dose. Designation of a specific person to monitor dose and to appropriately inform the interventionalist in a timely manner prevents this from occurring. This may be a technologist, nurse, or another individual.

For purposes of monitoring and post-procedure management, it is useful to consider the concept of a *substantial*[5] *radiation dose level* (SRDL) (Section 4.3.4.2) (Balter and Moses, 2007; Stecker *et al.*, 2009). As defined by Stecker *et al.* (2009), this is an appropriately-selected reference value used to trigger additional dose-management

[5]This was called "significant" in the referenced publications. The term has been changed to eliminate any confusion with statistical significance.

actions. There is no implication that a radiation level below an SRDL is absolutely safe or that a radiation level above an SRDL will always cause an injury. The numerical value for an SRDL will vary depending on both patient and procedural parameters. The interventionalist is periodically advised of the patient radiation level when it exceeds certain specified values and at regular intervals thereafter. It is useful to choose these values so that they are simple round numbers and so that three notifications, regardless of the dose metric used, indicate that an SRDL has been reached and patient follow-up is necessary (Stecker *et al.*, 2009). Suggested quantities and their SRDLs are shown in Table 4.7 (Stecker *et al.*, 2009).

The suggested values in Table 4.7 are based on Table 2.5 and are applicable to the normal range of patient radiosensitivities in the absence of aggravating physical or clinical factors. Additional information is presented in Appendices F.1 and F.2. ACR (2008b) takes a more conservative approach and recommends 3 Gy as the SRDL for $K_{a,r}$, but also allows different SRDL values when supported by the published literature.

The notification values and SRDLs presented in Table 4.7 should be reduced if known sensitizing factors are present. Specific attention should be given when high levels of radiation have been used in previous FGI procedures and when there is planned or prospective radiation therapy in the same anatomical region.

Procedures performed using biplane fluoroscopy systems are a special situation. The dose received from each plane should be considered independently when the fields do not overlap. When they do overlap, the doses are additive. If it is uncertain whether the fields overlap, it should be assumed that they do. Similarly, when

TABLE 4.7–*Suggested values for first and subsequent notifications and the SRDL.*

Dose Metric	First Notification	Subsequent Notifications (increments)	SRDL
$D_{skin,max}$	2 Gy	0.5 Gy	3 Gy
$K_{a,r}$	3 Gy	1 Gy	5 Gy[a]
P_{KA}	300 Gy cm^2 [b]	100 Gy cm^2 [b]	500 Gy cm^2 [b]
Fluoroscopy time	30 min	15 min	60 min

[a]See additional discussion concerning the value 5 Gy in Section 4.3.4.2.

[b]Assuming a 100 cm^2 field at the patient's skin. For other field sizes, the P_{KA} values should be adjusted proportionally to the actual procedural field size (*e.g.*, for a field size of 50 cm^2, the SRDL value for P_{KA} would be 250 Gy cm^2).

multiple fields are utilized for a procedure performed with a single-plane fluoroscopy system, the dose delivered to each field can be considered independently when field overlap does not occur.

Currently, skin dose estimates are not generally available in real time during the procedure. Most institutions in the United States can use $K_{a,r}$ for clinical dose monitoring. The quantity P_{KA} may also be used if $K_{a,r}$ is not available. Fluoroscopy time correlates poorly with other dose metrics (Fletcher *et al.*, 2002; Miller *et al.*, 2003b; 2004). In general, fluoroscopy time should be used with caution to monitor patient irradiation during interventional procedures (Stecker *et al.*, 2009). Automated dose documentation (Section 4.3.4.1) has the potential to track the entry field for each irradiation and thus support real-time skin-dose maps. At the present time, several research projects are directed toward this aim (Chugh *et al.*, 2004; den Boer *et al.*, 2001; Morrell and Rogers, 2006).

4.3.3.3 *Use of Radiation-Level Information During the Procedure.* When given a radiation-level notification, the interventionalist should consider the amount of radiation already received by the patient, the additional amount of radiation necessary to complete the procedure, and all relevant clinical factors in the continuing benefit-risk evaluation. A procedure should never be stopped just because an SRDL has been exceeded (Balter and Moses, 2007). It is understood that it is unlikely that a procedure will be stopped purely because of radiation dose concerns, as the clinical benefit of a successful procedure almost always exceeds any detriment to the patient due to radiation. Further, if the procedure is stopped before the desired clinical result has been reached, radiation risk has been incurred without any corresponding clinical benefit. The interventionalist is responsible for weighing the benefits and risks of proceeding, versus terminating, when an SRDL has been exceeded.

Any use of radiation should be clinically justifiable to the same extent as the use of contrast material and drugs. Failure to record radiation usage in the clinical report (Section 4.3.4.1) in the same way that the use of other agents is recorded might be considered a lack of justification of radiation use.

4.3.4 *Post-Procedure Management*

4.3.4.1 *Dose Documentation.* In 1994 and 1995, FDA recommended that information permitting estimation of the absorbed dose to the skin be recorded in the patient's medical record (FDA, 1994; 1995a). ICRP has also recommended recording patient radiation dose in the medical record for certain procedures (ICRP, 2000a). Monitoring and recording patient dose data for all procedures can be valuable for quality assurance purposes as well as for patient safety. Feedback to the interventionalist may help to optimize radiation doses overall (Vehmas, 1997). It is generally agreed that recording radiation dose in the medical record for all FGI procedures is essential and is the responsibility of the interventionalist (Hirshfeld *et al.*, 2004; Miller *et al.*, 2004; NCI, 2005).

Better means for documenting radiation use are beginning to appear. Digital Imaging and Communications in Medicine (DICOM, 2005) introduced the Radiation Dose Structured Report (RDSR). IEC (2007) provides a publically-available specification defining the contents of an RDSR for FGI equipment and IEC requires use of the specification in its latest FGI-equipment standard (IEC, 2010). RDSR provides dose and geometric information for each individual irradiation. The Radiation Exposure Monitoring Integration Profile from Integrating the Healthcare Enterprise (IHE, 2008) provides the tools needed for implementation of this technology. The Radiation Exposure Monitoring Integration Profile was successfully tested for interoperability in 2009. Continuing development of the Integration Profile, its baseline standards, and first commercial implementations occurred in 2010. This format is intended to support real-time skin-dose mapping.

RDSR combined with appropriate mathematical phantoms (Bolch *et al.*, 2009; Gu *et al.*, 2008; Lee *et al.*, 2010) should supply sufficient inputs to give modeling algorithms the ability to calculate skin and organ doses. Reports of such modeling are expected by 2011.

Recommendation 15

Patient dose data *shall* be recorded in the patient's medical record at the conclusion of each procedure. This *shall* include all of the following that are available from the system: $D_{skin,max}$, $K_{a,r}$, P_{KA}, fluoroscopy time, and number of fluorographic images.

Dosimetric information should be recorded in the patient's medical record as soon as practicable after the completion of the procedure. All available data should be recorded in those situations where complete data are unavailable. When deterministic effects are of concern, the most desirable dose estimate is $D_{skin,max}$. Ideally, this would include a skin dose map as a means for managing patient radiation dose. In the absence of skin dose estimates, $K_{a,r}$ is an acceptable substitute, but it does not correlate well with $D_{skin,max}$ in individual cases (Miller *et al.*, 2003b). Fluoroscopy time does not correlate with $D_{skin,max}$ (Fletcher *et al.*, 2002). Monitoring fluoroscopy time alone underestimates the risk of radiation-induced skin effects (O'Dea *et al.*, 1999). Fluoroscopy time and number of fluorographic images, used together, can provide a better guide to patient dose but are not themselves measures of dose. They do not provide sufficient information for dose calculations and are therefore suboptimal dose metrics. However, if none of the other metrics can be measured, fluoroscopy time and the number of fluorographic images, along with the patient's height and weight, can be used for recording patient irradiation until dosimetric means are available. It is unacceptable to record only fluoroscopy time if better estimators of radiation dose are available.

Radiation reactions in the lens of the eye have associated latent periods ranging from years to decades. Lens doses delivered during procedures where the eyes are in the useful x-ray beam for more than a few moments may be high enough to induce long-term effects. These procedures include many interventional procedures on the brain and face (Miller *et al.*, 2010a). Documentation of radiation levels will contribute to increased understanding of dose-time responses.

In those cases where the patient is pregnant, knowledge of the dose to the embryo and fetus may be of importance in managing the patient or the unborn child.

4.3.4.2 *Substantial Radiation Dose Level.* Suggested values for SRDLs are given in Table 4.7. These values are based on the dose conversion equations determined for a variety of noncardiac FGI procedures (Miller *et al.*, 2003b) and on the relationships between skin dose and skin effects described in Section 2.6.4. They are slightly less conservative than the highly conservative recommendations of ACR (2008b) and the National Cancer Institute (NCI, 2005), which recommend an SRDL (using the quantity $K_{a,r}$) of 3 Gy.

The values used in this Report are those recommended by the Society of Interventional Radiology and the Cardiovascular and Interventional Radiology Society of Europe (Stecker *et al.*, 2009) and are intended to trigger follow-up for a radiation level that might produce a clinically-relevant injury in an average patient. The purpose of a follow-up examination is to detect skin effects that may require further management or prolonged follow-up. A minor skin reaction, such as transient erythema, does not normally require follow-up and will not be detectable two to four weeks post-procedure, at the usual time that a patient is asked to perform self-examination. A trigger level lower than 5 Gy may be appropriate for patients undergoing a procedure where a substantial proportion of the delivered dose is likely to be directed to a single area of skin, such as TIPS creation, hepatic chemoembolization, and some neuroembolization procedures (Miller *et al.*, 2003b).

An SRDL should be adjusted if necessary for those types of FGI procedures where there is little or no beam movement and the reference point is near or inside the patient's skin. A qualified physicist can advise a facility on this matter. For example, where there is no beam motion and the reference point is at the entrance-skin surface, then $D_{skin,max} = D_{skin,e} \simeq 1.3\ K_{a.r}$, where $D_{skin,e}$ is the entrance-skin absorbed dose and the constant 1.3 accounts for backscattered radiation and other factors (*i.e.*, attenuation of the table top, ratio of the mass attenuation coefficients of skin to air) (Appendix D.1 and Figure D.3).

An SRDL should be low enough so that major tissue reactions are unlikely below this level. This is of increased importance when previously-irradiated skin is involved. The numerical value(s) chosen will depend on the nature of the procedure, patient sensitivity, and the patient's radiation history. The values used in ACR (2008b) and NCI (2005) guidelines are intended to prompt follow-up for a level that might result in even a minor reaction in average patients. These values may be used instead for all patients, according to local preferences (ACR, 2008b). It is important to remember that an SRDL in the context of these reports is simply a selected value that is used to trigger additional dose-management actions, and has no other meaning.

The Joint Commission recommends that doses from previous procedures to the same body area be summed over a 6 to 12 month period (TJC, 2006). Other recommendations include a follow-up period of six months (Miller *et al.*, 2010a). Biologically, the repair process in the skin is complete at approximately two months (Balter *et al.*, 2010). A six-month period is a reasonable compromise. This recommendation requires access to pertinent parts of the patient's previous radiation history prior to initiating the new procedure. Summation of dose over time increasingly overestimates the likelihood of skin reaction as the interval between procedures increases. Six months was selected as a working value based on a typical two-month period for cellular repopulation in the skin plus a very large empirical margin (see Appendix F.2 for the discussion of cellular kinetics). This is an appropriate default clinical recommendation in the absence of detailed patient-level radiobiological data.

It should also be noted that fluoroscopy/fluorography with a skin dose >15 Gy to a single skin field over a period of six months to 1 y, although extremely uncommon, is considered a reviewable sentinel event by The Joint Commission (TJC, 2006; 2007). Note that this definition of a sentinel event refers to skin dose. However actual skin dose is rarely measured during routine clinical procedures (Dauer *et al.*, 2009; Struelens *et al.*, 2005). A sentinel event does not always occur if $K_{a,r}$ for a patient exceeds 15 Gy during the integration period. Evaluation is made by reconstructing the skin dose from the procedure(s) using the best available information on such parameters as the orientations of the x-ray beam on the patient's skin surface and SSD for each beam port. Available direct skin dose measurements (Chida *et al.*, 2009b; Fletcher *et al.*, 2002; Rampado and Ropolo, 2005; Suzuki *et al.*, 2006; Tsapaki *et al.*, 2008; Vano *et al.*, 2001) are of value in this respect.

The Joint Commission has defined a sentinel event as an "unexpected" outcome (TJC, 2007), with the implication that a reviewable radiation overdose is "preventable" (TJC, 2006). In some circumstances, in order to achieve a life-preserving outcome, a planned intervention or series of interventions may require a sufficient dose of radiation to reach The Joint Commission's threshold for a sentinel event. This is especially the case if the patient has had multiple FGI procedures or radiation therapy in the recent past, with radiation delivered to the same area of skin (Balter and Miller, 2007). Examples of situations where it may be necessary to exceed a 15 Gy skin dose for multiple procedures include, among others, multiple PCIs for severe coronary artery disease and multiple neuroembolization procedures for the treatment of complex AVMs or multiple intracranial aneurysms.

4.3.4.3 *Interventionalist Responsibilities.* The interventionalist ensures that all available radiation metrics ($D_{skin,max}$, $K_{a,r}$, P_{KA}, fluoroscopy time and number of frames) are recorded in the medical record (Hirshfeld *et al.*, 2004; Miller *et al.*, 2004; NCI, 2005). In addition, the interventionalist writes an appropriate note in the patient's medical record if any of the values in the last column of Table 4.7 is exceeded, signifying that an SRDL has been exceeded and indicating the reason (Balter and Moses, 2007; Hirshfeld *et al.*, 2004; Stecker *et al.*, 2009). Notation in the medical record may also be appropriate even if these values are not exceeded, such as for patients on whom other procedures involving radiation exposure are planned or have already been performed within six months.

Recommendation 16

If a substantial radiation dose level (SRDL) (Table 4.7 and Section 4.3.4.2) is exceeded while performing an FGI procedure, the interventionalist *shall* place a note in the medical record, immediately after completing the procedure, that justifies the radiation dose level used.

4.3.4.4 *Patient Follow-Up*. Patients whose radiation level exceeded an SRDL (Table 4.7) should be clinically followed after the procedure (Hirshfeld *et al.*, 2004; NCI, 2005; Stecker *et al.*, 2009). Arrangements for radiation follow-up should be made before the patient leaves the facility. It may be desirable to perform follow-up for lower radiation doses in special situations, such as previous recent irradiation of the same anatomical region. If any SRDLs in Table 4.7 are exceeded during the performance of a procedure, the dose for additional procedures performed within the subsequent few months should be closely monitored, and generally should be considered additive to the dose already received (Stecker *et al.*, 2009).

Patient information and follow-up may not be necessary if fluoroscopy time exceeds the SRDL in Table 4.7 and if other dose metrics are measured and do not exceed their respective SRDLs. Examples of this situation include prolonged procedures performed on thin body parts or using very low fluoroscopic dose rates.

In addition to preprocedure informed consent, patients should be advised if they have undergone a procedure that has delivered an SRDL (Balter and Moses, 2007; FDA, 1994; Hirshfeld *et al.*, 2004; Miller *et al.*, 2004; NCI, 2005; Stecker *et al.*, 2009). A patient whose amount of radiation exceeded an SRDL should be given written radiation follow-up instructions in addition to their other discharge instructions. Sample discharge instructions are given in Appendix K.

The patient should understand the potential radiation effects that might result when a particular skin area has received a dose that might cause deterministic effects. This will help the patient understand any skin changes that may develop and to seek appropriate medical care. Patients, care givers, and responsible healthcare professionals should be made aware of the possible radiologic etiology of relevant signs and symptoms.

Recommendation 17

If an SRDL is exceeded for an FGI procedure, the patient and any caregivers *should* be informed, prior to discharge, about possible deterministic effects and recommended follow-up.

If fluoroscopy time exceeds the SRDL, but other measured dose metrics do not exceed the SRDL, patient information and follow-up *may not* be necessary.

The interventionalist is responsible for patient follow up for possible deterministic effects until the likelihood of a reaction has passed. The patient should be instructed to notify the interventionalist of the results of self-examination of the irradiated area (either positive or negative). Telephone contact can be sufficient if no tissue reactions are reported. Clinical follow-up is arranged if the examination is positive. All suspicious findings should be treated as a probable radiation effect unless an alternative diagnosis is established (Balter and Moses, 2007). If radiation has not been ruled out, it is essential to refer the patient to a physician experienced in managing radiation injuries (*i.e.*, injuries from radiation oncology) (Hymes *et al.*, 2006). Available skin dose information should also be provided to the treating physician. Because of the possibility of misdiagnosis, contact with the patient and other caregivers should be maintained to ensure that a possible radiogenic etiology is considered in subsequent patient care.

Recommendation 18

Follow-up for possible deterministic effects *shall* remain the responsibility of the interventionalist for at least 1 y after an FGI procedure. Follow-up *may* be performed by another health-care provider who remains in contact with the interventionalist.

All relevant signs and symptoms (Table 2.5) *should* be regarded as radiogenic unless an alternative diagnosis is established.

Clinical follow-up intended to diagnose radiogenic, visually disabling cataracts may not be practicable at lower radiation levels because of a dose-dependent latent period (*i.e.*, the higher the dose, the more rapidly cataracts develop), ranging from years to decades between irradiation and lens opacification (Worgul *et al.*, 1996).

A qualified physicist should review all patient reports of skin effects in order to evaluate the dosimetric aspects of the procedure and discuss these findings with the interventionalist. The qualified physicist may also assist in facilitating clinical follow-up as determined by the interventionalist. There may be other recommendations and/or requirements pertaining to patient follow-up according to the facility's policies.

4.3.5 *Radiation Dose-Management Processes*

The goals of radiation dose management differ between diagnostic radiography and FGI procedures. The goals for diagnostic radiography are minimizing stochastic risk to the population and providing appropriate image quality. Dose management for FGI procedures shares these goals, but has the additional goals of facilitating procedure completion and managing the risk of deterministic injuries.

4.3.5.1 *Diagnostic Radiography and Diagnostic Reference Levels.* The diagnostic reference level (DRL) process (ICRP, 2007a; 2007b) is a major quality tool for diagnostic radiography. The DRL process is designed to reduce the risk of stochastic effects by helping to manage the technical factors in image acquisition and processing. Only highly-standardized, high-volume diagnostic radiography examinations are evaluated (target procedures). Each imaging task is standardized with regard to the range of x-ray field sizes and locations, beam qualities, and other technical factors that affect the distribution of absorbed dose in the body, so that organ and tissue doses are generally proportional to the quantity selected for the DRL. The influence of patient size is removed by obtaining a single value for a standard phantom (*e.g.*, as typically done in the United States) or an average value for a sample of standard-sized patients (*e.g.*, as typically done in European countries).

The single value or average value for the target procedure is obtained from each of a large number of facilities. Most of the previous literature in the United States used the quantity entrance-skin exposure for this purpose. Currently, the quantity is usually incident air kerma ($K_{a,i}$). The Nationwide Evaluation of X-Ray Trends series is a good example of this process (Spelic *et al.*, 2010). The resultant data set has a tail at higher values but can usually be represented by a skewed normal distribution.

ICRP recommends that DRLs be determined by professional medical bodies, in conjunction with national health and radiation protection authorities (ICRP, 2007b). The same recommendation is made for what is called "guidance levels" in the IAEA's Basic Safety Standard (IAEA, 1996). The 75th percentile of the data set is often selected as the upper DRL for that procedure. In some cases, the 10th percentile is selected as the lower DRL.

A facility can evaluate its performance relative to the DRLs. A local value is obtained with the same protocol used by those facilities whose data determined the DRLs. If the local value is below the 10th percentile DRL, then a review of image quality is indicated. If the local value exceeds the 75th percentile DRL, a review of the appropriateness of the dose is indicated.

DRLs help avoid radiation dose to the patient that does not contribute to the medical-imaging task (ICRP, 2001; Wall and Shrimpton, 1998). They are a guide to what is achievable with current good practice, rather than optimum performance, and are neither limits nor thresholds that define competent performance of the operator or the equipment (IAEA, 1996). A value of the dosimetric quantity for an imaging task that is less than the DRL does not guarantee that the task is being performed optimally (Vano and Gonzalez, 2001). DRLs do not apply to individual patients or individual cases. However, the use of DRLs has yielded reduction in the levels of radiation observed for noninvasive x-ray-imaging procedures (Hart *et al.*, 2009).

4.3.5.2 *Fluoroscopy-Guided Interventional Procedures and Advisory Data*. A review of the amount of radiation delivered to patients from FGI procedures is an essential aspect of any performance improvement program, but the literal application of the DRL process used for diagnostic radiography to FGI procedures is not appropriate. As discussed below, the reasons are:

- standardized patients and procedures do not exist; and
- the DRL process is unable to detect and identify individual instances of high-dose procedures (those that are responsible for deterministic effects).

One aspect of the technical performance of FGI-procedure systems can be evaluated using a standard phantom and each facility's clinical protocol for the particular FGI procedure (*e.g.*, diagnostic coronary artery angiography) (CRCPD, 2009). The process is analogous to the DRL process for diagnostic radiography. The process yields a single air-kerma rate for each facility. The data from many facilities are combined to produce a distribution of air-kerma rates. The facility's single air-kerma rate can then be compared to both the 10th and 75th percentiles of the distribution and, if necessary, appropriate action is taken. However, this evaluation characterizes only the technical performance of the equipment; it supplies little information on equipment use and no information on patient mix or operator performance.

FGI procedures cannot be standardized in the same way as diagnostic radiography because of the great variations in patient anatomy, lesion characteristics, and disease severity. These factors cause wide variation in procedure complexity from one patient to another. Radiation dose is strongly affected by procedure complexity (Bernardi *et al.*, 2000; IAEA, 2009; Peterzol *et al.*, 2005; Vehmas, 1997). Direct application of the DRL methodology to FGI procedures would require multi-facility collection of a suitable metric (*e.g.*, P_{KA} or $K_{a,r}$) for a specific FGI procedure with standardized complexity performed by a standardized operator on a standard-size patient. It is impractical for an FGI-procedure facility to define such a standardized case.

Overall dosimetric performance for FGI procedures, incorporating the effects of equipment function, procedure protocols, and operator performance, can be characterized and analyzed. However, this requires a different process than that used for diagnostic radiography, and a more detailed presentation of the necessary data. In this Report, the full distribution of data used for evaluation of an FGI procedure is given the name advisory data set (ADS), to distinguish the nature of the data from the upper (or lower) single reference values used in the DRL process.

An ADS for an FGI procedure is generated by obtaining dosimetric data for a large number of cases for a specific procedure from each of many facilities. The resultant data set includes the data for all instances of the procedure at each facility, rather than a single data point from each facility. The ADS is usually well characterized by a lognormal distribution. A corresponding facility data set (FDS) consists of all of the dosimetric data for all of the cases of the specific procedure performed at an individual facility.

Because differences between the shapes of the dose distributions of the ADS and a FDS are potentially useful, the ADS should characterize the entire distribution, rather than just the 10th and 75th percentile values typically used for DRLs. Also, in order to provide a basis of comparison for facilities that use a locally-derived SRDL, a published ADS should indicate the percentage of cases of each procedure that exceed a specific radiation dose level. Ideally, percentages should be presented at 0.5 or 1 Gy intervals up to the maximum value observed in the ADS.

To illustrate the approach needed for FGI procedures, a sample ADS (for $K_{a,r}$) and a sample FDS (for $K_{a,r}$) for a particular facility are given in Tables 4.8 and 4.9. Table 4.8 presents the distribution of the two data sets as the percent of values in each $K_{a,r}$ bin. Table 4.9 presents the percentiles for the distributions of the two data sets. The numerical values in Tables 4.8 and 4.9 are only for the purpose of this illustration and cannot be used for any other purpose. The ADS and a FDS are presented graphically in Figures 4.2 and 4.3, respectively.

4.3.5.3 *Example of Using an Advisory Data Set to Evaluate a Facility.* The following example uses the quantity $K_{a,r}$ (the preferred quantity) to evaluate FGI-procedure dose utilization. The quantity P_{KA} could also be used to evaluate general dose performance, but would not be able to unambiguously identify the cases where a very-high skin dose may result in deterministic effects.

TABLE 4.8—*Distribution by $K_{a,r}$ bin for a sample ADS and a sample FDS for coronary angioplasty.*[a]

$K_{a,r}$ Range for Bin (Gy)	ADS (n = 2,546)		FDS (n = 1,697)	
	Percent in Bin	Cumulative Percent	Percent in Bin	Cumulative Percent
0.0 – 0.5	1.22	1.22	0.53	0.53
0.5 – 1.0	6.17	7.38	3.65	4.18
1.0 – 1.5	15.36	22.74	7.96	12.14
1.5 – 2.0	17.95	40.69	13.38	25.52
2.0 – 2.5	15.79	56.48	15.03	40.54
2.5 – 3.0	12.10	68.58	12.79	53.33
3.0 – 3.5	9.86	78.44	10.55	63.88
3.5 – 4.0	5.85	84.29	9.13	73.01
4.0 – 4.5	4.67	88.96	6.54	79.55
4.5 – 5.0	3.61	92.58	4.77	84.33
5.0 – 5.5	2.79	95.37	4.48	88.80
5.5 – 6.0	1.34	96.70	3.24	92.04
6.0 – 6.5	1.18	97.88	2.12	94.17
6.5 – 7.0	0.67	98.55	1.59	95.76
7.0 – 7.5	0.67	99.21	1.06	96.82
7.5 – 8.0	0.27	99.49	0.88	97.70
8.0 – 8.5	0.20	99.69	0.65	98.35
8.5 – 9.0	0.04	99.73	0.41	98.76
9.0 – 9.5	0.08	99.80	0.41	99.18
9.5 – 10.0	0.04	99.84	0.29	99.47
10.0 – 10.5	0.04	99.88	0.18	99.65
10.5 – 11.0	0.04	99.92	0.06	99.71
11.0 – 11.5	0.08	100.00	0.12	99.82
11.5 – 12.0	0.00	100.00	0.00	99.82
12.0 – 12.5	0.00	100.00	0.06	99.88
12.5 – 13.0	0.00	100.00	0.00	99.88
13.0 – 13.5	0.00	100.00	0.00	99.88
13.5 – 14.0	0.00	100.00	0.00	99.88
14.0 – 14.5	0.00	100.00	0.00	99.88
14.5 – 15.0	0.00	100.00	0.12	100.00
>15.0	0.00	100.00	0.00	100.00

[a]The numerical values in this table are only for the purpose of this illustration and cannot be used for any other purpose.

TABLE 4.9—*Percentiles for the distributions of the ADS and FDS for coronary angioplasty.*[a]

Percentile	$K_{a,r}$ (Gy) ADS	$K_{a,r}$ (Gy) FDS
1	0.5	0.6
5	0.9	1.1
10	1.1	1.4
20	1.4	1.8
25	1.6	2.0
30	1.7	2.2
40	2.0	2.5
50	2.3	2.9
60	2.6	3.3
70	3.1	3.8
75	3.3	4.1
80	3.6	4.5
90	4.7	5.7
95	5.4	6.8
99	7.3	9.3
Arithmetic mean	2.6 Gy	3.3 Gy
Percent of values >5 Gy	7.4 %	15.7 %

[a]The numerical values in this table are only for the purpose of this illustration and cannot be used for any other purpose.

An appropriate published ADS for the selected procedure is used as the starting point. Unfortunately, published ADSs for FGI procedures are sparse as of 2010 (Balter *et al.*, 2008; Bleeser *et al.*, 2008; Brambilla *et al.* 2004; IAEA, 2009; Miller *et al.*, 2009; Vano *et al.*, 2008a; 2009).

A facility judges its FGI-procedure dose performance in several steps. The first step is to compare the local SRDL to the ADS. The facility's local SRDL is a value taken from published recommendations in Table 4.7 or a locally determined value. The percentage of procedures in the ADS (Figure 4.2 and Table 4.8) that exceed this value can now be determined. This percentage will be used as part of the evaluation of a facility's performance.

The next step is to characterize the dose distribution for all cases of a specific procedure performed at the facility (Figure 4.3).

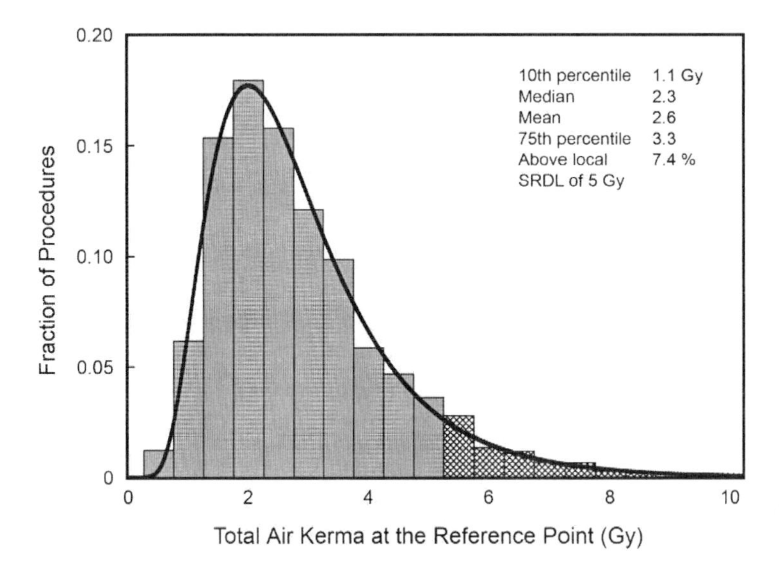

Fig. 4.2. Sample ADS for coronary angioplasty. Bins with $K_{a,r}$ values exceeding the facility's selected SRDL of 5 Gy are cross hatched. The percentage of procedures in the ADS that exceeded the facility's local SRDL is 7.4 %.

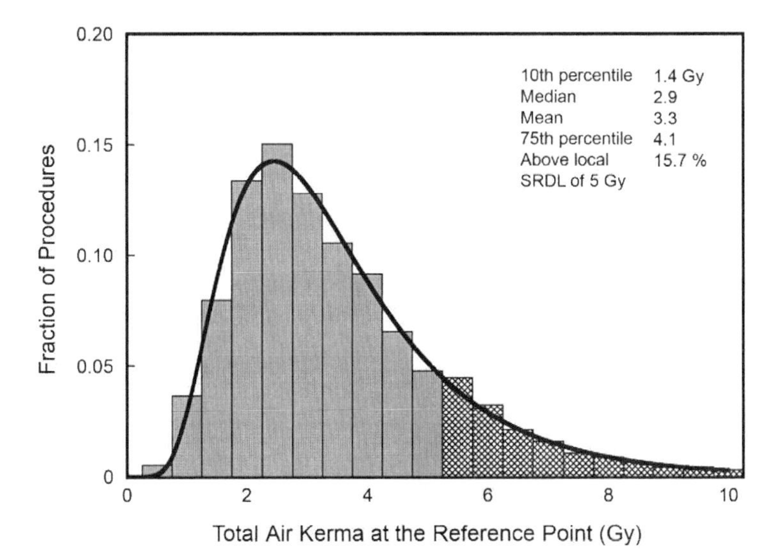

Fig. 4.3. Sample FDS of $K_{a,r}$ for the same FGI procedure shown in Figure 4.2. Bins with $K_{a,r}$ values exceeding the facility's selected SRDL of 5 Gy are cross hatched. The percentage of procedures in the FDS that exceeded the facility's local SRDL is 15.7 %.

Evaluation of subsets of these data sorted by procedure room and operator can be useful as well, as discussed below. The percentage of instances exceeding the local SRDL, and the median value of the entire FDS (and appropriate subsets) are calculated.

The local median should be compared with the 75th percentile of the ADS (Table 4.9). The average (mean) value should not be used because it can be strongly influenced by the high-dose tail of the distribution. An investigation should be undertaken if the local median exceeds the 75th percentile of the ADS. Investigations may also be desirable if the local median is below the 25th or above the 50th percentile (median) of the ADS. Too little radiation use might be attributable to incomplete procedures, inadequate image quality, or superior dose management. Too much radiation use might reflect poor equipment and settings, suboptimal procedure performance, or high clinical complexity. For the example shown in Table 4.9, the median $K_{a,r}$ of the FDS is 2.9 Gy. This is above the ADS median of 2.3 Gy but below the ADS 75th percentile of 3.3 Gy. The facility's overall performance is acceptable, but the reasons behind the additional radiation usage should be analyzed and understood. The additional radiation usage may be due to differences in clinical complexity or to other causes.

The next step is to determine the percentage of local facility cases that exceed the local SRDL. In this example, the local facility elects an SRDL of 5 Gy. A percentage for the FDS that is appreciably above or below that observed for the ADS should prompt investigation of clinical results and reported skin injuries.

It can be useful to perform the same analysis using a $K_{a,r}$ value of 3 Gy, in addition to the local SRDL. An FGI procedure is in the potentially-high radiation dose category if more than 5 % of instances of that procedure exceed a $K_{a,r}$ value of 3 Gy (Section 3). If fewer than 5 % of the instances of the procedure at the local facility exceed this value, then the procedure, as performed at the local facility, is not in the potentially-high radiation dose category. At the local facility, the procedure may be performed safely in a fluoroscopy suite that does not meet the requirements of IEC 60601-2-43 (IEC, 2010) (Table 3.1). Also, those procedures at the local facility that are not in the potentially-high radiation dose category may be audited less frequently than those that are in that category.

Lastly, the distribution of the FDS may be overlaid on the distribution of the ADS (Figure 4.4). Displacement or distortion of the FDS distribution relative to the ADS distribution may be due to differences in equipment, clinical complexity, or other factors.

The analysis can be extended to individual operators or to individual FGI-procedure rooms by comparing operator- or room-specific

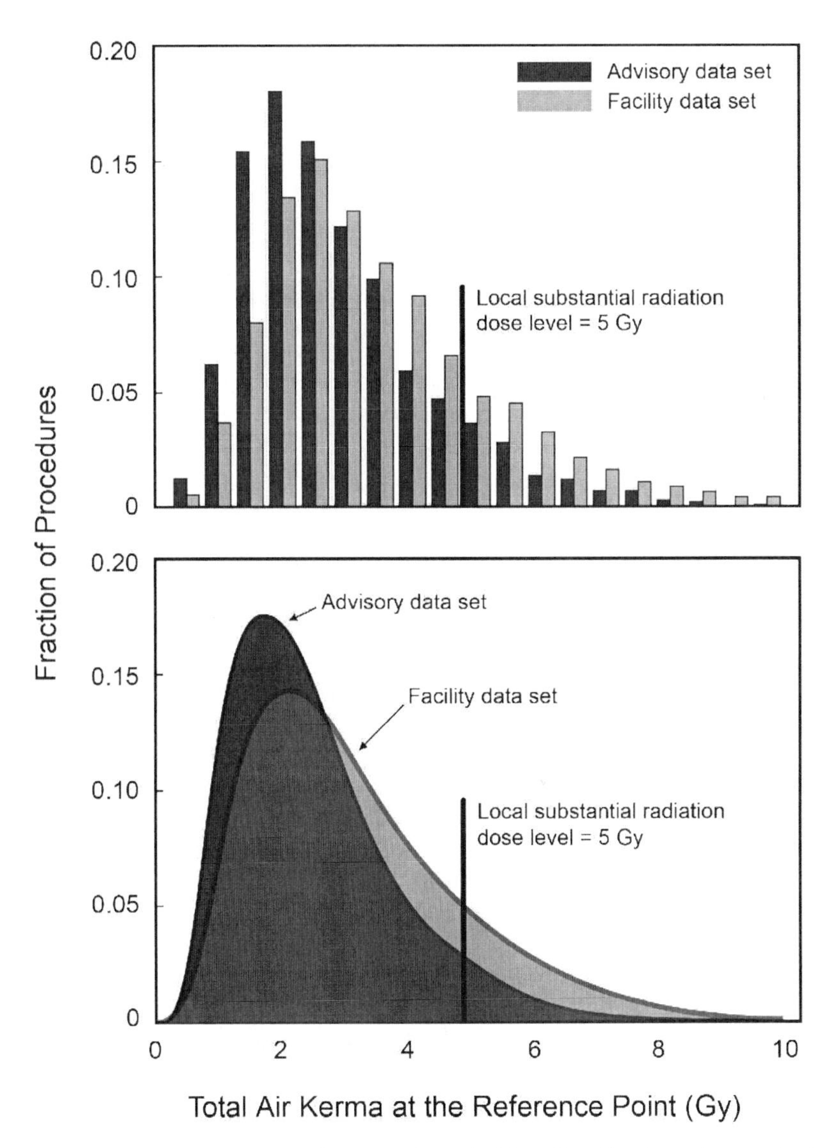

Fig. 4.4. Comparison of the FDS and ADS distributions (numerical comparisons are discussed in Section 4.3.5.3).

data to a published the ADS, the local FDS, or to other pooled distributions of data for multiple facilities (*e.g.*, Miller *et al.*, 2009). Care should be taken in such an analysis to account for statistical interactions (*e.g.*, statistical confounding between the operator and the FGI-procedure room).

Recommendation 19

Facilities *shall* have a process to review radiation doses for patients undergoing FGI procedures.

Advisory data based on measured dosimetric quantities (in particular P_{KA} or $K_{a,r}$ to manage overall performance, and $K_{a,r}$ to manage deterministic effects) *should* be used for quality assurance purposes.

4.3.5.4 *Further Research and Data Collection*. Further research in using advisory data for FGI procedures, particularly for noncardiac interventional procedures, is encouraged. Additional research on the analysis of dose distributions for local data is also needed.

Consideration should be given to adapting the FGI-procedure SRDL approach to other potentially-high radiation dose diagnostic imaging procedures such as encountered in CT-perfusion studies that can be quite variable in the dose delivered to the patient.

The collection of FGI-procedure ADSs should be a priority. International and national registries of dosimetric data can play an invaluable part in this process by providing a central data depository to evaluate patient doses from a variety of FGI procedures. Specialized registries for those cases with the potential for deterministic injury would permit the development of advisory data derived from a broad base of institutions and operators. An effort to develop the necessary infrastructure is underway, as described by FDA (2010).

5. Protection of Staff

Employers have a responsibility to protect their employees using a hierarchy-of-controls approach to radiation protection. This approach requires that controls be implemented, if possible, in the following order (more information is found in this Report in the respective sections noted):

- engineering (*i.e.*, structural and equipment shielding) (Sections 3.1, 3.2, and 4.2);
- safe work practices (*i.e.*, policies and procedures increasing safety either by changing the actual way the work is done or by adding a tool to help) (Sections 4.1 and 4.2);
- administration (*i.e.*, training and work assignments) (Sections 5.2, 5.4, 6.2, and 6.3); and
- personal protective equipment (aprons and/or personal shields) (Section 5.5).

Steps taken to manage (reduce) the patient dose to only the level necessary to perform the clinical task will also manage (reduce) the occupational dose to employees.

Radiation protection in occupational settings involving medical facilities has been the subject of previous detailed ICRP (1997) and NCRP (1989b; 1998; 2000a) publications that describe the principles of a multifaceted approach that includes facility and equipment design, organizational and management arrangements, training of staff, and operational radiation control practices. Given that all items listed above have been complied with to the greatest practical extent, this section of the Report concentrates on those aspects of radiation protection of FGI workers that require special attention because of a unique radiation environment in which the staff is required to be in proximity to the patient in order to accomplish the FGI procedures.

Radiation protection for interventionalists and other staff exposed to radiation has long been a concern. Overall, dose to radiation workers has decreased in the past 50 y (Kim *et al.*, 2008; Kumazawa *et al.*, 1984). However, FGI procedures are an exception to this trend. Radiation hazards include exposure to the hands of the interventionalist by the primary x-ray beam and exposure of the interventionalist and staff to scattered radiation from the patient and leakage radiation from the x-ray tube. The nature of FGI procedures causes several conditions that result in unavoidable occupational exposure. The interventionalist generally stands close to the patient to manipulate interventional devices. Close proximity and maintenance of a sterile field make use of lead shielding devices challenging. The clinical condition of the patient and use of anesthesia require that multiple healthcare workers be present in the procedure room. Plus, procedure complexity results in lengthy fluoroscopy times. This section will address issues related to staff radiation exposure, including occupational dose limits, expected dose levels, radiation effects that have been observed in radiation workers, and recommendations to reduce staff radiation exposure.

5.1 Review of Staff Exposure

There have been numerous publications that report staff radiation levels received in clinical practice for various types of interventional procedures. A review of these studies has been provided by Kim *et al.* (2008) for cardiac procedures. Reviews for noncardiac FGI procedures are included in Table 5.1. Values for E have been calculated as described in Section 5.7.3.

For FGI procedures, the mean neck/chest dose values per procedure range from as low as 3 μSv for diagnostic procedures, noncomplex interventions, and those procedures utilizing extensive dose-reduction techniques to as high as 660 μSv for interventional procedures with over-table x-ray tube equipment (Ramsdale *et al.*, 1990). Note that some procedures could result in annual values of E exceeding 20 mSv for a workload of 1,000 cases per year. Similarly, lens of the eye and hand doses are highest for interventional cardiology and interventional radiology procedures, particularly when the over-table x-ray tube is used. For both lens of the eye and hand exposure, several studies indicate that the annual equivalent dose limits for these tissues (Section 5.7.1) would be exceeded for a workload of 1,000 procedures in a year if eye and hand protection are not used.

5.2 Essential Radiation Protection Knowledge

Most FGI procedures cannot be conducted without exposing the participating staff members to a radiation field. Each staff member's irradiation is strongly influenced by the individual's actions during a procedure. Appropriate use of radiation reduction techniques and protective equipment can have a marked effect on occupational dose

(Kim and Miller, 2009; Martin, 2009). Appropriate education in radiation protection is the best way to avoid unnecessary exposure. Appropriate education has been shown to improve compliance with occupational dose-monitoring requirements among trainees (Kim *et al.*, 2010).

Organizing educational programs for individuals who are routinely in the FGI-procedure room during procedures is relatively simple. Providing live instruction for individuals occasionally in the room (*e.g.*, anesthesiologists) can be difficult. Web-based training can be a viable alternative for individuals who are only occasionally in the room.

Recommendation 20

Each individual present in an FGI-procedure room while a procedure is in progress *shall* have appropriate radiation protection training.

Conversely, the individuals accountable for radiation protection programs in facilities and regulatory agencies can best discharge their responsibilities if they have a sufficient understanding of clinical interventional procedures to evaluate worker performance and patient safety. This should include a basic understanding of the nonradiation risks experienced by both interventional workers and patients.

5.3 Pregnant or Potentially-Pregnant Workers

Specific restrictions apply to the occupational exposure of pregnant women. In the United States, NCRP (1993) recommends a 0.5 mSv monthly limit (equivalent dose) for the embryo and fetus (excluding medical and natural background radiation) once the pregnancy is declared. In the United States, workers who do not wish to declare their pregnancy are not required to do so (NRC, 1999). ICRP recommends that the standard of protection for the embryo and fetus should be broadly comparable to that provided for members of the general public (ICRP, 1991a; 2007a), and recommends that after a worker has declared her pregnancy, her working conditions should ensure that the additional dose to the embryo and fetus does not exceed 1 mSv during the remainder of the pregnancy (ICRP, 2007a).

Very few individuals working in the interventional environment accumulate as much as 1 mSv (E) in a year beneath a radiation protective apron (*i.e.*, as measured by a personal dosimeter under the apron). The shielding provided by a standard protective lead apron

TABLE 5.1—*Operator exposure during noncardiac interventional procedures.*

Procedure Type	Mean Dose per Procedure (μSv)					Reference
	Neck or Chest, Outside Apron	Inside (under) Apron	Effective Dose	Lens of the Eye	Hand	
Biliary stent/drainage	660		31	310	1,290	Ramsdale *et al.* (1990)
Biliary stent/drainage					670	Stratakis *et al.* (2006)
Biliary stent/drainage, PTHC,[a] nephrostomy					920	Whitby and Martin (2005)
Cerebral angiography	11		0.52			Marshall *et al.* (1995)
Cerebral angiography /embolization				13.6	19.3	Marshall *et al.* (1995)
Cerebral angiography /embolization	32		1.5			Layton *et al.* (2006)
Cerebral embolization	25		1.2			Marshall *et al.* (1995)
Complex percutaneous nephrostomy				18.5	163	Vehmas (1997)
Dialysis access graft intervention					780	Stavas *et al.* (2006)
Endovascular surgery aneurysm repair	293	32	23		398	Lipsitz *et al.* (2000)
Endoscopic retrograde cholangiopancreatography	450		21	550	640	Buls *et al.* (2002)

Femoral angiograms					50	Whitby and Martin (2005)
Femoral angioplasty					210	Whitby and Martin (2005)
Femoral embolization					140	Whitby and Martin (2005)
Femoral stent placement					510	Whitby and Martin (2005)
Mixed interventional and diagnostic radiology	80	5	3.8		269	Williams (1997)
Mixed interventional radiology	50	0.9	1.7			Marx *et al.* (1992)
Mixed interventional radiology	325		15	296	396	Vano *et al.* (1998a)
PCNL[a]				68	48	Hellawell *et al.* (2005)
PCNL[a]	270		13	320	520	Ramsdale *et al.* (1990)
Percutaneous drainage				12.5	84	Vehmas (1997)
Spine injections	3	<0.1	0.14	4	7	Botwin *et al.* (2002)
TIPS[a]	40 – 240		1.9 – 11			Zweers *et al.* (1998)
TIPS,[a] transhepatic liver biopsy					630	Whitby and Martin (2005)
Vertebroplasty	220		10	84	453	Harstall *et al.* (2005)
Vertebroplasty					1,280	Kallmes *et al.* (2003)
Vertebroplasty					422	Synowitz and Kiwit (2006)

[a]PTHC = percutaneous transhepatic cholangiogram
PCNL = percutaneous nephrolithotomy
TIPS = transjugular intrahepatic portosystemic shunt

is sufficient to protect the embryo and fetus for typical exposure to staff involved in interventional procedures (Wagner and Hayman, 1982). Therefore, pregnant individuals involved in FGI procedures generally do not need to limit their time in the procedure room to remain below the dose limit for the embryo and fetus.

Conformance to the dose limit is demonstrated using a single personal dosimeter worn under the apron by the pregnant worker at waist level, from the date the pregnancy is declared until delivery. This dosimeter overestimates actual dose to the embryo and fetus because radiation attenuation by the mother's tissues is not considered. The dosimeter should be evaluated monthly. Electronic dosimeters can be used to provide rapid access to data (Balter and Lamont, 2002). In those centers where a two-dosimeter system is used, workers who may become pregnant should place the dosimeter that is worn under the apron at waist level. Data from that dosimeter provides an estimate of the equivalent dose to the embryo and fetus from conception to declaration of pregnancy. When that dosimeter shows an average value for personal dose equivalent [$H_p(10)$] of <0.1 mSv per month, the equivalent dose to the embryo and fetus would be in conformity with ICRP and NCRP recommendations. This is because the dose to the embryo and fetus can be approximated as one-half of the $H_p(10)$ for the dosimeter under the protective apron (Faulkner and Marshall, 1993; Trout, 1977).

ICRP (2000b) noted that a dosimeter worn by a diagnostic radiology worker (typically at the collar level without a protective apron) may overestimate the actual dose to the embryo and fetus by about a factor of 10 or more due to attenuation by the body tissue overlying the uterus. If the dosimeter is worn outside a protective lead apron, the measured value is likely to be about 100 times higher than the actual dose to the embryo and fetus because of additional attenuation by the apron.

5.4 Radiation Sources

Sources of radiation exposure for the interventionalist and staff include the primary x-ray beam, x-ray tube leakage radiation, and scattered radiation. For both fluoroscopy and CT, the x-ray beam is required by regulation to be closely collimated to the image receptor (FDA, 2009a; 2009b), so that primary radiation outside of the visible image is minimal. For certain procedures where the hands of the interventionalist directly manipulate devices in the primary beam, care should be taken to avoid direct hand exposure. Examples of these types of procedures include fluoroscopy for spinal injections, vertebroplasty, biliary drainage and stent placement, dialysis access graft intervention, and CT-guided biopsy.

Secondary radiation in fluoroscopy and CT suites consists primarily of leakage radiation from the x-ray tube and scattered radiation from the volume of patient tissue in the primary beam. Leakage radiation levels are regulated to a maximum of 0.88 mGy (air kerma) in an hour at 1 m for maximum operating kilovolt peak and continuous milliampere (FDA, 2009c; IEC, 2008). At typical operating techniques, the leakage rate is lower than this level, in the range of 0.001 to 0.01 mGy h^{-1} at 1 m. Scattered radiation levels are generally substantially higher. During fluoroscopic imaging, scattered radiation air-kerma rates adjacent to the patient range from 1 to 10 mGy h^{-1} (Balter, 2001b; Schueler *et al.*, 2006). For CTGI procedures operated in the continuous acquisition mode, scattered radiation levels generally range from 10 to 30 mGy h^{-1} in the area adjacent to the imaging slice and CT gantry (Nawfel *et al.*, 2000). Secondary radiation levels fall off rapidly with increasing distance from the radiation source, roughly following the inverse-square law, where the air-kerma rate decreases in proportion to the inverse of the distance from the source squared. For example, when the distance from the source is doubled, the air-kerma rate is decreased by a factor of four.

Generally, staff doses correlate with patient dose levels. Specifically, scattered radiation levels during fluoroscopy have been found to be proportional to P_{KA} for a given distance from the central ray (Marshall and Faulkner, 1992; Schueler *et al.*, 2006; Servomaa and Karppinen, 2001; Vano *et al.*, 1998c; Williams, 1996; 1997). For a typical position of the interventionalist at 0.75 m from the central ray, the ratio of scattered radiation to P_{KA} is 5 to 10 μGy $(Gy\ cm^2)^{-1}$ (Servomaa and Karppinen, 2001).

There are many techniques that interventionalists and staff involved in FGI procedures can use to minimize their occupational radiation exposure. When these techniques are used together, the dose can be significantly reduced. The first step in staff dose reduction is to employ patient dose-reduction methods and equipment settings, as described in Section 3. Since air-kerma rates incident on the patient's skin increase substantially for large patients and steeper angulations, staff should be aware that scattered radiation levels are also much higher under these conditions.

Secondary (scattered plus leakage) radiation distributions surrounding the patient are not uniformly distributed in time and space. Sample distributions are shown in Figure 5.1. Actual isodose curves for individual fluoroscopes and the associated measuring conditions are found in the fluoroscope's instructions for use (IEC, 2000; 2010). Backscatter from the patient dominates the secondary radiation distribution. Leakage from the x-ray tube housing provides only a small contribution because interventional fluoroscopes typically

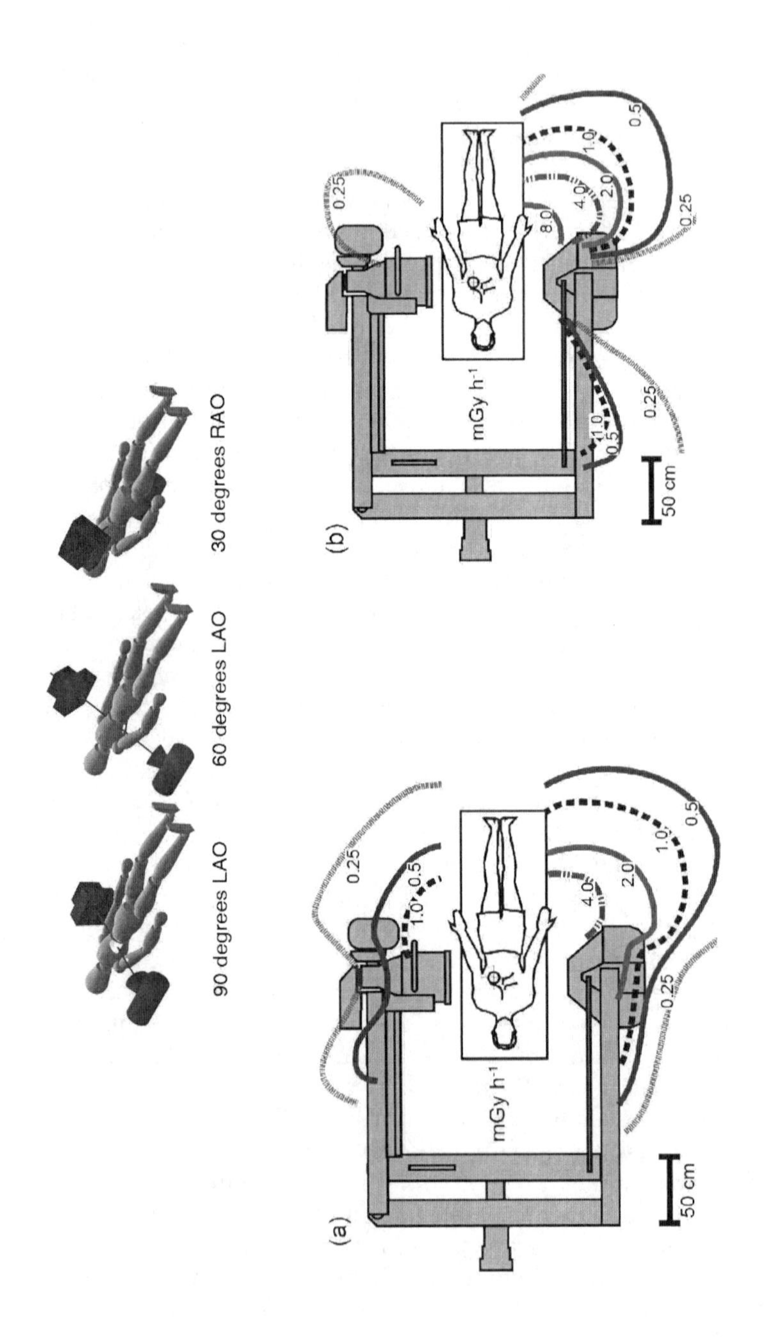
90 degrees LAO
60 degrees LAO
30 degrees RAO
(a)
mGy h⁻¹
0.25
0.5
1.0
2.0
4.0
50 cm
(b)
mGy h⁻¹
0.25
0.5
1.0
2.0
4.0
8.0
50 cm

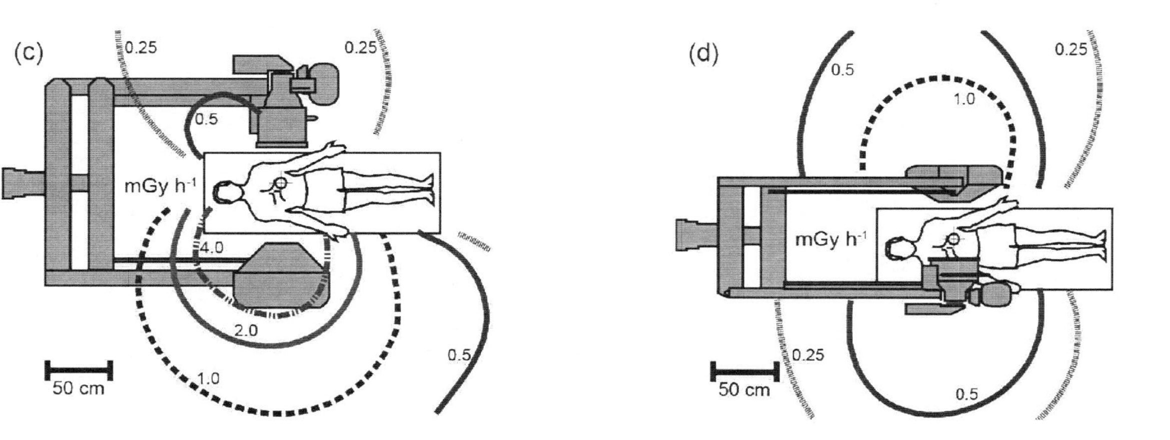

Fig. 5.1. Sample scattered-radiation isodose curves (air-kerma rates) measured 100 and 150 cm above the floor.: (a) 90 degrees LAO, 150 cm; (b) 90 degrees LAO, 100 cm; (c) 60 degrees LAO, 100 cm; and (d) 30 degrees RAO, 100 cm. Actual isodose curves for individual fluoroscopes and the associated measuring conditions are found in the fluoroscope's instructions for use in (IEC, 2000; 2010).

operate well below the maximum-rated kilovolts of the x-ray tube (NCRP, 2004). Interventionalists typically work on the patient's right side and are usually positioned between the patient's hips and knees.

In general, scattered radiation is most intense on the entrance-beam side of the patient (Balter, 2001b; Schueler *et al.*, 2006, Stratakis *et al.*, 2006). When using a C-arm in a lateral angulated projection, staff should position themselves on the image-receptor side of the patient, if possible. For a frontal projection, positioning the x-ray tube under the table will place higher scattered radiation areas toward the floor so that the head and neck of the interventionalist receive less radiation. Figure 5.1a depicts a horizontal plane corresponding to eye level (150 cm above the floor) with the fluoroscopy system in position for a 90 degree left anterior oblique (LAO) projection. The entry port for the x-ray beam is on the patient's right side (interventionalist's side of the patient). There is minimal attenuation by the patient's tissues between the entry port and the interventionalist's eyes. Figure 5.1b is the horizontal plane corresponding to waist level (100 cm above the floor). Scattered radiation air-kerma rates on the patient's right side are higher because the entry port is essentially in this plane. The scattered radiation air-kerma rates on the patient's left side are approximately a factor of 10 less than on the right side. The shadow cast by the x-ray tube is evidence of low leakage compared to scattered radiation. Figure 5.1c depicts a horizontal plane 100 cm above the floor with the beam rotated to a 60 degree LAO projection. The beam is underneath and on the right side of the patient. The patient's tissues provide some attenuation of the backscatter from the entry port. Figure 5.1d depicts a horizontal plane with the fluoroscopy equipment in position for a 30 degree right anterior oblique (RAO) projection. The beam is now closer to vertical and on the left side of the patient. The scattered radiation air-kerma rate on the patient's right has decreased substantially.

When the x-ray beam enters the patient from below the table, the intense backscatter is directed toward the floor and away from the interventionalist's eyes. Cross table (lateral) views may be clinically unavoidable and therefore present additional eye hazards. Over-table x-ray tube positions increase the eye hazard and should be avoided.

Information about the secondary radiation distribution for a specific fluoroscopy system can be obtained by referring to the isodose maps (expressed in air kerma) provided by the manufacturer as a requirement of IEC standards (IEC, 2000; 2010). This information is usually provided in the user manual supplied with the equipment.

5.5 Personal Radiation Protective Equipment

Employers are responsible for ensuring that radiation protective equipment is available to each staff member who, as part of their job responsibilities, is exposed to radiation. All personnel present in an FGI-procedure room who are not positioned behind a radiation barrier should wear a radiation protective garment when the x-ray beam is activated. Workers may be required to wear additional radiation protective equipment (*e.g.*, protective gloves, protective glasses) that they have been provided with, as appropriate, in order to ensure that their occupational radiation exposure is consistent with the ALARA principle.

5.5.1 *Lead and Other Attenuating Materials (lead-equivalent thickness)*

Radiation protective shielding, such as lead aprons or structural barriers, is intended to attenuate radiation fields to a small fraction of the incident intensity of the radiation field. Lead and other radiation-attenuating materials are used in permanent and moveable barriers in FGI-procedure rooms and in personal radiation protective garments (*e.g.*, an apron) worn by staff members during FGI procedures.

Aprons should provide the desired protection at an acceptable weight, because the apron weight itself can pose a substantial ergonomic risk to its wearer (Klein *et al.*, 2009). Apron weight can be reduced by using a lesser thickness of lead or by replacing lead, completely or partially, with a combination of one or more nonlead materials that have the same or better attenuation for the scattered radiation from fluoroscopic beams. Replacing lead with other materials also eliminates concern for the chemical toxicity of lead when aprons are discarded.

Aprons that use lead are specified in terms of the lead thickness. Aprons that use nonlead or composite materials are specified in terms of the thickness of lead that would afford the same attenuation as the materials used (called the lead-equivalent thickness). The lead-equivalent thickness is determined by physical measurements for an appropriate reference x-ray spectrum. However, the accuracy of the specification depends on how closely the reference x-ray spectrum simulates the scattered x-ray spectrum encountered in the FGI-procedure room (Section 5.5.2).

The adequacy of any radiation protective apron can only be assessed if the individual always wears a personal dosimeter under the apron (located at the waist or chest level). The personal-dosimeter reading includes contributions from transmission of scattered radiation through the apron, x rays emitted from photon interactions in the apron, backscatter from the wearer, and radiation reaching the personal dosimeter without passing through the apron.

5.5.2 *Protective Garments*

Numerous surveys have noted an increased incidence of back pain in interventionalists who routinely wear protective aprons (Section 2.4.3.2.1). This has resulted in the introduction of protective materials for aprons that are lightweight lead composites or lead-free materials that incorporate proprietary combinations of barium, tungsten, tin and antimony. Different manufacturers supply aprons using various materials or combinations of materials. In general, aprons made from these materials provide the same attenuation as an equivalent thickness of lead for conventional fluoroscopic beams at less weight (Hubbert *et al.*, 1993; Yaffe *et al.*, 1991). However, studies have shown that there are significant differences in the attenuation properties of aprons of similar nominal lead-equivalent thicknesses (Christodoulou *et al.*, 2003; Finnerty and Brennan, 2005) and the attenuation varies as a function of x-ray beam quality in a different manner than lead (Finnerty and Brennan, 2005). It is most important that lead equivalence of protective aprons be characterized for broad-beam conditions with the specific beam quality and energy range expected during use. Protective aprons should be tested to ensure that the attenuation properties are adequate for the radiation beam qualities used by the wearer. International standards (IEC, 1994; 1998) have been published that could be used as a resource for establishing such testing methods and acceptance limits. Standardized methods for measuring the x-ray transmission of aprons, and acceptance limits for the transmission of aprons of nominal lead equivalence at specific beam qualities should be defined (Christodoulou *et al.*, 2003; Eder *et al.*, 2010; Schlattl *et al.*, 2007).

Weight reduction can be accomplished by selecting an apron with thinner lead thickness or one appropriately constructed using other materials. Though 0.5 mm lead aprons are currently considered the standard, a thinner lead apron may provide adequate protection. Based on the calculation of E from dual personal dosimeters, a 0.3 mm lead apron will result in a value of E that is only moderately higher (7 to 16 %) than a 0.5 mm lead apron (NCRP, 1995b). If a lead-equivalent thickness <0.5 mm is used, personal dosimeters located at both the collar level outside and above the apron and at the waist or chest level under the apron are recommended. Monthly dose monitoring can also be implemented to ensure that staff members who use garments with <0.5 mm lead-equivalent thickness continue to maintain an occupational dose below the required dose limits. With these precautions in place, it is quite possible to provide adequate protection with a 0.35 mm or less lead-equivalent thickness.

Staff members in FGI- and CTGI-procedure rooms should wear appropriate protective garments (aprons). A 0.5 mm lead-equivalent garment attenuates over 95 % of incident radiation, protecting body organs and lowering the value of E to the wearer (McCaffrey *et al.*, 2007). Staff members who may have their back to the patient during exposure should wear a wraparound-style apron. Proper fit is required to ensure the apron provides the necessary protection with minimum ergonomic hazard. All individuals who routinely participate in interventional procedures should be provided with custom-fitted protective garments. This is essential for those staff members who are routinely within 1 m of the patient, where scattered radiation levels are highest. The apron should fit snugly around the arms and not fall too low at the neckline. This is especially important for female staff members, for protection of breast tissue. Proper apron sizing is also critical if the apron design includes two layers that are overlapped to meet the lead-equivalent thickness specification.

Some individuals, due to their physical status, cannot tolerate the weight of 0.5 mm lead-equivalent protective aprons. These individuals should be fitted with protective aprons of lesser lead-equivalent thickness, if the readings for the personal dosimeter worn under the lighter-weight apron remain within an acceptable range.

Some individuals are able to tolerate protective aprons with lead-equivalent thicknesses exceeding 0.5 mm. Adding additional lead thickness will add significant weight while yielding small gains in dose reduction. If the reading for an individual's personal dosimeter worn under the apron is already at or close to the minimum detectable level, this additional weight is not accompanied by a useful improvement in radiation protection.

Lead protective garments should be maintained in a state of good repair such that the article of protective clothing can provide the maximum shielding benefit for which it was designed. They should be stored in a manner conducive to the protection of the garment from possible damage that would reduce its effectiveness as a radiation shield. Lead aprons should be properly placed on designated hangers and should not be folded, creased or crumpled in any way. Newer lead aprons that are made from lead composites or lead-free materials are less likely to be affected by folding and creasing.

All lead aprons, gloves, and other protective clothing should be inspected before they are put into service (Glaze *et al.*, 1984) and then periodically reinspected to determine that they provide the shielding benefit for which they were designed. Inspection procedures, the type and frequency of inspection and the criteria for removing these radiation protective devices from service should be established in institutional policy (Lambert and McKeon, 2001; Michel and Zorn, 2002). A combination of visual, physical and fluoroscopic inspection can be employed to ensure the integrity of the garments. Consideration should be given to minimizing the exposure of inspectors by minimizing unnecessary fluoroscopy.

An institutional policy might require each user to visually inspect the apron for any sign of damage prior to each use. Any suspected compromise of the shielding effectiveness of the device would require that it be removed from service pending fluoroscopic inspection, repair or replacement. Items which fail any of the inspection criteria developed in the policy would be removed from service immediately. Users should consult their state regulations for any additional requirements that might apply.

5.5.3 *Thyroid Shields*

In younger workers, the thyroid gland is relatively sensitive to radiation-induced cancer. However, the cancer-incidence risk is strongly dependent on age-at-irradiation, with very little risk after 30 y of age for males and 40 y of age for females (NA/NRC, 2006). For younger workers, wearing a thyroid collar and a protective apron reduces E to ~50 % of the effective dose achieved by wearing a protective apron alone (Martin, 2009; von Boetticher *et al.*, 2009). The use of thyroid collars (or protective aprons with thyroid coverage) is recommended for younger workers and for all personnel whose personal-dosimeter readings at the collar level (unshielded) exceed 4 mSv [$H_p(10)$] in a month (Wagner and Archer, 2004).

5.5.4 *Protective Glasses*

As discussed in Section 2.6.4.2, the threshold for radiation-induced cataracts may be lower than the currently accepted value of 2 to 5 Gy for fractionated exposure. The annual equivalent dose limit of 150 mSv (Section 5.7.1) is based on the currently accepted value. Until current dose-limit values are reassessed, it is prudent to regard eye exposure in much the same way as whole-body exposure (*i.e.*, ensure exposures are consistent with the ALARA principle). It is known that high eye doses are possible for interventionalists performing FGI procedures (Ciraj-Bjelac *et al.*, 2010; Vano *et al.*, 2008b) (Table 5.1) and cases of injury to the lens of the eye from occupational exposure have been reported, specifically for an over-table x-ray tube configuration (Vano *et al.*, 1998b). Leaded eye protection is available and has been shown to provide a reduction in eye dose (Moore *et al.*, 1980). The frontal attenuation of 0.25 mm lead-equivalent glass is ~90 %. Standard glass lenses also provide some dose reduction (30 to 35 %), though plastic lenses attenuate only a minimal amount of radiation (Marshall *et al.*, 1992).

The extent of radiation attenuation by eyeglass lenses is not an adequate descriptor, by itself, of the effectiveness of protective eyewear. For maximum effectiveness, radiation protective eyewear should intercept as many as possible of the scattered photons that are directed at the radiation worker's eyes. During interventional procedures, the interventionalist and others in the room normally turn their heads away from the primary beam to view the fluoroscopy monitor. This results in exposure of the eyes to scattered radiation from the side (Figure 4.1) Therefore, protective eyewear should provide shielding for side exposure, using either side shields or a wraparound design. Leaded glasses attenuate the scattered radiation by a factor of two or three (Moore *et al.*, 1980). Proper fit is also necessary to ensure that lenses and side shields adequately protect the eye and minimize exposure (Schueler *et al.*, 2009). Proper fit is also important to minimize discomfort from the weight of the eyewear. (Protective eyewear is considerably heavier than normal eyeglasses.) The net effect of protective eyeglasses is dependent on the design of the glasses, the nature of the clinical procedure, and the wearer's work habits.

Leaded eye protection is required for those who perform FGI procedures. Preferably, this will be provided by ceiling-suspended shields, as these devices protect the entire head, and not just the eyes. However, there are many FGI procedures where it is not practical to use ceiling-suspended shields, as they interfere with the operator's ability to perform the procedure (Miller *et al.*, 2010b). The facility should have a policy which details the circumstances under which the use of leaded eyewear (typically leaded eyeglasses) is required. As a minimum, eye protection is always required for individuals routinely performing potentially high-dose procedures. The facility's policy should also state the means by which those who may not be able to wear leaded eyeglasses (*i.e.*, the inability to have adequate prescription eyewear manufactured) may be exempted from the requirement.

When leaded eyewear use is required, the facility is responsible for providing each person who is required to use this protective equipment with individually fitted eyewear that is sized and adjusted to the individual's face, has the correct optical prescription for the individual, provides appropriate radiation protection, and is comfortable to use.

5.5.5 *Protective Gloves*

Flexible, sterile, radiation-attenuating surgical gloves are available to reduce interventionalist hand exposure. An earlier report recommended that protective gloves be worn in high-exposure situations (NCRP, 2000a). However, there are several factors that could lead to higher hand doses for interventionalists when these gloves are used (Miller *et al.*, 2010b). Just as with special tools that allow for increased distance between the hands of the interventionalist and the primary x-ray beam, the reduction in tactile feedback from radiation-attenuating surgical gloves may lead to an increase in fluoroscopy time or CT exposure time for delicate procedures. With or without added protection, the hands should not be placed in the primary x-ray beam, except for those rare occasions when it is essential for the safety and care of the patient. This should be done for the shortest possible time.

Attenuating surgical gloves have been shown to provide only a minimal reduction in hand dose for FGI and CTGI procedures (Damilakis *et al.*, 1995; Nickoloff *et al.*, 2000; Wagner and Mulhern, 1996). They may therefore produce a false sense of security in the wearer, and lead to higher dose levels (Miller *et al.*, 2010b). In addition, if hands in attenuating surgical gloves are placed in the primary beam during fluoroscopy, and intercept the automatic brightness control sensing area, the air-kerma rate incident on the patient's skin will automatically increase, resulting in an increase in dose to the interventionalist's hands and to the patient.

For FGI procedures where the hands of the interventionalist are required to be located close to the exposed patient area, careful management of the x-ray beam can be used to limit hand exposure. Proper collimation is critical. For frontal projections, the x-ray tube should be placed under the patient. The interventionalist should work on the exit-beam side of the patient and the beam should be collimated to the area of interest. For lateral projections, the beam height and collimation should be adjusted so that the primary beam is confined to the patient, thus preventing direct x-ray exposure to the hands of the interventionalist. Attenuating surgical gloves may be used to provide a small degree of protection when hands are exposed only to scattered radiation, but the use of these gloves does not permit an interventionalist to safely place the hands in the primary beam.

Recommendation 21

Each individual present in an FGI-procedure room while a procedure is in progress *shall* be provided with and *shall* use appropriate radiation protective equipment.

5.5.6 *Movable Shielding*

Various types of movable lead shields have been developed to provide added protection for interventionalists and ancillary staff (Figure 5.2). These devices are typically 0.5 to 2 mm lead equivalence, and provide nearly complete protection in the shield's shadow. Lower extremity shields on wheels or mounted to the table provide excellent protection against high scattered radiation levels and are generally unobtrusive.

Large transparent mobile shields are useful for ancillary personnel who do not need to be next to the patient but are required to remain within the procedure room. They are also useful for interventionalists and other staff to step behind during high-dose-rate digital image-acquisition series. If ancillary personnel can remain completely behind a protective barrier at all times during the procedure, there is generally no need for them to also wear a lead apron.

Ceiling-mounted shields are designed to shield the upper body of the interventionalist, with transparent upper sections and contoured lower sections that are positioned around the patient. This type of shield is particularly useful for cardiac and neurointerventional procedures, where the imaged area is separated from the catheter entrance site (where the interventionalist's hands are located). Movable lead shields should be placed to block the most intense source of scattered radiation, the patient x-ray beam entrance port. Whenever possible, the shield should be placed close to the patient entrance port, resulting in a larger protective shadow (Figure 5.3).

Another type of mobile shield is a "radiation protective cabin" which includes a transparent leaded plastic upper section, armholes for patient access, and wheels to position the cabin adjacent to the patient table. One of many such devices was reported by Dragusin *et al.* (2007). These devices are intended to protect the interventionalist. Spine and abdominal procedures present special challenges because the interventionalist is required to manipulate interventional devices in or near the irradiated area. Several reports have described various types of sterile shielding drapes placed on the patient that can result in substantial interventionalist dose reduction for these FGI procedures (Cohen *et al.*, 2007; Dromi *et al.*, 2006; King *et al.*, 2002; Kruger and Faciszewski, 2003) and for CT-guided procedures (Nawfel *et al.*, 2000; Neeman *et al.*, 2006). Use of these disposable protective drapes should be considered for complex procedures and procedures where the operator's hands must be near the radiation field (*e.g.*, management of dialysis fistulas and grafts, biliary and genitourinary interventions) (Miller *et al.*, 2010b). These drapes should not be visible in the fluoroscopic image. If they are, the result will be an increase in patient dose.

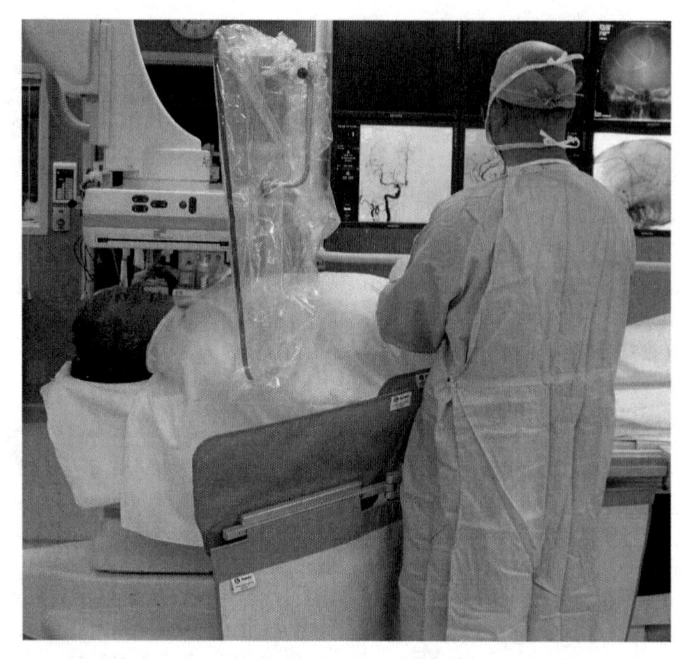

Fig. 5.2. Optimum use of movable radiation protection. The ceiling-mounted shield is perpendicular to and in contact with the patient. A floor shield (seen in front of the operator) increases the shadowed zone (photo taken circa 2010 at Mayo Clinic, Rochester, Minnesota).

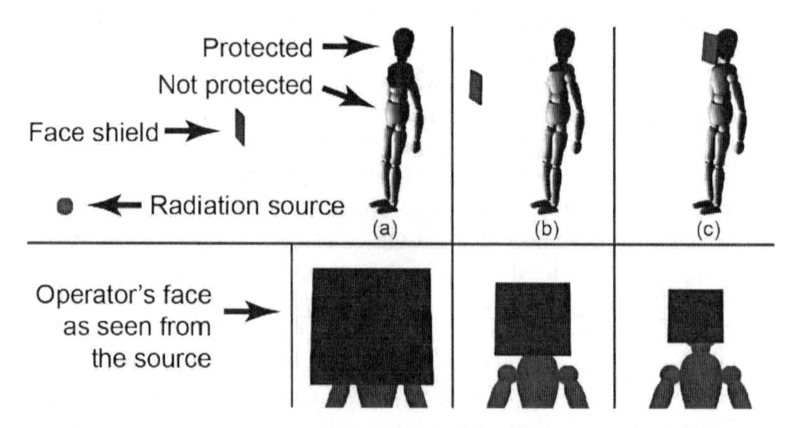

Fig. 5.3. A radiation face shield placed close to the source of radiation [location (a)] protects more of the operator's body than the same shield placed close to the operator [locations (b) and (c)]. In the figure, protected areas of the body are dark, irradiated areas are light. Corresponding views of the operator's face as seen from the radiation source are shown below locations (a), (b), and (c) of the face shield.

5.6 Situational Awareness

Proper management and awareness of radiation use is the key to safety. Before initiating exposure, the interventionalist should make sure all assisting personnel are properly protected and facing the patient, if wraparound-style lead aprons are not worn. When several staff members are present in the procedure room, a verbal warning to everyone is recommended. Assisting staff should make sure they are aware of the specific exposure indicators for the imaging equipment in use, whether they are warning lights or tones. The presence of images on the display monitor does not necessarily indicate radiation use, since either live or stored images may be visible. It is not sufficient to listen for x-ray tube or generator sounds during exposure since many newer model systems operate nearly silently.

Assisting personnel should leave the procedure room whenever their presence is not required. This is especially true during acquisition of recorded images when scattered radiation rates are much higher than during fluoroscopy.

All reasonable precautions should be used to prevent irradiation of individuals other than the patient by the primary beam. However, in some situations it is not possible to completely avoid irradiation of a portion of a healthcare worker's body by the primary entrance or exit beam. An example of necessary irradiation is the performance of cardiopulmonary resuscitation (CPR) in emergency situations; the alternative may be death of the patient. In other situations, the interventionalist's fingers might be exposed to the direct beam when inserting a medical device into the patient at the site of irradiation.

Increasing one's distance from the irradiated tissue volume is a simple way to decrease scattered radiation exposure. Staff members who do not need to be adjacent to the patient should step back. This is particularly important during high-dose-rate image acquisition. When over-table x-ray tube fluoroscopy systems are used, the light field indicator can indicate to the interventionalist where the radiation field is located on the patient's surface. Interventionalists who need to be close to the patient should incorporate additional tools and methods to allow for increased distance. These include using power injectors instead of hand injection for DSA (Hayashi *et al.*, 1998; Layton *et al.*, 2006) and using bone cement injection devices for vertebroplasty procedures (Komemushi *et al.*, 2005). Various needle holders have been developed specifically for CTGI procedures (Irie *et al.*, 2001a; 2001b; Kato *et al.*, 1996). The objective of each of these devices is to increase the distance between the interventionalist's hands and the primary x-ray beam, thereby reducing both hand dose and body dose. However, some reports have noted that some devices cause a reduction in tactile feedback, leading to increased fluoroscopy time (Kallmes *et al.*, 2003). It is important that individual interventionalists thoroughly evaluate the interventional tools in their own practice to ensure that they do not lead to increased procedure difficulty and increased fluoroscopy time.

5.7 Personal Monitoring and Risk Assessment

The purpose of a personal monitoring program is to help ensure that doses received by workers do not exceed the occupational dose limits, and to help implement the ALARA principle. The occupational dose limits recommended by both ICRP (1991a; 2007a) and NCRP (1993) have been set so that the stochastic risks of an occupational radiation dose *at the dose limit* are comparable with the overall occupational risks incurred by a worker in a "safe" occupation (*e.g.*, retail clerk, office worker).

Most monitored healthcare workers spend little if any time in a radiation environment. For 2006, the data in NCRP Report No. 160 (NCRP, 2009) indicate that 71 % of all such workers had no measurable reading on their personal dosimeters (*i.e.*, below the minimum detectable level) and an additional 24 % had readings that were between the minimum detectable level and 1 mSv y^{-1}. Workers involved in FGI procedures are in the small minority of monitored healthcare workers who may experience higher levels of radiation exposure while delivering healthcare. Appropriately modified personal monitoring policies are needed for this group of workers.

As of 2010, the risk experienced by interventionalists performing FGI procedures in comparison to a control cohort of nonoccupationally-irradiated physicians is under investigation by the Radiation Epidemiology Branch of the National Cancer Institute. The total U.S. pool of interventionalists (current and former) is ~50,000. There are ~150,000 additional physicians in various control groups. The levels of E received by most interventionalists (when personal-dosimeter readings are properly interpreted) are almost always substantially <10 mSv y^{-1}.

5.7.1 *Occupational Dose Limits*

Current NCRP (1993) recommendations for dose limits are given in Table 5.2. These dose limits are intended to prevent the occurrence of clinically-significant deterministic radiation effects and to achieve a stochastic radiation risk comparable to the overall risk of working in a "safe occupation." The linear nonthreshold dose-response model assumes that risk increases proportionally with dose. This implies that occupational doses just below the limit

TABLE 5.2—*Recommended dose limits for occupational exposure (NCRP, 1993).*

Dose Quantity	Dose Limit
Effective dose	
• Annual	50 mSv
• Cumulative	10 mSv × age (y)
Equivalent dose[a] *(annual)*	
• Lens of the eye	150 mSv
• Skin, hands and feet	500 mSv
Embryo and fetus, equivalent dose[a] *(monthly), once pregnancy is known*	0.5 mSv

[a]For x-ray fields, the equivalent dose in millisievert is numerically equal to the mean absorbed dose in milligray because the radiation weighting factor is assigned a value of unity for low linear energy transfer radiation (ICRP, 1991a; 2007a; NCRP, 1993).

are not intrinsically safe. It also implies that levels just above the limit do not inevitably lead to injury.

NCRP (1993) also states "The Council believes that most occupations and industries involving ionizing radiation can adhere to this risk level without undue difficulty, and many can operate at a much lower risk level due to the relatively low occupational exposure required to accomplish the work. There may exist, however, unusual occupational situations in which the worker population cannot carry out required functions under the recommended annual limits and, in which, risks other than radiation may be much larger than normal. Such special circumstances may require special limits and a different basis of comparison. For example, the radiation risks in these circumstances could be compared with those from the less safe industries such as public utilities and transportation, construction, agriculture and mining. In these situations, the main concern also would be for the accumulated exposure of individuals over their working lifetime and thus a career limit would be of special importance." However, it would be extremely rare to identify an interventionalist who uses optimum radiation protective practices and has actually exceeded the recommended dose limits.

As previously discussed, the nonradiation aspects of FGI procedures expose workers to greater risk than would be experienced working in a safe occupation. Nevertheless, physicians and staff performing FGI procedures have a societal obligation to care for their patients in this environment. Almost all interventionalists, and virtually all other workers involved in FGI procedures, receive occupational radiation doses well below the dose limits. However, neither relative occupational risks nor societal obligations justify ignoring dose limits. Most state regulations require exclusion of individuals from radiation work when their dose, determined *accurately*, exceeds regulatory limits.

On rare occasions, in order to save a patient's life or to prevent severe and irreparable injury to a patient, it may be necessary for the interventionalist to be exposed to a radiation dose (from that specific procedure) that when added to the cumulative dose received thus far in the year would exceed an occupational dose limit. Table 5.3 describes situations where this may be necessary. Rarely, assisting staff members may also be exposed to radiation doses that exceed an occupational dose limit. This might occur, for example if cardiac arrest occurs while a patient is undergoing an FGI procedure to reopen a narrowed or blocked coronary artery. It may be necessary for a staff member to perform CPR while the interventionalist attempts to complete the procedure. In this event, the staff member's hands will be in the primary beam during CPR, and the staff member's body will be very close to the primary beam. Direct exposure of the staff member's hands and scattered radiation to the staff member's body may lead to occupational doses that exceed dose limits. However, under these circumstances, failure to perform CPR would inevitably lead to the patient's death within minutes.

Recommendation 22

Policies and procedures *should* be in place so that in the event of a time-critical urgent or emergent situation, as defined in this Report (Table 5.3), advanced provision exists for exceeding an annual occupational dose limit.

The planned special exposure provisions in NRC (1998) describe similar circumstances within the scope of that NRC regulation. The major difference is one of time urgency. The need for an individual planned special exposure under NRC (1998) is usually known far enough in advance to allow time for approvals to be sought and to provide the training required in that regulation. When time permits, these preresponse steps would also be desirable for an urgent

TABLE 5.3—*Conditions under which an annual occupational dose limit for an interventionalist or assisting staff member may be exceeded.*

- there is a need for an urgent or emergent FGI procedure; *and*
- no qualified individual is readily available whose occupational dose would be below the annual dose limit after the procedure; *and*
- there is no time to locate and bring to the site another qualified individual; *and*
- there is no time to move the patient or the patient is too ill to be moved; *and*
- the NCRP lifetime cumulative dose limit is not exceeded by each such individual (Table 5.2).

or emergent FGI procedure. However, the necessary immediate response for many urgent or emergent FGI procedures does not allow time for the preresponse steps envisioned in NRC (1998). In these situations, this Report recommends that the FGI-procedure facility establish the necessary policies and procedures in advance. As soon as practicable after the procedure, estimates of the dose received by individuals falling under this recommendation should be appropriately reviewed by the facility.

5.7.2 *Dose-Monitoring Policy Considerations*

The vast majority of healthcare workers have little or no measurable reading on their personal dosimeters (NCRP, 2009). Even if the value of E is highly overestimated, the net occupational exposure of these individuals is small. However, the conservative models for converting personal-dosimeter readings to occupational dose often greatly overestimate the levels of E experienced by interventionalists.

NCRP recommends that all workers who are likely to receive an annual level of E exceeding 1 mSv y^{-1} be monitored (NCRP, 2000a). Individuals whose duties require them to be routinely in an FGI-procedure room during the procedures are likely to be included in this group. They should not be directly compared to the majority of healthcare workers, who are seldom if ever exposed occupationally. A department performing FGI procedures might be protecting its workers optimally even if the number and percentage of investigations are higher than in other hospital departments.

Accurate assessment of workers' occupational radiation dose can only be obtained if the workers always wear their personal dosimeters while working, wear them correctly, and their dosimeter readings are interpreted appropriately. Assessments are confounded if an individual simultaneously works in more than one facility. This work pattern is not uncommon. Physicians who work in several facilities without an employee-employer relationship at any of them are a particular challenge. As of 2010, IAEA and the European Union are investigating the feasibility of a central radiation worker database. A national database would be of great value in the United States as well. In its absence, individuals who work in more than one facility should track their own cumulative radiation dose to ensure that they do not place themselves at unnecessary risk.

5.7.3 *Operational Dose Monitoring*

Various types of radiation monitors are available. NCRP (2000a) provides detailed descriptions of several types of dose-monitoring devices. The radiation monitors and monitoring services should comply with the National Voluntary Laboratory Accreditation Program (NIST, 2008).

Worker irradiation in the FGI-procedure environment is highly nonuniform (Chida *et al.*, 2009a). Major factors affecting the dose levels actually experienced by the worker are the personal methods of working and the use of personal radiation protective equipment. A worker in the FGI-procedure environment may wear as many as three personal dosimeters (*i.e.*, on the torso, at the neck, on the hand). However, these devices indicate only the radiation level received by the device. None of these dosimeters directly measures the value of E or H_E received by the worker. The actual values require accounting for the attenuation of the radiation due to the use of protective equipment by individual workers.

Recommendation 23

Determinations of occupational doses *shall* take into account the personal protective equipment used by each individual in the FGI-procedure environment in order to properly assess compliance with occupational dose limits.

Two different methods for positioning personal dosimeters on staff wearing protective aprons are used at present in the United States. These are a single dosimeter worn at the neck outside and above the protective apron, and dual dosimeters, one worn under the protective apron at the waist or on the chest and the other worn outside and above the apron at the neck. ICRP (2000a) recommended that staff performing FGI procedures wear two dosimeters, one under the apron and one at collar level above the protective apron.

The preferred two-dosimeter method provides:

- the best estimate of E or H_E for comparison with the dose limit for stochastic effects (NCRP, 1995b);
- a better indication (from the dosimeter worn under the protective apron at the waist or on the chest) of the shielding provided by the protective apron; and
- an estimate of the dose to the lens of the eye from the dosimeter worn outside and above the apron at the neck.

A single dosimeter worn at neck level outside protective garments provides:

- a less precise estimate of E or H_E for comparison with the dose limit for stochastic effects (NCRP, 1995b); and
- an estimate of the dose to the lens of the eye from the dosimeter worn outside and above the apron at the neck (Kim *et al.*, 2008).

A single dosimeter worn under radiation protective garments is unacceptable because it provides no information about eye irradiation, and would not account for contributions to E or H_E from irradiation of organs or tissues not covered by a protective garment (*e.g.*, the thyroid if no thyroid shield is worn).

Recommendation 24

Two personal dosimeters, one worn under the protective apron and a second worn at neck level above protective garments, are preferred and *should* be used in the FGI-procedure environment.

A single personal dosimeter worn at neck level above protective garments *may* be used in the FGI-procedure environment.

A single personal dosimeter worn under the protective apron *shall not* be used in the FGI-procedure environment.

The method used at a facility may be specified by state or local regulations. The location on the body where personal dosimeters are worn should be standardized by the facility. This will facilitate comparison between workers. With any method, consistency is essential. When dual dosimeters are used, they should be worn in the specified locations and clearly labeled to avoid accidental exchange. When a single dosimeter is used, it should always be placed outside and above the apron so that exposure to the head and neck is monitored. Also, since there is a large difference between the exposure to the left and right side of the interventionalist (Balter *et al.*, 1978), it is important that the personal dosimeter at the neck be worn in the same location to avoid artificial variations in dose estimates from one monitoring period to another (Vano *et al.*, 1998c). While a single collar dosimeter has the virtue of simplicity, for FGI procedures it will produce large overestimates of E or H_E because the attenuation of the protective apron is not taken into account (NCRP, 1995b; Siiskonen *et al.*, 2007). Therefore, a policy of assigning the unmodified highest measured personal-dosimeter reading located outside and above a protective apron as the E or H_E value of record is inappropriate for any workers wearing protective aprons during the conduct of FGI procedures.

For comparison with dose limits, several calculation models have been used to estimate the regulatory dose quantity when a protective apron is worn. These can yield widely differing estimates of the regulatory dose quantity (Chida *et al.*, 2009a; Jarvinen *et al.*, 2008). As with single- or dual-dosimeter conventions, state or local regulatory agencies frequently specify the quantity (H_E or E)[6] and the calculation method. NCRP Report No. 122 provides a summary of several methods along with an estimation of their ability to estimate H_E or E. The methods discussed in that report are shown in Table 5.4 (NCRP, 1995b).

Calculation models do not yield the same numerical value for E and H_E because of the differences in the organs and tissues included in and the w_T values used in the E and H_E formulations (NCRP, 1995b). Significant overestimates will occur if results using H_E calculations are inappropriately interpreted as E. NCRP recommended E limits are given in Table 5.2. Until federal and state radiation protection regulations are revised to express dose limits in E, as recommended by NCRP, instead of H_E, one should be cognizant of which quantity is being used in a given situation.

Monitoring of hand dose is recommended if high hand exposure is expected. If hand exposure is not known or a new procedure is implemented, exposure should be monitored for a trial period of several months. Monitoring is recommended for workers who may receive an equivalent dose over 50 mSv y^{-1} to their hands (Whitby and Martin, 2005). Ring dosimeters are the most practical monitoring method. The sensitive area of the ring dosimeter should be

[6]NCRP (1993) radiation protection recommendations express dose limits as the quantity E. However, in the United States, current federal and state radiation protection implementing regulations often still include dose limits expressed as the quantity H_E. The quantity E was adopted by ICRP (1991a) and superseded the previous quantity H_E adopted by ICRP (1977).

TABLE 5.4—*Recommended methods of estimating E (or H_E) from the results of personal dosimeters when protective aprons are worn (NCRP, 1995b).*[a]

Dose Quantity Being Estimated	Dosimeters Worn	Calculation Formula[b]	Resulting Estimate
H_E	Single, neck[c]	$H_N / 5.6$	<3 H_E
H_E	Dual, waist or chest[d] and neck[c]	$1.5\ H_W + 0.04\ H_N$	0.97 to 1.72 H_E
E	Single, neck[c]	$H_N / 21$	<3.4 E
E	Dual, waist or chest[d] and neck[c]	$0.5\ H_W + 0.025\ H_N$	1.06 to 2.03 E

[a]Staff performing or assisting FGI procedures may also wear protective thyroid shields in addition to protective aprons (Section 5.5.2). These recommended methods do not account for the additional reduction in H_E and E from the use of thyroid shields.

[b]H_N = neck-dosimeter personal dose equivalent [$H_p(10)$] for strongly-penetrating radiation at a depth of 10 mm

H_W = waist- or chest-dosimeter $H_p(10)$

[c]At the neck level, outside and above the apron.

[d]At the waist or chest level, inside (under) the apron.

pointed towards the exposure source: inward when direct hand exposure is not expected and for under-table x-ray tube configurations, outward for over-table x-ray tube configurations where direct hand exposure may occur.

Equivalent dose to the lens of the eye is usually inferred from a personal dosimeter placed elsewhere on the worker's body. The preferred locations are either at the collar level outside any radiation protective garments or near the eyes.

The reading on a collar dosimeter is likely to be somewhat higher than the actual dose to the lens of the eye (Kim *et al.*, 2008). Measurements can be performed to define a correction factor if needed. The collar-dosimeter reading should be directly used in the absence of such a measured correction.

Over-table x-ray systems result in more scattered radiation to the upper body of workers performing FGI procedures than do under-table x-ray systems. Opacities in the lens of the eye have been reported with over-table x-ray systems (Vano *et al.*, 1998b). When over-table x-ray equipment is used, monitoring of the dose to the lens of the eye is required.

When protective eyewear is worn it reduces exposure to the lens of the eye. Useful attenuation depends on the size and shape of the device as well as on the working conditions of the wearer. This is seldom as high as the nominal attenuation of the protective eyewear (Moore *et al.*, 1980; Schueler *et al.*, 2009).

Recommendation 25

Monitoring of equivalent dose to the lens of the eye *should* be performed with a personal dosimeter placed either at the collar level outside any radiation protective garment or near the eyes.

5.7.4 *Investigation of Personal-Dosimeter Readings*

A personal dose-monitoring program should include periodic review of measured worker doses. This review not only ensures that dose limits are not exceeded, but also evaluates whether the dose received is at a level expected for that worker's particular job responsibilities.

At any point in time, there are expected ranges of occupational dose for different job assignments. For example, evidence of occupational dose is expected for interventionalists; no occupational dose is expected for full-time chest radiographers. Doses to individuals outside the expected dose ranges (either above or below) should be investigated, as would sudden changes in an individual's dosimeter readings.

Interventionalists and others performing FGI procedures are expected to have evidence of occupational dose on personal dosimeters worn outside of radiation protective garments. The use of these garments, including radiation protective eyewear, is intended to reduce the dose to the person (or organ) to a level below that indicated by the dosimeter. Investigation levels for these individuals should be based on job assignments and the nature of the radiation protective equipment actually used. Current practice is to investigate only "high" readings. There is no consequence for having minimum readings. Optimization of radiation protection is critically dependent on all workers routinely and consistently wearing their assigned personal dosimeters. Unfortunately, workers who perform FGI procedures often choose to not wear their dosimeters, in order to avoid investigations (Marx *et al.*, 1992). Individuals who do not wear assigned dosimeters are not in compliance with most regulatory requirements. This could cause difficulties when a facility is inspected. Not wearing a personal dosimeter is more than a regulatory compliance issue; it also leads to poor occupational dose control and increases the potential for unnecessary health effects in staff. The failure to wear personal dosimeters should not be minimized.

A facility performing FGI procedures might establish two special groups of radiation workers. These are "scrubbed individuals" and "circulating individuals." The first group is comprised of those individuals who are required to be within arm's length of the patient when the beam is on. They typically wear sterile gowns when working, hence the name "scrubbed." The second group is comprised of those individuals who are required to be in the procedure room when the beam is on, but usually work at a greater distance from the patient. Different expected dose ranges should be developed for each group. There is no need to investigate personal-dosimeter readings that are within the expected dose range. Dosimeter readings outside of the expected dose range (either too high or too low) should be investigated. There is the possibility that as radiation protection improves the upper limits of the expected dose ranges will decrease.

Too low a dosimeter reading should prompt a formal investigation. For example, collar-dosimeter readings that are <25 % of the average reading for that worker or worker group should be investigated to determine if the assigned dosimeter is being worn appropriately. In one publication, almost half of the interventionalists studied indicated they rarely or never wore personal dosimeters, presumably to avoid scrutiny of exposure practices (Marx *et al.*, 1992).

Recommendation 26

Investigations *should* occur if personal-dosimeter readings for an individual are substantially *above* or *below* the expected range for that individual's duties.

5.7.5 *Investigation of High Occupational Dose*

WHO recommends investigation when the monthly level reaches 0.5 mSv E, 5 mSv (equivalent dose) to the lens of the eye, or 15 mSv (equivalent dose) to the hands or extremities (WHO, 2000). The radiation safety officer or a qualified physicist should contact the worker directly to determine the cause of the unusual level and to make suggestions about how to keep the worker's level consistent with the ALARA principle.

Personal-dosimeter results for workers in FGI-procedure laboratories can be expected to be higher than for most other hospital workers. Most other hospital workers are expected to have minimal occupational radiation exposure. Using the same investigation criteria for both groups leads to nonproductive investigations of interventional radiologists and, often, to their reduced compliance with personal-dosimeter use. ICRP Publication 103 (ICRP, 2007a) discusses how this situation may be avoided, by considering both the need for optimization of radiation protection and the avoidance of arbitrary design or operational targets, and states: "The use of prescriptive requirements should always be carefully justified. In any event, they should never be regarded as an alternative to the process of optimizing protection. It is not satisfactory to set design or operational limits or targets as an arbitrary fraction of the dose limit, regardless of the particular nature of the plant and the operations."

Investigation of a high personal-dosimeter value begins with a check of the validity of the personal-dosimeter result. Potential sources of invalid personal-dosimeter results include wearing of designated under- and over-apron dosimeters in the wrong location, wearing of a different worker's dosimeter, and dosimeter storage in a location where it is exposed to radiation. If an invalid result is suspected, the result for the individual's next monitoring period should be reviewed to ensure the problem has been corrected.

If the dosimeters have been stored and worn correctly, the worker will be asked if there was a change in work habits that could explain the increase in radiation exposure. Typical inquiries would be:

- Was a new type of procedure initiated during the monitoring period?

- Were procedure techniques or equipment settings modified?
- If so, did these new methods require increased patient dose or closer proximity to the patient?
- Did procedure workload or complexity increase?

Sometimes, a temporary cause is found. If this is the case, personal-dosimeter levels should return to usual levels during the next monitoring period, when workload returns to normal, equipment settings are corrected, or there is additional experience with a new procedure or technique. The individual's personal-dosimeter result for the next monitoring period should be reviewed to confirm that levels have returned to the expected range.

If the cause is not thought to be temporary, or if no cause can be identified, the individual's working habits should be observed during a series of representative procedures. The observer could be a qualified physicist or a physician colleague with knowledge of radiation protection principles and the operation of the specific imaging equipment being used. The observer should pay close attention to equipment settings (particularly those that affect patient dose and P_{KA}), the worker's proximity to the patient, and the use of equipment mounted shields and personal protective devices. While individual workers may be able to assess their own working habits, an external observer has a different perspective and can point out otherwise unrecognized practices that result in high-exposure levels.

Once the cause(s) of high personal-dosimeter levels have been identified, and changes to work practices implemented, it can be helpful for the individual to wear a real-time dosimeter to provide frequent feedback of radiation dose levels. With adequate cooperation and attention to dose-reduction principles, forced limitation of workload to ensure compliance with dose limits is generally not needed.

6. Administrative and Regulatory Considerations

The goal of administrators and regulators is the efficient delivery of safe and effective procedures to patients while minimizing all worker hazards. Achieving this goal requires a balanced evaluation of all relevant risks and their mitigation whenever practicable. Although the scope of this Report is limited to radiological risk, its recommendations should be used in the context of an all-hazards risk mitigation program. Significant nonradiation safety topics, such as the management of mechanical, electrical and biological hazards in the FGI-procedure environment, are beyond the scope of this Report.

Interventionalists are the key to all fluoroscopy safety management programs. Ultimately, their knowledge, skill and experience determine the amount of radiation used during the procedure, the distribution of radiation in the patient, and staff irradiation. The interventionalist performing or personally supervising a procedure should possess both fundamental knowledge of radiation management and machine-specific training in order to optimally-control radiation utilization. Optimal patient radiation management also includes the participation of other knowledgeable staff members before, during, and after a procedure. Continuing and adequate communication between all members of the interventional team is necessary.

Recommendation 27

Every person who operates FGI equipment or supervises the use of FGI equipment *shall* have current training in the safe use of that specific equipment.

Levels of supervision for trainees performing medical procedures are defined by the Centers for Medicare and Medicaid Services (CMMS, 2002) and are reproduced in Table 6.1. For safety reasons, it is inappropriate to perform FGI procedures under

TABLE 6.1—*Levels of supervision for diagnostic x-ray tests, diagnostic laboratory tests and other diagnostic tests as defined by the Centers for Medicare and Medicaid Services (CMMS, 2002).*

- *General supervision* means the procedure is finished under the physician's overall direction and control, but the physician's presence is not required during the performance of the procedure. Under general supervision, the training of the nonphysician personnel who actually perform the diagnostic procedure and the maintenance of the necessary equipment and supplies are the continuing responsibility of the physician.
- *Direct supervision* in the office setting means the physician is required to be present in the office suite and immediately available to furnish assistance and direction throughout the performance of the procedure. It does not mean that the physician is required to be present in the room when the procedure is performed.
- *Personal supervision* means a physician is required to be in attendance in the room during the performance of the procedure.

general supervision. Direct and personal supervision levels are appropriate in defined circumstances such as formal training programs. The appropriate level of supervision is dependent on the nature of the procedure and the skill set of the trainee.

Occupational exposure in the interventional environment is not completely controllable by engineering design (*e.g.*, structural shielding, remote control instruments). Interventionalists, by minimizing patient irradiation, will simultaneously minimize the radiation fields to which staff involved in FGI procedures are exposed. All workers, including interventionalists, have a responsibility to use appropriate personal radiation protective equipment and personal dosimeters. All workers also have an obligation to communicate with each other promptly and adequately when necessary for their own safety or the safety of other workers.

The technical optimization of radiation use can only be achieved when well-maintained and appropriate equipment is used for procedures. Adequate resources should be provided to meet this goal. The proportional cost of equipment and its maintenance is seldom more than a few percent of the total costs of performing an interventional procedure (Section 6.4.1).

Regulatory, accreditation and professional organizations provide the oversight and guidance needed to meet accepted safety standards and standards of care. The requirements of such bodies are accomplished through a facility's quality program.

6.1 Human Resources

Individuals play many different roles in the support and delivery of FGI procedures. The job titles held by these individuals vary from location to location depending on facility policy and the constraints of the relevant regulatory bodies. For the purposes of this Report, the roles played by all of these individuals are grouped into one of the categories shown in Table 6.2 and discussed below.

6.1.1 *Principals (physicians and other authorized independent operators)*

Principals are interventionalists who are legally responsible for conducting or supervising individual procedures. (The professional titles of the other authorized independent operators vary from state to state and are likely to change over time.) Principals have immediate and continuing responsibilities for both patient and worker safety. These individuals are expected to identify safety deficiencies and to have the authority to implement appropriate improvements. One individual is usually the principal of record for a given procedure. Trainees in principal roles have similar safety responsibilities. Supervisors are required to be fully qualified to perform any procedure done under their supervision and to immediately supersede a trainee whenever necessary.

6.1.2 *Routine Direct-Support Staff*

Routine direct-support staff refers to those individuals who are routinely in the procedure room or the control room while a procedure is in progress. Their duty is to assist the principal in the performance of the procedure within the limits prescribed by their licenses or other credentials. Depending on local policies and regulations this may or may not include the performance of designated portions of a procedure (*e.g.*, obtaining vascular access). Typical job titles in this group include physician extenders, nurses, cardiovascular technologists, and x-ray technologists.

These individuals should be knowledgeable about the conduct of the procedure. Their main task is to facilitate the procedure. They are also expected to provide appropriate communications to the principal and to each other on matters of patient and staff safety.

6.1.3 *Occasional Direct-Support Staff*

This group includes individuals such as anesthesiologists, respiratory therapists, perfusionists, and others who are needed to attend the patient during selected procedures. All such individuals should have sufficient knowledge about FGI procedures and radiation protection to optimize their own safety.

6.1.4 *Indirect-Support Staff*

Examples of indirect-support staff include radiation safety officers, medical physicists, medical health physicists, service engineers, and biomedical engineers. Indirect-support staff should be encouraged to be an integral part of the team by regularly being present during procedures and actively providing constructive practical advice to the team to minimize radiation to the patient, physician and staff. In any event, these individuals should witness clinical procedures from time to time so that they can maximize their contributions to the optimal staff procedures and equipment configuration. Radiation protection training in the FGI-procedure environment is an important safety component.

6.1.5 *General-Support Staff*

Examples of general-support staff include nurses' aides, patient transporters, material managers, and housekeepers. These individuals are likely to spend part of their time in an FGI-procedure room. Radiation safety training is directed toward worker personal safety whether or not these individuals are formally classified as radiation workers. For example, knowing when x rays are being produced is critical.

6.1.6 *Administrative and Management Staff*

Examples include facility leadership such as departmental managers, risk management officers, and senior administrators. The facility medical board along with its officers and key staff are also included in this group. These individuals should have sufficient understanding of both the clinical and radiation protection aspects of FGI procedures so that they can provide human and material resources needed for optimized safety.

6.1.7 *Visitors to the Procedure Room*

FGI procedures can be observed by a variety of visitors, including referring physicians, students, industrial personnel, and others. No special precautions are needed if these individuals are only present in uncontrolled areas properly shielded for full occupancy. Visitors entering procedure rooms or other controlled areas need sufficient knowledge to take appropriate radiation safety precautions. In addition, facilities may need to provide personal dosimeters to such individuals.

TABLE 6.2—*Categories of individuals and their roles in support and delivery of FGI procedures.*

Category	Brief Description	Example	Directly Influences Radiation Safety for	
			Patient	Staff
Principal	Individual medically and legally permitted to perform a procedure and responsible for the patient being treated; usually performs procedures or directly supervises the performance of procedures by other authorized individuals	Physician	Yes	Yes
Routine direct support	In the procedure room or adjacent control room during the performance of most procedures	Nurse, technologist	Yes[a]	Yes
Occasional direct support	In the procedure room or adjacent control room during the performance of selected procedures	Anesthesiologist	No	Personal safety only
Indirect support	Responsible for technical items related to imaging equipment performance and radiation safety	Qualified physicists, radiation safety officers, service engineers	Yes	Yes

General support	Usually in the procedure room before or after procedures	Housekeeping	No	Personal safety only
Administrative	Responsible for human and material resources necessary for procedures	Administrator	Clinical privileges	Infrastructure
Visitors	Individuals with no duties in the procedure room	Technical representatives, students, visiting physicians	No	Personal safety only

[a]Expected to keep principals informed regarding patient positioning and dose.

6.2 Education and Training

Students and other healthcare trainees should receive basic radiation management training before observing or participating in FGI procedures.

6.2.1 *Physicians and Other Authorized Independent Operators*

Clinical training and experience is not an acceptable substitute for formal training in radiation management.

Interventionalists who perform FGI or other procedures with the potential for high patient doses require additional knowledge and training beyond that necessary for interventionalists whose practice is limited to low-dose FGI procedures. However, training all interventionalists to this level of understanding may increase overall patient and worker safety. As recommended by the American College of Cardiology, the American Heart Association (AHA), the Heart Rhythm Society, and the Society for Cardiovascular Angiography and Interventions, interventionalists should have knowledge of x-ray production and imaging, and advanced radiation biology, safety and protection, as well as knowledge of basic radiation physics and safety (Hirshfeld *et al.*, 2004). Some medical specialties (*e.g.*, pain management, interventional cardiology) require most or all of this training as part of a residency or fellowship, and candidates may be examined on this knowledge as part of the board certification process. Physicians in other medical specialties may or may not have received training or been examined on this subject matter. It is suggested that radiation management training should be integrated into all relevant residency and fellowship programs.

Interventionalists should possess both fundamental knowledge of radiation management and machine-specific training in order to control radiation utilization optimally (Hirshfeld *et al.*, 2004). Setting up fluoroscopy equipment for an FGI procedure requires not only knowledge about radiation management, but also training on how to set up a particular machine to apply that knowledge (Wagner, 2007). Each person who operates or directs the operation of fluoroscopy equipment should be trained in the safe use of that specific equipment (ACR, 2008b). Major areas of importance include mechanical operation of the equipment, radiation protection and management, and aspects of image handling and processing. Initial training should include didactic training, hands-on training, and clinical operation under a preceptor physician.

Training should be conducted or verified on arrival at a new facility and periodically thereafter as part of the reprivileging process.

Records of the training should be kept according to facility policy. Training need not be performed at or by the medical facility, provided that the facility determines that external training meets these requirements and was sufficiently recent, and the facility obtains written certification of successful completion of the training.

Didactic training in radiation management is a course of instruction in radiation technology and safety that meets guidelines established by a responsible authority. It should include the following topics:

- physics of x-ray production and interaction;
- technology of fluoroscopy machines, including modes of operation;
- characteristics of image quality and technical factors affecting image quality in fluoroscopy;
- dosimetric quantities and units;
- health effects of radiation;
- principles of radiation protection in fluoroscopy;
- applicable federal, state and local regulations and requirements; and
- techniques for minimizing dose to the patient and staff.

For interventionalists who perform potentially-high radiation dose procedures, adequate coverage for *each* of these topics will typically require 30 to 60 min (EC, 2000; Hirshfeld *et al.*, 2004; ICRP, 2000a). Thus, a minimum of a full day of formal training is needed. A condensed presentation may be adequate for other individuals. The required training may be supplied by a combination of formal lectures, supervised web-based training, and reading. Professional societies and clinical facilities have the responsibility to specify acceptable sources and versions of training materials. This phase of training should include successfully completing a written examination.

Interventionalists who have completed the initial didactic training, have passed a specific examination on this radiation management knowledge, and have had refresher training on a continuing basis since then, may use that in lieu of repeating the didactic portion of the training, at the discretion of the facility. Alternatively, newly arrived interventionalists who meet these criteria may be permitted to complete a written examination without repeating the training at the discretion of the facility.

Hands-on training means training on the operation of each fluoroscope that is to be used clinically (or an essentially similar fluoroscope), including the use of controls, activation of various modes of operation, and radiation dose displays. This training is conducted using the actual fluoroscope, or an essentially similar unit, that will be used clinically. This phase of training may include demonstrations of the effect of different modes of operation on the dose rate to a simulated patient and may include demonstration of the dose rates at various locations in the vicinity of the fluoroscope. An understanding of the mechanical motions of the equipment and the function of the various controls will minimize impact damage to equipment and injury to personnel. Training should include how to turn on a system and configure it for clinical use, along with key first-level responses (*e.g.*, recovery from a power interruption, configuration for CPR).

An interventionalist should have a fundamental knowledge of patient and staff safety in the interventional environment as well as specific knowledge of normal and emergency operation of each specific fluoroscopy unit for which clinical privileges are granted (Table 6.3). Hands-on training should be successfully completed for each model of fluoroscopy equipment before clinical privileges to use that model are granted. In particular, adequate training should be completed by all users when a new fluoroscope is installed in a facility.

Clinical operation under a preceptor physician means operation of the fluoroscope for clinical purposes under the direct supervision of a preceptor physician privileged to use and experienced in the operation of the device. Clinical operation is intended to demonstrate that the candidate manages radiation appropriately. Completion of this phase of training should include written attestation, signed by the preceptor physician, that the individual has achieved a sufficient level of competency.

Periodic radiation management refresher training should include clinically-appropriate content (*e.g.*, a journal club review of literature on radiation injuries resulting from FGI procedures). At the facility's discretion, refresher training might also include hands-on operation and clinical operation under a preceptor.

6.2.2 *Routine and Occasional Direct-Support Staff*

The major radiation safety difference between routine and occasional direct-support staff categories is the time that these individuals spend in the procedure room. Individuals in the routine direct-support category are frequently in the procedure room while x rays are being produced. Individuals in the occasional category are in the room less frequently and often have fewer responsibilities regarding patient radiation dose monitoring.

TABLE 6.3—*Essential safety-related equipment knowledge for physicians and other independent operators.*

- The management of fire safety or any other emergency situation that might require emergency evacuation of the patient and staff from the procedure room should be understood. How does one verify the urgency of an alarm? What is the best route of escape? Who is in command?
- The location of facility emergency shutdown switches and their intended use. Switch(s) typically are provided that shutdown all power to the imaging equipment. How to reset emergency shutdown switches after activation.
- The location of emergency shutdown switches that are part of the imaging equipment and the functionality that each switch provides. How to reset emergency shutdown switches after activation.
- The availability and scope of emergency power for patient life support equipment and for the imaging equipment. Is emergency power to the imaging equipment provided to allow the procedure to continue, to allow momentary operation of only fluoroscopy to allow safe egress of devices from the patient's body, or not at all?
- When emergency power is available, how to manage patient care when the power source switches from normal to emergency mode or vice versa.
- The required steps to minimize the time to recover emergency fluoroscopy after equipment shutdown whether due to loss of power or equipment malfunction (*e.g.*, equipment shutdown, equipment lockup).
- Safety-related mechanical aspects of potential collisions between gantry components and the patient's body. What types of sensors, if any, are provided to warn of collision? If collision occurs, what steps are necessary to back off the equipment from the patient's body?
- Maximum load rating of the patient support.
- Any limitations of the patient position on the patient support to allow safe emergency resuscitation of the patient. For example, does the patient's torso need to be placed directly above the pedestal of the patient support prior to emergency resuscitation.
- The safety features of the power contrast injector and its proper use.

Personnel assisting an interventionalist during procedures (physician extenders, radiologic technologists, nurses, and others) should be trained in radiation management and safety relative to their responsibilities in procedures. Patient safety is improved when these workers are trained in patient radiation safety and patient radiation dose monitoring. Members of the routine direct-support staff are expected to assist principals in tracking patient radiation dose during procedures.

Their training can also include those aspects of crew resource management (Grogan *et al.*, 2004; Musson and Helmreich, 2004; Oriol, 2006) relevant to effective communication of patient and staff safety items while procedures are in progress (*e.g.*, communicating the current radiation dose level, providing recommendations for dose-saving measures).

Direct-support staff responsible for the set-up, programming, and operation of fluoroscopy equipment should have completed both technical and radiation management training. Technical training should include use of controls and features of the equipment and various modes of operation, particularly those that affect dose and dose rate. Because system configuration is increasingly task specific, these individuals should have adequate understanding of the clinical goals and performance of interventional procedures.

Completion of necessary training is documented by appropriate current credentials. This training should enable the detection of unsafe or potentially-unsafe conditions affecting either patients or staff.

6.2.3 *Indirect-Support Staff*

The qualifications of technical support individuals (*e.g.*, radiation safety officers, medical physicists, medical health physicists, service engineers) should be evaluated. Facilities may accept appropriate licenses, board certifications, or equivalent credentials as evidence of competence.

The ability of these individuals to contribute to radiation safety during FGI procedures is enhanced if they have appropriate knowledge about the medical and technical-clinical aspects of FGI procedures. Therefore, training on the basic concepts of patient care and procedure protocol should be included. Observation of one or more clinical procedures, followed by discussion with the principal, is also part of this process.

6.2.4 *General-Support Staff*

General-support staff are usually in a fluoroscopy room only before or after a procedure. Occasionally, their services are needed while an FGI procedure is in progress. These individuals should be trained appropriately in the relevant aspects of personal radiation protection. In particular, they should receive sufficient training to recognize when x rays are being generated and to avoid inadvertently causing x rays to be produced (*e.g.*, stepping on the fluoroscopy pedal).

6.2.5 *Administrative and Management Staff*

Managers should be able to evaluate their facility's interventional program. This requires sufficient understanding of how clinical practice, image quality, radiation management and quality assurance make essential contributions to safe and effective FGI procedures. This knowledge facilitates prudent judgments about the need to purchase, upgrade or replace equipment or facilities. It also contributes to a full understanding of the necessary qualifications and roles of all individuals performing or assisting in procedures.

6.2.6 *Visitors*

Visitors should be able to look after their own safety in the procedure room. Training can be as simple as identifying "safe" areas or extended to include appropriate aspects of personal radiation protection.

6.3 Radiation Safety Credentials and Clinical Privileges

Every facility has both the ethical and the legal responsibility to ensure that all aspects of healthcare are provided by qualified individuals. This goal is accomplished by the granting of procedure-specific clinical privileges to physicians and other independent operators by the facility's medical director. Similar specific privileges are granted to others by the facility's management. Credentials are the documents and other evidence of competence presented with the request for clinical privileges. The minimum required credentials are compliance with the state health code and state licensure. Additional credentials are often required before clinical privileges are granted. This section discusses the additional credentials needed for FGI procedures.

6.3.1 *Physicians and Other Independent Operators (interventionalists)*

Each person who operates or directs the operation of FGI-procedure equipment should be privileged to perform FGI procedures by the medical facility. The credentials needed for granting of these clinical privileges should be contingent upon successful completion of training (Section 6.2.1). Maintenance of clinical privileges should be contingent upon successful completion of periodic refresher training and on compliance with regulatory and facility requirements for the safe use of the fluoroscopy equipment. Radiation safety credentials should include documentation of successful completion of appropriate initial and refresher training. FGI-procedure privileges should not be granted to individuals who are not included in the facility's personal radiation-monitoring program. Clinical experience by itself is not an acceptable radiation safety credential. Policies and procedures should ensure that only specifically-trained individuals are privileged to perform or supervise procedures.

Recommendation 28

An FGI procedure *shall* be performed or supervised only by a physician or other medical professional with fluoroscopic and clinical privileges appropriate to the specific procedure.

6.3.2 *Routine Direct-Support Staff*

Many states require radiologic technologists obtain American Registry of Radiologic Technologists certification. If not required, this credential is recommended for technologists assisting with FGI procedures, along with maintenance of current American Registry of Radiologic Technologists registration which include compliance with rules, ethics, and continuing-education requirements. For cardiovascular technologists or cardiovascular invasive specialists, certification with Cardiovascular Credentialing International is recommended, along with a current renewal of certification. Appropriate credentials for physician extenders and nurses include academic transcripts, state licenses or registration, board certification, and current continuing medical education courses.

In addition, all routine direct-support staff should be privileged to assist with FGI procedures. As with physicians, these privileges require compliance with state licensure regulations and successful completion of initial and refresher training as described in Section 6.2.2.

6.3.3 *Medical Physicists, Medical Health Physicists, Radiation Safety Officers*

Some states limit the practice of these professions to individuals possessing the relevant state-mandated credentials. In the absence of legal requirements, necessary services should be provided by a qualified physicist. Facilities should also ensure that these professionals have adequate specific knowledge of the clinical and technical aspects of FGI procedures.

6.4 Purchase, Configuration, and Technical Quality Control

6.4.1 *Equipment Economics Overview*

For the purposes of this section, consider the following hypothetical example for a major fluoroscopy system with a purchase price of one million dollars. Such a system is expected to have a useful lifetime of 5 to 10 y. An all-inclusive service contract typically costs 10 % of the purchase price per year. Thus, the cost of the equipment and its service is between $200,000 and $300,000 per year. A major FGI-procedure room is expected to accommodate at least 1,000 procedures per year. Given this model, the fluoroscopy equipment cost is $200 to $300 per procedure. Other direct costs of an individual procedure include contrast material, disposables, drugs, and medical devices. Interventionalist and support staff costs also should be considered. Personnel and supply costs can easily average several thousand dollars per procedure. Thus, without factoring in the list of less direct costs (*e.g.*, room construction), it can be seen that the fluoroscope contributes a very small percentage to the cost of performing most FGI procedures.

Purchasing minimally-acceptable equipment and limited service contracts can significantly reduce initial expenses. However, there will be little impact on the total cost of running the procedure room. The long-term nonfinancial costs of reducing equipment costs might include fewer clinical options, less advanced imaging and radiation dose management, and possibly lower equipment reliability and availability.

6.4.2 *Initial Activities*

The resources needed to perform interventional procedures are expected to comply with applicable regulations and accreditation standards. Guidelines from professional societies such as the American Association of Physicists in Medicine, the American College of Cardiology, ACR, AHA, the Society of Interventional Radiology, the Society for Cardiovascular Angiography and Interventions, and the Cardiovascular and Interventional Radiological Society of Europe also provide inputs regarding minimum and optimum requirements. These guidelines reflect the clinical expertise of the members of these societies.

Recommendation 29

Standards and guidelines provided by professional societies *shall* be considered when establishing radiation-related resources, and quality and performance requirements.

Modern fluoroscopes are complex instruments. In addition to hardware, clinical performance is strongly determined by the installed software and operational setting configuration. Manufacturers do their best to deliver systems that comply with both their own specifications and with regulatory requirements such as from FDA (2009a). The systems are preconfigured to deliver the manufacturer's understanding of typical clinical requirements. Configurations are typically changed to meet specific local requirements at the time of installation. Participation of interventionalists and qualified physicists in the purchase process and in the establishment of local configurations helps to optimize the dosimetric and performance properties of the system.

Recommendation 30

Interventionalists and qualified physicists *should* participate in the process for purchase and configuration of new fluoroscopes and fluoroscopy facilities.

Once equipment is installed, its safety and performance should be tested by a qualified physicist prior to first clinical use. Two major series of tests are performed at this time. The first, acceptance testing, verifies that the equipment complies with the manufacturer's specifications, the purchase contract, and applicable local regulations. The purchase contract may contain additional technical requirements unique to the clinical application. The second, commissioning, verifies that the equipment is properly configured for its clinical application at that facility. Additional discussion of equipment acquisition and acceptance testing is found in Appendix L.

Recommendation 31

A qualified physicist *shall* perform acceptance and commissioning tests before first clinical use of new, newly-installed, or newly-repaired fluoroscopy equipment, and *shall* perform subsequent periodic tests as part of a technical quality-control program.

6.4.3 *Technical Quality-Control Program*

A technical quality-control (TQC) program consists of periodic equipment testing, which is also referred to as acceptability or consistency testing. Initial TQC tests should be done as part of acceptance and commissioning using TQC tools and procedures. These tests should be repeated at appropriate intervals. An appropriate subset of the tests should also be performed after repairs or configuration changes that have the possibility of altering radiation output or x-ray beam confinement. Periodic testing of each fluoroscope should be performed by a qualified physicist to validate safety and performance milestones.

An acceptability test, as defined by the European Union, includes a judgment of the acceptability (fitness for use) of a system after it is tested (EU, 1996). Based on equipment characteristics and testing, a given fluoroscope may be acceptable for all proposed clinical procedures, for only a restricted set of procedures, or unusable for any clinical purpose. Servicing the equipment often lessens or removes such restrictions. However, some equipment may have to be replaced to safely accommodate the desired range of procedures.

The evaluation of fluoroscopes used for interventional purposes should include additional tests beyond those used to evaluate general purpose fluoroscopes. A full list of the required testing is beyond the scope of this Report. Two examples illustrate specific testing for FGI procedures:

- For each expected clinical mode of operation (including DSA and/or cinefluorography), patient (phantom) entrance dose rates should be evaluated with the system at maximum output at a measuring point that corresponds to the shortest SSD expected in clinical practice. Doing this always requires the system to be set at its maximum source-to-image-receptor distance (SID), and may require measurements at the end of the x-ray tube's spacer.
- Many fluoroscopes have added or built-in dosimetric systems. Periodic testing should be performed to validate the accuracy of these instruments.

Additional information on equipment testing and references can be found in Appendix L.

7. Summary and Conclusions

Fluoroscopically-guided interventional (FGI) procedures encompass a wide range of organ systems, procedure types, and physician specialties. Due to the substantial patient and societal benefits of FGI procedures, their volume has grown considerably in the last few decades. However, FGI procedures also include risks, including the potential for high patient radiation dose. Generally, the nonradiation risks of interventional procedures and the risks of alternative therapy, such as open surgery, are substantially greater than the radiation risk. Similarly, risks to personnel performing FGI procedures include both radiation risks and nonradiation risks, such as infectious disease transmission and chronic orthopedic problems. It is important to consider appropriate management of all occupational risks so that efforts to reduce radiation risk do not disproportionally increase nonradiation risks (Appendices B and C).

Though infrequent, severe skin injuries to patients have occurred as a result of potentially-high radiation dose procedures. The severity of skin and subcutaneous tissue injury depends on the skin dose, the portion of the patient's body exposed, and other patient sensitivity factors. An increase in the probability of cancer is also a radiation effect, though the risk level is relatively low for the typical older adult patient undergoing an FGI procedure. However, cancer risk is a more important consideration when justifying procedures in children and young adults. Another tissue at notable risk from radiation exposure is the lens of the eye. Though previous studies indicated that lens opacification required a relatively-high threshold dose, new findings suggest the threshold dose may be lower than previously believed. Each of these radiation effects (Appendices E, F and G) also apply to staff irradiation. To evaluate the radiation exposure from FGI procedures, this Report uses and extends the definitions of radiation quantities found in ICRU Report 74 (ICRU, 2005) (Appendix D).

To manage radiation doses for both patients and staff in FGI procedures, this Report provides specifications for fluoroscopic equipment intended for use in potentially-high radiation dose procedures. A potentially-high radiation dose procedure is defined in this Report as one where $K_{a,r}$ exceeds 3 Gy, or P_{KA} exceeds

300 Gy cm^2, for more than 5 % of cases of that procedure. The specifications for such equipment include the use of dose monitors, variable rate pulsed fluoroscopy, automatic spectral filtration, and table- and ceiling-mounted shielding (Table 3.1 and Appendix H). Additional specifications apply for FGI equipment intended for use with pediatric patients (Appendix I).

Patient radiation dose management requires active review, monitoring and recording. Prior to an FGI procedure, a review of the patient's radiation history is necessary. For potentially-high dose procedures or when significant previous radiation exposure of the patient has occurred, obtaining informed consent from the patient is recommended (Appendix J). During the procedure, radiation dose should be continuously monitored, with adjustment of imaging technique to manage skin dose as needed. After the procedure, the patient's dose should always be recorded in the medical record. In instances where an SRDL (Table 4.7) has occurred, additional actions are necessary. These actions include providing justification for the radiation dose used in the patient's medical record, informing the patient about possible skin effects, and providing follow-up (Appendix K). An SRDL is defined in this Report as a $D_{\mathrm{skin,max}}$ exceeding 3 Gy, or $K_{\mathrm{a,r}}$ exceeding 5 Gy, or P_{KA} exceeding 500 Gy cm^2, or a fluoroscopy time exceeding 60 min.

Exposure levels of personnel present in the procedure room during fluoroscopy depend predominantly on their individual actions. Therefore, radiation protection training is a critical component in management of staff radiation risk. In addition, staff members are to be provided with and utilize protective aprons, thyroid shields, leaded eyewear, and movable shields as appropriate to their exposure level and typical position within the FGI-procedure room. In order to provide the most accurate assessment of occupational radiation risk, personal monitoring with both an over-apron collar dosimeter and under-apron waist or chest dosimeter is recommended, along with calculation of E (or H_{E}) using the formulas provided in NCRP Report No. 122 (NCRP, 1995b). Given the recent data suggesting that a threshold for cataract formation may be lower than previously believed, personal monitoring of dose to the lens of the eye is recommended, using either the collar dosimeter or a separate eye-dose dosimeter. Finally, monitoring compliance among staff is recommended by setting both low and high investigational dose levels based on an individual's expected level of exposure.

Administrators have specific responsibilities regarding the performance of FGI procedures in their facility. They should ensure that FGI-equipment operators receive initial and refresher training in the function of the specific FGI equipment they use, along with clinical instruction under a trained physician. Other support staff, managers and visitors should also receive appropriate safety training. All interventionalists who perform FGI procedures are required to have both fluoroscopic and clinical privileges for these procedures. When purchasing FGI equipment, administrators should ensure that interventionalists and qualified physicists participate in the equipment and site selection processes, and should also ensure that appropriate acceptance testing of new and modified equipment is performed (Appendix L).

Listings of specific practical actions that will reduce unnecessary radiation exposure to both patient and staff during FGI procedures are provided in Section 4.3.3.1 and in Appendix M.

The goal of FGI procedures is to safely and effectively achieve desired clinical results at an acceptable cost. The FGI equipment is one of the many tools needed for this purpose. Even with the best FGI equipment, the desired results cannot be obtained without appropriately trained interventionalists and support staff. Conversely, even the best FGI-procedure team may be limited by suboptimal equipment. As is often the case, ensuring optimum human resources is the greater challenge.

Appendix A

Benefit-Cost Analyses

A.1 Diagnostic Procedures

A cost analysis of percutaneous abdominal biopsy for tissue diagnosis in 439 patients with an abdominal mass compared the costs of percutaneous biopsy as the first procedure, followed by surgical biopsy if the percutaneous biopsy was not diagnostic, with the estimated costs of surgical biopsy alone for the same patients (Silverman *et al.*, 1998). Percutaneous biopsy as the first procedure resulted in a savings of $1,376,622 for the entire group; an average savings of $3,136 per patient (in 1995 U.S. dollars).

In a retrospective analysis of 114 patients with a solitary pulmonary nodule <3 cm in diameter, whose subsequent outcome was known, Baldwin *et al.* (2002) demonstrated that percutaneous biopsy of the nodule made a cost-effective contribution to clinical management, primarily by reducing the number of unnecessary operations and by increasing agreement between physicians on the need for surgery.

FGI diagnostic procedures are not always the most cost-effective. Fraser-Hill *et al.* (1992) compared the cost-effectiveness of fluoroscopically-guided percutaneous biopsy, CT-guided biopsy, and open surgical biopsy for musculoskeletal lesions. For infections, suspected metastatic foci and a number of other types of lesions, both fluoroscopically-guided percutaneous biopsy and CT-guided biopsy were cost-effective as compared to open surgical biopsy. For suspected primary tumors, on the other hand, percutaneous biopsy and surgical biopsy were about equally cost-effective. Note, however, that this analysis did not consider the relative morbidity of the various procedures or the costs for treatment of complications.

Kliewer *et al.* (1999) compared the relative cost of percutaneous liver biopsy using CT guidance to percutaneous liver biopsy using ultrasound guidance. The variables used to construct the benefit-cost model were obtained from a series of 437 liver biopsies.

Ultrasound guidance was substantially more economical than CT guidance. Ultrasound guidance also has the additional advantage of not using ionizing radiation, but does not always provide adequate visualization of the target lesion.

A.2 Therapeutic Procedures

A.2.1 *Percutaneous Coronary Intervention*

A benefit-cost analysis of percutaneous coronary intervention (PCI) is complex because the procedure is performed for a variety of clinical indications (*e.g.*, chronic stable angina pectoris, unstable angina pectoris, and acute myocardial infarction). The most appropriate comparison may be with either intensive medical therapy or surgical coronary artery bypass grafting (CABG), depending on the particular indication.

Clinical decision making is heavily influenced by the location and severity of coronary artery stenoses as defined by coronary angiography. Minor lesions and diffuse disease not amenable to either angioplasty or bypass are treated with medication. Left main coronary artery disease and complex multi-vessel disease are most often treated surgically. Patients with more localized coronary artery disease, those with acute myocardial infarction and those with prior surgical bypass are treated with PCI when possible.

Randomized trials comparing PCI to other forms of treatment have usually enrolled only a small portion of the patients screened because of restrictive inclusion criteria that only permit enrollment of patients who are suitable candidates for both PCI and the alternative treatment under current clinical guidelines. Extrapolation of data from these trials to a "real world" situation should be done with caution.

According to recent data from AHA, the mean charges for CABG are $85,653 and the mean charges for PCI are $44,110. In-hospital mortality is 2.1 and 0.8 %, respectively (AHA, 2008). It should be noted, however, that bypass patients are usually older and more ill than those receiving PCI.

While there are obvious differences in procedure-related costs and risk, the durability of the procedure also affects cost over time. The frequency of recurrent stenosis at the angioplasty site, ultimately requiring further hospitalization and additional procedures, should also be included in the economic analysis. The Randomized Intervention Treatment of Angina (RITA-1) study compared balloon angioplasty without stent placement to CABG and found the initial cost of the latter to be twice that of the former (Henderson *et al.*, 1998). However, by 5 y of follow-up the need for repeat procedures was six times higher in the angioplasty group and there was no significant difference in cost between the two strategies (Henderson *et al.*, 1998). The Angina with Extremely Serious Operative Mortality Evaluation (AWESOME) trial randomized medically-refractory angina pectoris patients either to PCI using uncoated stents (with no drug-eluting capability) or to CABG. A PCI was less costly and as effective as surgery over a 5 y follow-up (Stroupe *et al.*, 2006). The introduction of drug-eluting stents in 2003 has had significant economic impact. These stents are markedly more effective than uncoated stents in reducing restenosis. The rate of subsequent PCI procedures for revascularization is reduced from ~20 to ~5 % as compared to uncoated stents, but drug-eluting stents are three times as expensive as uncoated stents (Holmes *et al.*, 2004).

The cost-effectiveness of the routine use of drug-eluting stents has been questioned (Kuukasjarvi *et al.*, 2007). The COURAGE (Clinical Outcomes Utilizing Revascularization and Aggressive Drug Evaluation) study, reported in 2007, compared optimal medical management to the combination of optimal medical management and PCI using uncoated stents in patients with clinically-stable coronary artery disease. The need for subsequent revascularization procedures was 10 % higher in the medically-treated group, but there were no significant differences in death rates or myocardial infarction rates over the 4.6 y median follow-up (Boden *et al.*, 2007).

The overall sense of these published trials is that, as compared to medical therapy, PCI is cost-effective for both acute ST-segment elevation myocardial infarction and acute non-ST segment myocardial infarction. In patients with acute or chronic angina without very-high-risk anatomy, PCI can be cost-effective compared to CABG. In this patient group, if symptoms are controlled with medical therapy, PCI does not decrease myocardial infarction or increase survival as compared with medical therapy. However, when clinical symptoms are not well controlled with medical therapy or the patient cannot tolerate medical therapy, PCI is indicated. There remains some ambiguity as to the role of PCI with drug-eluting stents in patients with very-high-risk coronary disease (*i.e.*, left main coronary artery stenosis and multi-vessel disease), especially in diabetic patients (Brinker, 2008). A number of large-scale randomized controlled trials specifically designed to address this issue are ongoing.

A.2.2 *Uterine Fibroid Embolization*

The benefits and risks of uterine fibroid embolization (UFE) and hysterectomy, alternative treatments for patients with fibroid tumors of the uterus, are discussed in more detail in Appendix B.2.3.1. In terms of benefit-cost, the interventional procedure (*i.e.*, UFE) is superior. Edwards *et al.* (2007) conducted a randomized, controlled trial of UFE and hysterectomy in 106 patients in the United Kingdom. At 1 y follow-up, total costs for UFE were significantly lower than for hysterectomy, with a mean difference of £951 per patient ($1,712 at the exchange rate of £1 = $1.80 prevailing at the time of the study). Beinfeld *et al.* (2004) developed a decision model to compare the costs and effectiveness of UFE and hysterectomy. UFE was equally or more effective than hysterectomy and less expensive ($6,916 versus $7,847 per patient, in 1999 U.S. dollars) across a wide range of assumptions about the costs and effectiveness of the two procedures. A cost analysis performed as part of a randomized trial of UFE versus hysterectomy concluded that UFE had significantly-lower overall costs and significantly-lower direct medical in-hospital costs (Volkers *et al.*, 2008).

A.2.3 *Endovascular Aneurysm Repair*

The aorta is the large artery that carries oxygenated blood from the left side of the heart to the body. It is a muscular-walled artery that usually withstands the pulsations of the heart's blood flow and is flexible enough to help push the blood along its course. In older patients, the aorta can suffer degeneration of the elastic fibers in the wall, leading to abnormal dilation. This dilated portion of the artery is called an aneurysm. Left untreated, aortic aneurysms can rupture, with very-high mortality. The risk of rupture increases with increasing aneurysm diameter. There is increasing clinical interest in screening patients for aneurysms with ultrasound or CT, as most patients are asymptomatic until the aneurysm ruptures (Kim *et al.*, 2007).

Other than controlling hypertension and avoiding smoking, there is no medical management for aortic aneurysms. For years, patients with aneurysms approaching a size where the risk of rupture becomes substantial were treated with an open surgical procedure. A graft made of a Dacron® (Unifi Manufacturing, Inc., Greensboro, North Carolina) tube was used to exclude the aneurysm. Endovascular aneurysm repair (EVAR) has recently become an important alternative to open surgical repair. EVAR is the placement of an endograft (a flexible cylinder of metal mesh covered in a flexible plastic cloth). Unlike open repair, which requires an incision big enough to visualize the entire aneurysm, endovascular repair is performed through a small incision in the groin at the leg crease. The endograft is introduced through a hard plastic tube (catheter) placed temporarily into the common femoral artery in the groin. Using fluoroscopic guidance, the device is slid into position through the catheter in a compressed form, and then deployed inside the aneurysm. The device springs open to make contact with normal artery wall above and below the aneurysm. In most cases, the endograft excludes flow into the aneurysm sac. DSA is used to help guide placement and to avoid covering any important branches of the aorta.

EVAR is an important alternative to open surgical repair because patients usually spend less time in the hospital and have a quicker recovery. EVAR also provides a treatment option for those considered high risk for surgery. Not every aneurysm can be treated using EVAR; there are anatomic constraints that can make EVAR difficult or impossible. Patients with a known aneurysm undergo a CT scan with intravenously-administered contrast material to determine whether or not they are a candidate for EVAR.

Initial interest in EVAR was due to its minimally-invasive nature, which offered patients a quicker recovery. Shortcomings with EVAR, such as continued filling of the aneurysm sac (endoleak) and the need for reintervention to stop endoleaks make it clear that EVAR is not yet a replacement for surgery.

Aneurysms can occur anywhere along the length of the aorta, from the heart to the lower abdomen. For accurate comparisons between open surgery and EVAR, the efficacy and costs of treatment of aneurysms in the chest [thoracic aortic aneurysm (TAA)] and the abdomen [abdominal aortic aneurysm (AAA)] are considered separately. Cost analysis for EVAR includes:

- price of the initial procedure (EVAR is more expensive than open repair, despite the shorter hospital stay, due to the cost of the devices used);
- follow-up (EVAR is more expensive due to the need for serial follow-up CTs);
- cost of reinterventions (more common with EVAR); and
- cost of morbidity associated with the procedure (generally much higher with surgery).

AAA repair is the more common of the two procedures and has been managed by EVAR the longest. In a randomized trial of 1,082 patients conducted in the United Kingdom, EVAR cost 33 % more after 4 y of follow-up (Greenhalgh *et al.*, 2005). Similar results have been found in other, less complete, trials conducted in the United States (Sternbergh and Money, 2000).

Experience with EVAR for TAA is less extensive. Hospital costs associated with EVAR are lower than for open repair due to the much shorter hospital stay and decreased procedure morbidity with EVAR (Glade *et al.*, 2005).

The high costs associated with EVAR for AAA and TAA are mostly due to the high costs of the devices used in AAA and TAA repair. For AAA, device costs are not completely offset by the shorter hospital stay for EVAR, but for TAA they probably are. Despite the device costs, EVAR usage continues to grow, due to reduced procedure-related morbidity and reduced mortality, particularly in high-risk patients (Greenhalgh *et al.*, 2005). As the cost of these devices decreases, and methods to prevent reintervention improve, EVAR will continue to grow as an important alternative to surgery.

Appendix B

Benefit-Risk Analysis

B.1 Diagnostic Procedures

FGI diagnostic procedures consist primarily of percutaneous biopsies and angiography. Percutaneous biopsy is currently performed on almost all organ systems. Conventional catheter angiography was a mainstay for diagnosis of vascular disease and some tumors for decades, but more recently has largely been replaced by CT angiography and magnetic resonance angiography, which are less invasive and require only administration of a contrast agent through a standard intravenous line. Catheter angiography is now used most frequently for evaluation of the heart and coronary arteries (Table 2.2). Only a sample discussion of the benefits and risks for percutaneous biopsy is provided in Appendix B.1.

Up until the 1970s and early 1980s, tissue biopsy of nonpalpable lesions was generally done by surgeons, either as part of a larger surgical procedure or as a separate surgical procedure. Fluoroscopically-guided needle biopsy was introduced in the late 1930s, but technological limitations in fluoroscopy equipment limited its usefulness for several decades (Hopper, 1995). In the 1970s, the introduction of CT, specialized ultrasound biopsy transducers and 22-gauge (0.7 mm diameter) "skinny" needles made it possible to perform image-guided needle biopsy of virtually every organ in the body with very low risk (Hopper, 1995). Percutaneous, image-guided biopsy is now the standard method for obtaining tissue diagnoses and identifying infectious organisms. In most patients, it is less invasive, less risky, and less expensive than surgical biopsy, and equally likely to be diagnostic (Adler *et al.*, 1983; Logrono *et al.*, 1998).

Percutaneous biopsy can be performed with a wide range of needle types and sizes, using guidance from ultrasound, fluoroscopy or CT. Biopsy needles may be as small as 25 gauge (0.5 mm) for aspiration of cells for cytological examination or as large as 11 gauge (3.05 mm) for excisional core biopsies (Gazelle and Haaga, 1991;

Jensen *et al.*, 1997). These needles are available in a wide variety of tip configurations as well. Needle choice depends on the lesion location and texture, as well as on interventionalist preference. The choice of imaging modality is based on the size, location and depth from the skin surface of the lesion to be biopsied; the ability to visualize the lesion with each of the imaging modalities; and on the experience and preference of the interventionalist.

Ultrasound is attractive for guiding biopsies because it can provide real-time images in multiple planes, does not use ionizing radiation, and is relatively inexpensive compared to CT (Kliewer *et al.*, 1999). It is frequently used for biopsy of abdominal and pelvic masses, especially those in the liver, kidney and retroperitoneum. It is also very useful for biopsy of relatively-superficial soft-tissue masses, such as those in the thyroid and breast. It is less useful for guiding biopsies when the target is relatively deep or obscured by overlying air, bowel gas, or bone.

Fluoroscopy is commonly used to guide percutaneous biopsy of pulmonary and pleural masses and bone lesions. Pulmonary and pleural lesions are visible because they are outlined by air in the lung; bone lesions are visible because of the attenuation differences between bone and soft tissue. Fluoroscopy provides real-time imaging guidance, but requires ionizing radiation.

Fluoroscopy can also be used to guide transvenous biopsy of the liver and kidney. This approach is often chosen in patients with bleeding or clotting disorders and diffuse abnormalities of the liver or kidney because of the decreased risk of bleeding into the abdomen or retroperitoneum when the transvenous route is used. The most common route is transjugular. The catheter and biopsy needle are introduced into the jugular vein in the neck, and guided into a hepatic vein for liver biopsy or into the renal vein for kidney biopsy. Adequate biopsy specimens can be obtained in 95 to 98 % of procedures in adults and children (Cluzel *et al.*, 2000; Smith *et al.*, 2003).

CT is used primarily to guide biopsy of small or deep lesions in the chest, abdomen and pelvis that are not seen well with ultrasound or fluoroscopy. CT provides high-resolution images and the ability to visualize bowel and bone. In a prospective series of 1,000 consecutive CT-guided biopsies performed in all regions of the body from 1973 through 1983, sensitivity was 91.8 %, specificity was 98.9 %, the positive predictive value was 99.7 %, and the negative predictive value was 73.3 % (Welch *et al.*, 1989). There were 11 complications (1.1 %) and no deaths. In a series of 1,015 percutaneous lung biopsies performed with 22 gauge needles, a diagnosis was established from the biopsy specimen in 94.6 % (960/1,015) of biopsies (Johnston, 1984). In a series of 510 percutaneous

CT-guided liver biopsies, sensitivity was 94 % and overall accuracy was 92 % (Luning *et al.*, 1991). Bommer *et al.* (1997), in an evaluation of percutaneous needle biopsy of bone lesions, predominantly obtained with CT guidance, reported that an adequate specimen was obtained from 385 (86 %) of 450 biopsies. For these specimens, the sensitivity was 95.7 %, the specificity was 99.3 %, the positive predictive value was 99.6 %, and the negative predictive value was 93.6 %.

When ultrasound guidance is not appropriate, CT guidance is effective for percutaneous biopsies in children as well as adults. The procedure can usually be performed with intravenous sedation rather than general anesthesia (Cahill *et al.*, 2004; Klose *et al.*, 1990). In a series of 75 CT-guided lung biopsies in pediatric patients, 64 biopsies (85 %) were diagnostic, with one complication (1.3 %) (Cahill *et al.*, 2004). In another series of 44 CT-guided biopsies performed at 41 various sites in the skull, chest and abdomen of 39 pediatric patients, sensitivity and specificity were both 100 % for the diagnosis of malignancy (Klose *et al.*, 1990). Classification of benign or inflammatory disease was correct in 24 of 26 biopsies (92 %). There was one complication (2 %) as a result of the 44 biopsies. General anesthesia and surgery were avoided in all but two patients.

CT fluoroscopy (CT performed either in a continuous or intermittent mode) (Carlson *et al.*, 2001) is used for needle guidance during drainage of fluid collections; spinal pain management procedures; and percutaneous needle biopsy in the chest, spine, abdomen and pelvis. The principal advantage of CT fluoroscopy over conventional CT guidance is the ability to use real-time monitoring to access lesions that move within the body as a result of patient breathing or other motion. Two methods of CT fluoroscopy are used: continuous, real-time guidance during biopsy needle placement (*i.e.*, continuous fluoroscopy technique) or intermittent imaging of the needle between needle manipulations (*i.e.*, quick-check technique) (Paulson *et al.*, 2001). Although some reports indicate that CT fluoroscopy is superior to conventional CT guidance in that it permits higher success rates and shorter procedure times (Gianfelice *et al.*, 2000), other reports indicate no statistically-significant difference in these indices (Silverman *et al.*, 1993).

B.2 Therapeutic Procedures

FGI therapeutic procedures are performed in a number of different organ systems and regions of the body (Table 2.1). This section briefly describes several relatively common interventions performed in a variety of different organ systems and regions of the body.

B.2.1 *Interventions in the Central Nervous System*

FGI procedures in the central nervous system are preformed in the brain and spinal cord, primarily for treatment of tumors and vascular abnormalities (aneurysms and AVMs).

An intracranial aneurysm, also called a cerebral or brain aneurysm, is an abnormal outward bulging of one of the arteries in the brain. It is estimated that as many as one out of every 15 people in the United States will develop a brain aneurysm during their lifetime. Brain aneurysms are sometimes discovered when they rupture and cause bleeding into or around the brain. This bleeding is called a subarachnoid hemorrhage. A ruptured brain aneurysm is a serious condition; as many as 50 % of patients with a ruptured aneurysm die within the first 30 d. If a patient survives the initial brain aneurysm rupture, the main goal of treatment is to block the flow of blood into the aneurysm in order to prevent another episode of bleeding. Sometimes, unruptured brain aneurysms are treated to prevent rupture, depending on the size, shape and location of the aneurysm (Wiebers *et al.*, 2003).

Traditionally, treatment of both ruptured and unruptured aneurysms was accomplished with invasive surgery. This surgical procedure, called craniotomy and clipping, required temporarily removing a piece of the skull and placing a metal clip on the outside of the aneurysm to stop blood from entering the aneurysm. In the 1990s, a minimally-invasive treatment option was developed to treat brain aneurysms from the inside, through the artery. This technique, known as endovascular embolization or aneurysm coiling, is an FGI procedure that does not require the removal of a portion of the skull. The procedure is performed by interventional neuroradiologists, endovascular neurosurgeons, and interventional neurologists. Microcatheters are carefully navigated with fluoroscopic assistance through the arterial tree and into the aneurysm. Small platinum coils are then placed in the aneurysm through the microcatheter until there is no longer any blood flow into the aneurysm.

While there are still circumstances when an aneurysm is best treated with traditional craniotomy and clipping, many aneurysms are now treated with the less invasive coiling technique. Molyneux *et al.* (2002), the International Subarachnoid Aneurysm Trial, is an international randomized trial that examined the outcomes of patients treated with either neurosurgical clipping or endovascular coiling for ruptured aneurysms. It enrolled 2,143 patients. This study demonstrated a significantly better outcome at 1 y after treatment in those patients treated with endovascular coiling compared to those treated with neurosurgical clipping. At 1 y, the relative and absolute risk of death or inability to live independently were reduced by 22.6 and 6.9 %, respectively, in those patients who underwent endovascular coiling. Although no multi-center randomized clinical trials comparing endovascular coiling and neurosurgical clipping of unruptured aneurysms have been conducted, retrospective studies have found that endovascular coiling is safer, with shorter hospital stays and shorter recovery times compared with surgery. In a large, retrospective study, the average hospital stay for a patient treated with endovascular coiling was 4.6 d compared with 9.6 d in patients treated with neurosurgical clipping (Johnston *et al.*, 1999). This difference in length of hospital stay has likely increased as more experience has been gained with the coiling procedure. The majority of patients who undergo elective endovascular coiling now require only an overnight stay in the hospital.

In a separate study, investigators found that a significantly-increased recovery time was required for neurosurgical clipping (1 y) compared to endovascular coiling (27 d). Furthermore, the rate of new symptoms or disability was significantly higher in the clipping group (34 %) versus the coiling group (8 %) (Johnston *et al.*, 2000). In adult patients, the major advantages of endovascular embolization greatly outweigh any radiation-related risk from the FGI procedure. Stochastic risk is of greater concern in pediatric patients, but is still outweighed by the considerable morbidity and mortality of surgery or no therapy (Thierry-Chef *et al.* 2006). Patient radiation dose from the procedure is presented in Section 4.1.2 and Table 4.3.

B.2.2 *Interventions in the Heart*

Cardiac interventions are performed for the treatment of arterial lesions, dysrhythmias due to abnormal electrical conduction in the heart, valvular lesions, and congenital heart disease. PCIs for treatment of coronary artery disease and EP procedures for treatment of dysrhythmias are described below.

B.2.2.1 *Percutaneous Coronary Interventions.* There are three main arteries in the heart of most individuals that supply blood to the heart muscle. These are the right coronary artery and the two branches of the left coronary artery: the circumflex and the left anterior descending coronary arteries. The latter two branches arise from a single trunk called the left main coronary artery. In ~15 % of the population, the left main coronary artery trifurcates with the middle branch called the ramus intermedius. This vessel supplies the high lateral wall of the heart and is an anatomic variant of either the high marginal branch of the circumflex artery or the high diagonal branch of the left anterior descending artery.

Selective imaging of the coronary arteries using fluoroscopy and cineangiography was first performed in 1959.

Atherosclerosis causes narrowing or blockage (stenosis) of various arteries throughout the body, including the heart. Until 25 y ago, the cardiac catheterization laboratory was primarily a diagnostic facility, used in patients with chest pain (angina pectoris) and/or positive stress testing to identify coronary artery disease. Approximately 15 to 30 % of the entire adult population studied with coronary angiography had normal coronary arteries. This reflected limitations in the specificity of the clinical and noninvasive test criteria used for patient selection. However, as the sensitivity and specificity of noninvasive studies has improved, this percentage of normal studies has progressively declined.

In the past, patients with coronary artery disease had two therapy options, medical therapy or CABG. A third option became available in 1977 when the first PTCA was performed, using a balloon catheter to dilate a narrowed segment of a coronary artery. Refinements in both diagnostic and interventional equipment and devices have occurred since. Percutaneous coronary intervention (PCI) was initially restricted to patients with normal cardiac function who had a single, noncalcified lesion in the proximal portion of one coronary artery and who were suitable candidates for CABG. However, PCI is now performed as preferred therapy in many groups of patients and routinely performed in patients who are not candidates for CABG (King *et al.*, 2008; Smith *et al.*, 2006). Worldwide, the number of PCIs has grown rapidly (Section 2.2.3).

PCI with balloon dilation alone had serious complication rates, due both to acute vessel closure during the procedure, requiring emergent CABG in nearly 5 % of cases, and restenosis of the artery within the first six months after PCI. Restenosis is caused by the development of scar tissue during the reparative process after PCI and occurs in over 30 % of balloon angioplasty cases.

Stents, small metal scaffolds mounted on a balloon catheter and implanted on the inside surface of the coronary artery, were introduced in the early 1990s. Their use has reduced the emergent CABG rate to 0.4 % (King *et al.*, 2008). In-hospital death rates after PCI are currently 0.4 to 1.9 %, including those patients who present with cardiovascular collapse, called "cardiogenic shock." The rate of myocardial infarction (or heart attack) after PCI is 0.4 to 4.9 %. Overall success rates for PCI are currently 91 to 92 % (King *et al.*, 2008). Bare metal stents have clinical restenosis rates of 15 %. Drug-eluting stents have anti-proliferative drugs bound to a polymer coating on the stent. These drug-eluting stents have demonstrated a further improvement in clinical restenosis rates, to 5 %, accompanied by a small increase in the incidence of late thrombosis (development of blood clots within the stent) as compared to bare metal stents. The incidence of late thrombosis in patients treated with drug-eluting stents is <0.5 % (Smith *et al.*, 2006).

Radiation doses from PCI are presented in Table 4.2; the radiation doses and radiation risk from this procedure are discussed in Section 4.1.1.

There are few absolute contraindications to PCI. Those patients considered at the highest risk for PCI include patients with stenosis of the left main coronary artery who are surgical candidates, the presence of diffusely diseased coronary arteries, or a single remaining conduit for myocardial circulation. These patients should undergo PCI only after a lengthy discussion with both patient and surgeon. Left main and multi-vessel PCI is more frequently performed, and is a reasonable alternative to CABG in selected patients (Serruys *et al.*, 2009). However, CABG remains the preferred therapy for many of these patients, particularly those with diabetes (King *et al.*, 2008; Smith *et al.*, 2006).

Therapy for coronary artery disease has two primary goals or endpoints. The first is symptom relief and improvement of quality of life. The second is to reduce cardiac mortality. In general, both PCI and CABG are effective for symptom relief. Except for certain high-risk situations (*i.e.*, left main coronary artery disease, triple vessel disease, acute coronary syndrome) there is little evidence of a reduction in cardiac mortality for interventions as compared to intensive medical therapy alone. The choice of therapy is based on patient factors such as symptom presentation (stable versus unstable angina); extent of symptoms (Canadian Cardiovascular Society classification of functional status); the presence or absence of other medical conditions (*e.g.*, diabetes, chronic kidney failure); and cardiac diagnostic findings such as the location and extent of arterial disease, ventricular function, and disease of the cardiac valves. Medication intolerance, loss of productivity from delayed return to work, the need for repeat procedures, pain and suffering post-procedure, and morbidity and mortality are all taken into consideration in choosing among the strategies of medical (*i.e.*, drug) therapy, PCI, or surgical revascularization (Patel *et al.*, 2009).

Limited data are available comparing medical therapy to PCI in stable patients (*i.e.*, those patients with slowly progressive narrowing of the coronary arteries) (King *et al.*, 2008). The Clinical Outcomes Utilizing Revascularization and Aggressive Drug Evaluation (COURAGE) trial was designed to determine whether PCI in the stent era, coupled with optimal medical therapy, reduces the risk of death and nonfatal myocardial infarction in patients with stable coronary artery disease (and without high-risk coronary artery anatomy), as compared to medical therapy alone. The trial included 2,287 patients randomized to PCI with medical therapy versus medical therapy alone. No difference was observed between the two groups in the primary endpoints of death and nonfatal myocardial infarction, but relief of angina was significantly improved in the PCI group (Boden *et al.*, 2007). In patients presenting with acute coronary syndrome (*i.e.*, a ruptured plaque in the coronary arteries resulting in unstable angina or myocardial infarction), studies have demonstrated improved prognoses in patients treated with early PCI versus medical therapy (Anderson *et al.*, 2007).

Acute myocardial infarction is typically due to occlusion of a coronary artery by a blood clot that has formed in an area of preexisting stenosis. The preferred therapy is PCI if the artery can be opened within 90 min of presentation to the emergency department or first medical contact (Antman *et al.*, 2008). Medical therapy in conjunction with PCI (facilitated PCI) or instead of PCI (thrombolytic therapy with "clot-busting" drugs) has also been studied. Facilitated PCI has not been established as beneficial compared to primary PCI alone. Intravenous thrombolytic therapy has been a mainstay for reestablishing arterial patency (reperfusion) for over 25 y, but is less effective than PCI as primary therapy within the first 90 min after medical contact (Antman *et al.*, 2008). When PCI is not available within 90 min, thrombolytic therapy should still be considered. Due to the delays in operating room availability and patient preparation, CABG is reserved for the most critically ill patients when PCI is not an option or has been unsuccessful.

Numerous studies over the years have compared PCI to CABG in patients with complex coronary artery disease. A 2007 review analyzed 23 randomized controlled trials of PCI versus CABG, with a total of 9,963 patients (Bravata *et al.*, 2007). The early studies, begun in the 1980s and early 1990s, enrolled patients treated with angioplasty alone. Bare metal stents were included in clinical trials in the mid 1990s. Only one small study of drug-eluting stents has been completed as of 2008. Approximately 25 % of the study population has been female. Patients with prior CABG, patients older than 75 y, and patients with poor ventricular function have typically been excluded.

Several conclusions can be drawn from this overview of PCI versus CABG (Bravata *et al.*, 2007). Procedural survival is high for both PCI and CABG. Fifty-nine of 5,019 (1.2 %) PCI patients died compared to 85 of 4,944 (1.7 %) in the CABG group. Eleven of the 23 studies reported survival data at 1 and 5 y. There was no difference between the two groups. Relief from angina was more common in the CABG group than the PCI group at 1 and 5 y [75 % (PCI) versus 84 % (CABG) and 79 % (PCI) versus 84 % (CABG), respectively]. The need for repeat revascularization was higher in the PCI group in comparison to the CABG group at 1 and 5 y (23 % at 1 y and 33 % at 5 y). Quality-of-life issues were often related to recurrent angina. Quality of life was slightly better for the CABG group overall, but the difference was not significant in most trials. The cost per patient was lower in the PCI group initially. However, due to the more frequent need for repeat revascularization in PCI patients, over the long term there was only a modest cost reduction for the PCI group as compared to CABG. Finally, most subsets of patients had similar clinical outcomes at 5 y, with the exception of diabetic patients. In one of the trials (performed prior to the availability of coronary artery stents), the survival rate at 5 y for nondiabetic patients was 91.1 % for PCI and 91.1 % for CABG (Bravata *et al.*, 2007). In the same trial the 5 y survival rate for diabetic patients was 65.5 % in the PCI group and 80.5 % in the CABG group.

Though the randomized controlled trial is the most definitive method for comparing therapies, large clinical registries (*i.e.*, regional and referral center databases from Canada, New England, New York State, Duke University, and the Cleveland Clinic) provide useful insights into practice patterns (Bravata *et al.*, 2007). These registries demonstrate striking differences between patients referred for PCI versus CABG. This suggests that relatively few patients would be eligible for randomization, because few would be reasonable candidates for both PCI and CABG. This calls into question the clinical relevance of published randomized controlled trials. Interventional cardiologists most commonly treat patients with symptomatic but less extensive disease, while patients with complex coronary artery disease are referred for CABG (Patel *et al.*, 2009).

B.2.2.2 *Electrophysiology Procedures*. The normal cardiac rhythm is generated by electrical impulses that arise in the sinus node in the wall of the right atrium. The sinus node acts as a physiologic pacemaker that is immediately responsive to the body's need for cardiac output. These impulses are conducted by components of the cardiac conduction system to the remainder of the heart. Abnormal rhythms (dysrhythmias), which may be either very slow or very fast, can be due to abnormalities of electrical impulse generation or conduction. Dysrhythmias may be treated with drugs, surgery, or interventional procedures. The interventional procedures are termed electrophysiology (EP) procedures. They are performed with fluoroscopic guidance, either for device placement or for ablation of abnormal cardiac conduction pathways. As implantable devices advanced beyond simple pacemakers, many procedures moved from the operating room to the cardiac catheterization/EP laboratory where they can be performed with intravenous sedation, low morbidity and mortality, and a short hospital stay. EP cardiologists became primarily involved in lead placement and device programming (Perisinakis *et al.*, 2005). Finally, as specific pathways causing arrhythmias were identified in the heart, EP therapeutic procedures became available to ablate these pathways using fluoroscopically-guided catheters, instead of open surgical techniques.

Device implantation is commonly performed in the cardiac catheterization laboratory/EP suite. This is the most suitable place for pacemaker implantation, as it provides a fixed C-arm fluoroscopy unit suitable for guiding complex placement of pacemaker leads within the moving heart and an environment that is optimally suited to cardiac imaging. The EP suite also provides diagnostic electrophysiologic monitoring equipment for assessment of complex dysrhythmias, a sterile environment for programming of complex devices, and patient holding areas that are dedicated to patients undergoing catheter-based procedures. Guidelines for pacemaker indications were revised in 2008 (Epstein *et al.*, 2008).

One device implanted frequently is a pacemaker. Patients with an enlarged heart, poorly functioning heart muscle, and congestive heart failure often have abnormal conduction of electrical impulses within the heart, such as prolonged conduction between the atria and the ventricles and prolonged conduction within the ventricles. Prolonged ventricular conduction, with the left and right ventricles beating asynchronously, occurs in one-third of patients with advanced congestive heart failure and has been associated with worsening heart failure, sudden cardiac death (SCD), and increased total mortality (Epstein *et al.*, 2008; Perisinakis *et al.*, 2005).

To resynchronize heart contraction, percutaneous pacemaker leads are placed in the right ventricle and the left ventricle through the coronary sinus, a vein at the base of the heart. This allows multi-site ventricular pacing which can improve ventricular systolic function, reduce medical costs, improve heart valve function, improve functional capacity, and decrease mortality (Epstein *et al.*, 2008; Perisinakis *et al.*, 2005). This technique of biventricular pacing, known as cardiac resynchronous therapy, was shown in the Comparison of Medical Therapy, Pacing, and Defibrillation in Heart Failure (COMPANION) trial to reduce overall mortality by 36 % compared to medical therapy in patients with advanced heart failure (Bristow *et al.*, 2004). No similar surgical alternatives are available.

Another important device is an implantable cardioverter defibrillator. This is used to treat patients at risk for SCD which is caused by the abrupt onset of cardiac dysrhythmias incapable of sustaining life. Untreated, these dysrhythmias are followed within minutes by loss of consciousness and death. These patients seldom survive a dysrhythmic event if it occurs outside the hospital. If the patient does survive, there is a 30 to 50 % likelihood of recurrence within 2 y of the initial SCD event. Medical therapy alone is not sufficient to prevent recurrent SCD (Zipes *et al.*, 2006).

An implantable cardioverter defibrillator is a small device implanted into a patient that is designed to diagnose life-threatening arrhythmias and to deliver a small electrical shock to the heart to restore normal contraction. The patients at highest risk for SCD are those with poor ventricular function, known as cardiomyopathy, which can be the result of previous heart attacks, viral infections, abnormal heart valves, toxins, or other causes. When compared to patients receiving medical therapy alone, patients who receive an implantable cardioverter defibrillator have 23 to 55 % lower mortality, with the improvement in survival due almost entirely to a reduction in SCD (Zipes *et al.*, 2006).

The cardiac catheterization laboratory/EP suite is also used to perform ablation therapy. Two major advances over the past two decades permit percutanous therapy of dysrhythmias. Three-dimensional mapping identifies the abnormal conduction pathways and sites of origin for electrical impulses within the heart that are responsible for symptomatic, life-threatening dysrhythmias. Advances in catheter technology permit the delivery of RF energy to specific areas of the heart to damage or ablate these abnormalities and eliminate the arrhythmia.

Catheter-based, nonsurgical ablation of conduction abnormalities in the ventricles is performed to treat ventricular tachycardia, a life-threatening arrhythmia. This is most frequently accomplished using RF energy, but may also be done with catheter-delivered electrical shock, microwaves, or lasers. The success rate for catheter-based ventricular tachycardia ablation is variable and depends on the cause of the ventricular tachycardia (Zipes *et al.*, 2006). Surgical alternatives to catheter-based ablation require major cardiac surgery. The short- and long-term success rates of surgical therapy are derived from older literature. Few reports are available to assess benefit-risk in the current era (Zipes *et al.*, 2006). Surgical ablative techniques have a <10 % recurrence rate, but increased morbidity and mortality as compared to catheter-based techniques. Due to the invasive nature of the procedure, operative ablation is now most commonly performed only when other cardiac surgery is also necessary (CABG or heart valve surgery) (Zipes *et al.*, 2006).

A number of dysrhythmias that arise above the level of the ventricles (supraventricular dysrhythmias) can cause significant symptoms, and on occasion can be life threatening. Wolff-Parkinson-White syndrome involves an accessory conduction pathway. Medical therapy is moderately effective in decreasing symptoms, but does not decrease SCD. Until the development of catheter-based ablation therapy, open-heart surgery was required to destroy the accessory pathway. Catheter ablation using RF energy began in 1982 and currently has a >90 % success rate and a 3 % complication rate (Marine, 2007). When medications are not tolerated or are ineffective, RF ablation is performed and has a 97 % success rate, a 5 % lifetime recurrence rate, and a complication rate (heart block requiring pacemaker therapy) of 0.5 to 1 % (Marine, 2007).

Atrial fibrillation (AF) is the most frequent supraventricular arrhythmia. Normal heart rhythm can be maintained using anti-dysrhythmic drugs in one-half to two-thirds of patients, but these drugs may affect quality of life and reduce left ventricular function. The Atrial Fibrillation Follow-Up Investigation of Rhythm Management Study addressed this issue, comparing therapies in two patient populations over 65 y of age (*i.e.*, treatment of AF by controlling heart rate with drugs versus treatment of AF by restoring normal heart rhythm) (Wyse *et al.*, 2002). Though no difference in major bleeding, death, stroke, or quality of life was seen between the two groups, restoring normal heart rhythm with anti-dysrhythmic therapy or catheter ablation is recommended in symptomatic individuals in whom the dysrhythmia interferes with regular activities.

The surgical technique for AF ablation, the Maze procedure, is based on the concept of dividing the atria into sufficiently small segments so that no one area of the atria can maintain AF. Several centers have reported long-term maintenance of sinus rhythm in 75 to 95 % of patients. Because of the complexity of the operation and the potential complications of open-heart surgery, the Maze procedure is typically performed only when open-heart surgery is also required for CABG or cardiac-valve replacement.

Haissaguerre *et al.* (1998) identified the pulmonary veins as the site of origin for 95 % of the atrial ectopic beats that initiate AF. A worldwide survey of catheter ablation for AF reported that 4,550 of 8,745 patients (52 %) were maintained in a normal heart rhythm without the need for drug therapy. With the addition of drug therapy to catheter ablation, 76 % of patients achieved a normal heart rhythm (Cappato *et al.*, 2005). Though success rates vary significantly from study to study, with published studies reporting 37 to

95 % success rates, current success rates are around 70 %. The major complication rate of 6 % includes stroke, vessel perforation, and pulmonary-vein stenosis (Marine, 2007).

B.2.3 *Interventions in the Abdomen and Pelvis*

A wide variety of interventions are performed on the various organs in the abdomen and pelvis (Table 2.1). This section describes two of these procedures: uterine fibroid embolization (UFE) and treatment of abdominal aortic aneurysms (AAAs) with stent grafts.

B.2.3.1 *Uterine Fibroid Embolization.* FGI procedures may present advantages over the corresponding open surgical procedures, even if they are less likely to be successful. Uterine fibroid embolization (UFE) is an interventional radiology procedure developed in the 1990s to treat benign smooth-muscle tumors of the uterus. These tumors, properly referred to as leiomyomas, are commonly called fibroids. Fibroids are the most common tumor of the uterus, and are present in 20 to 40 % of women 35 y of age and older (Murase *et al.*, 1999). Most of these women are asymptomatic. However, uterine fibroids can cause heavy, painful or prolonged menstrual bleeding, sometimes severe enough to result in anemia. If large enough, they can also cause pelvic pain, pressure on the bowel leading to constipation, and pressure on the bladder leading to a constant urge to urinate.

Hormonal therapy is often used as the initial treatment option for symptomatic patients, but for many women this is not sufficient. The traditional surgical therapy for uterine fibroids is hysterectomy, a procedure in which the uterus is removed, often along with the ovaries and fallopian tubes. UFE is an alternative FGI therapeutic procedure performed by interventional radiologists. A catheter is guided, using fluoroscopy, through the arterial system and sequentially into each of the two uterine arteries (left and right) which supply blood to the uterus. Arteriograms of these arteries are obtained. Fluoroscopy is used to monitor the injection of small particles into these arteries. The particles block arterial blood flow to the fibroids, and the fibroids subsequently shrink. UFE preserves the uterus, and requires only a 5 mm skin incision.

UFE has become popular as an alternative to hysterectomy because it avoids the need for major surgery in most patients, and it permits the woman to keep her uterus. In the Kaiser Permanente Northern California system, hysterectomy rates decreased significantly and UFE rates increased significantly between 1997 and 2003 while the combined rate for all surgical and interventional therapies remained constant (Jacobson *et al.*, 2007). UFE yields marked improvement in health-related quality of life (Bucek *et al.*, 2006; Hehenkamp *et al.*, 2008). It has a lower clinical success rate for symptom relief (80 to 95 %) than the surgical equivalent, hysterectomy (100 %), but it also demonstrates a lower incidence of major complications (3.9 % versus 12 %) (Spies *et al.*, 2004). The length of hospital stay (mean 0.83 d versus 2.3 d) and the length of time lost from work (mean 10.7 d versus 32.5 d) are both significantly shorter for UFE than for hysterectomy (Pron *et al.*, 2003; Spies *et al.*, 2004). In a randomized, controlled study, Edwards *et al.* (2007) found that the length of hospital stay (median 1 d versus 5 d), the lengths of time until resumption of various normal activities of daily living, and the length of time until return to work (median 20 d versus 62 d) were significantly shorter for UFE than for hysterectomy. These advantages compensate for the lower clinical success rate. They also far outweigh any radiation-related risk of the FGI procedure. Radiation doses for this procedure are presented in Table 4.3. Using a quality-adjusted life-years model, O'Brien and van der Putten (2008) quantified the benefits and risks of UFE as compared to hysterectomy and myomectomy (a surgical procedure where only the fibroid is removed). They estimated the deterministic risk as "very small" and the stochastic risk, using the quality-adjusted life-years model, as equivalent to <1 d of life lost. They concluded that "the increased risk due to radiation dose delivered is small when compared with the benefits."

B.2.3.2 *Endovascular Aneurysm Repair.* AAA and TAA are defined as enlargement of these portions of the aorta to a diameter two to three times normal. The incidence of aortic aneurysms increases with advancing age. For AAA the incidence is 21 per 100,000 person-years and for TAA the incidence is 6 to 10 per 100,000 person-years (Katzen *et al.*, 2005). When relatively small, these aneurysms are generally benign, but as they enlarge they are increasingly likely to rupture. In most cases, rupture of an aortic aneurysm is rapidly fatal due to massive hemorrhage. Prevention of rupture has driven intervention in this disease. When aneurysms reach a diameter where the risk of treatment is less than the risk of rupture within a year, an intervention is indicated. For TAA this diameter is generally accepted as 6 cm, and for AAA it is 5 cm. (There are exceptions such as rapidly growing aneurysms and symptomatic aneurysms.) Due to the significant morbidity and mortality from open repair, an endovascular option using stent grafts was developed to prevent blood flow in the aneurysm sac and effectively exclude the aneurysm from the circulation. For endovascular aneurysm repair (EVAR) the aneurysm diameter at which the benefit and risk cross are still being defined. The size thresholds defined for open repair are used for EVAR. A growing fraction of TAA and AAA are treated using endografts. This is largely due to the less invasive nature of the procedure and the resultant lower morbidity, mortality, and hospital length of stay.

Comparing open repair to EVAR is difficult because the complications are different. The overall risk associated with EVAR for TAA is less than for open surgery, particularly in terms of post-operative pneumonia, paraplegia and death (Corbillon *et al.*, 2008; Dillavou and Makaroun, 2008; Glade *et al.*, 2005). The risks associated with EVAR of AAA are lower than open repair in the initial post-operative period, but several studies have failed to demonstrate long-term survival benefit over open surgical repair (Rutherford, 2006). This is due in large part to associated diseases such as cardiac or cerebral vascular disease in these patients. In patients with lung, kidney, or heart disease who are high risk for open surgery, EVAR shows benefit (Jean-Baptiste *et al.*, 2007).

EVAR is currently problematic in aneurysms that extend to involve critical aortic branch vessels such as the arteries that supply the brain, the intestine, and the kidneys. At present, aneurysms involving these structures are treated with either open surgical repair or combined surgical/EVAR procedures. Over the next decade, new technology (branched endografts and other solutions) should allow these types of aneurysms to be treated with EVAR (Chuter *et al.*, 2003).

B.2.4 *Pediatric Interventional Radiology*

Pediatric interventional radiology is a relatively new but rapidly growing specialty. Although the interventional techniques are similar to that of adult practice, the wide range in age and size of patients in pediatric interventional radiology often necessitate significant modification of technique. In particular, the use of ultrasound is much more prevalent in pediatric practice and has revolutionized minimally-invasive intervention in children. Greater radiosensitivity and a greater risk of contrast-induced nephrotoxicity also require meticulous attention to radiation and contrast dose (Heran *et al.*, 2010; Strauss and Kaste, 2006). Appropriate sedation is a necessity. Standards of practice are currently being established by the Pediatric Interventional Radiology Subcommittee of the Society of Interventional Radiology (Heran *et al.*, 2010; Roebuck, 2009).

B.2.4.1 *Central Venous Access.* The combined use of ultrasound-guided venous access and fluoroscopically-guided catheter placement allows safe and accurate placement of central venous catheters, with high rates of success and low rates of complications, even in small children (Donaldson, 2006; Peynircioglu *et al.*, 2007). The types of access devices placed include peripherally-inserted central catheters, tunneled central venous catheters, and subcutaneously-implanted venous ports.

B.2.4.2 *Angiography and Vascular Intervention.* Improvements in noninvasive imaging modalities, such as CT and magnetic resonance angiography, have decreased the requirement for diagnostic catheter angiography (Heran *et al.*, 2010, Roebuck, 2009). However, advancement in technology, including the development of microcatheters and microcoils, has also increased the number and types of endovascular procedures that can be performed in children (Christensen, 2001; McLaren and Roebuck, 2003).

B.2.4.3 *Treatment of Vascular Anomalies.* The treatment of vascular anomalies has evolved significantly over the last few years. Percutaneous sclerotherapy is now the mainstay of treatment of venous and lymphatic malformations (Burrows *et al.*, 2008; Legiehn and Heran, 2008). The lesion is initially accessed percutaneously with sonographic guidance, followed by contrast injection into the lesion with fluoroscopy. Presclerotherapy venography is also performed for venous lesions. Finally, the sclerosant opacified with contrast is injected under direct fluoroscopic guidance. Endovascular embolization plays a key role in the treatment of AVMs, either as primary therapy or in combination with surgery (Wu and Orbach, 2009). The nidus of the malformation is accessed with microcatheters followed by injection of liquid embolic agents. AVMs may also be accessed percutaneously and, under fluoroscopic guidance, directly injected with sclerosants such as ethanol.

B.2.4.4 *Biopsies and Drainages.* Due to the good sonographic visualization of the abdomen in children, the vast majority of solid organ biopsies are performed under sonographic guidance, with fluoroscopy and CT reserved for lung and bone biopsies (Garrett *et al.*, 2005; Sebire and Roebuck, 2006). Ultrasound is also the primary modality for access of chest or abdominal collections, as well as the renal pelvis, with access also confirmed with the use of fluoroscopy. The cavity is initially accessed with a thin gauge needle. Following confirmation of accurate positioning, the access needle is exchanged for a pigtail catheter over a guidewire. In children, the most common cause of an abdominal collection is perforated appendicitis and initial percutaneous drainage followed by interval appendectomy is now often the standard of care (Roach *et al.*, 2007).

B.2.4.5 *Gastrointestinal Intervention*. Image-guided gastrostomy and gastrojejunostomy procedures offer a minimally-invasive method of obtaining enteric access, with high success and low complication rates (Lewis *et al.*, 2008; Sy *et al.*, 2008). Following fluoroscopically-guided percutaneous puncture, the catheter can be placed *via* an antegrade or retrograde approach. A similar technique can also be employed to place a cecostomy tube for antegrade enema administration in children with fecal incontinence (Chait *et al.*, 2003). Other gastrointestinal and biliary interventions include esophageal dilatation, percutaneous transhepatic cholangiography and cholecystography (Connolly, 2003).

Appendix C

Medical Risks to Interventionalists and Staff

Interventionalists can be classified as surgeons for the purpose of estimating their medical occupational risk, because scalpels and needles are used routinely during percutaneous interventions. Hypodermic needles, suture needles, other hollow-bore needles, and scalpels were together responsible for 72 % of reported percutaneous injuries to hospital workers from 1995 to 2000 (NIOSH, 2004).

Makary *et al.* (2007) surveyed surgeons in training at 17 institutions and found that by their final year of training, 99 % of these surgical residents had had a needlestick injury. Fifth-year (chief) residents had sustained a mean of 7.7 needlesticks during training. Of note, residents failed to report 51 % (297/578) of their most recent injuries. Fifty-three percent of residents had sustained a needlestick injury while performing a procedure on a patient with a history of infection with human immunodeficiency virus (HIV), hepatitis B (HBV), hepatitis C (HCV), or injection drug use.

Pruss-Ustun *et al.* (2005) estimate that worldwide, in the year 2000, healthcare workers may have contracted 16,000 HCV, 66,000 HBV, and 1,000 HIV infections due to occupational exposure as a result of percutaneous injuries. The fraction of HCV, HBV, and HIV infections in healthcare workers attributable to occupational exposure were 39, 37, and 4.4 %, respectively. Sepkowitz and Eisenberg (2005) estimate that the U.S. healthcare worker death rate from infection-related causes (HBV, HCV, HIV and tuberculosis) in 2002 was 13 to 42 per million.

Data on healthcare worker conversion rates for HBV, HCV and HIV exposure are available from the Centers for Disease Control and Prevention (CDC, 2001). The HBV virus is highly contagious. An unvaccinated healthcare worker who sustains a needlestick injury from a patient with HBV has a 22 to 62 % risk of developing clinical hepatitis. HBV infection can also occur as a result of direct or indirect blood and body fluid exposures, with transfer of the virus into skin scratches, abrasions or burns, or onto mucosal surfaces. Because of the high risk of HBV infection in healthcare workers, routine vaccination against HBV has been recommended since the early 1980s (CDC, 2001). The combination of vaccination and the use of universal precautions to prevent exposure to blood and other body fluids has reduced the estimated annual number of HBV infections among healthcare workers in the United States by 96 %, from 10,721 cases in 1983 to 384 cases in 1999 (NIOSH, 2004). Post-exposure prophylaxis with HBV immune globulin or HBV vaccine is estimated to reduce the likelihood of infection by 75 % (CDC, 2001).

HCV is less likely to be transmitted as a result of occupational exposure, but there is no effective vaccine to prevent infection and no effective post-exposure prophylaxis is available. The average incidence of seroconversion after accidental percutaneous exposure is 1.8 % (CDC, 2001). In the United States, 50 to 150 instances of transmission of HCV from an infected patient to a healthcare worker can be expected annually, and three to eight healthcare workers can be expected to die annually of liver disease due to occupationally-transmitted HCV (Sepkowitz and Eisenberg, 2005).

The risk of transmission of HIV after occupational exposure is relatively low. The average risk of transmission after percutaneous exposure to HIV-infected blood is 0.3 and 0.09 % after mucous membrane exposure (CDC, 2001). There is no vaccine, but post-exposure prophylaxis is available and recommended because of the life-threatening nature of the disease. The drugs used are not innocuous, nearly 50 % of individuals receiving post-exposure prophylaxis experience adverse effects, and 33 % stop the four-week drug regimen early because of these effects (CDC, 2001).

Infectious disease may also be transmitted to healthcare workers by other routes. Transmission of pulmonary tuberculosis is a classic example of the inhalation route of infection. The risk of serious illness and death has increased with the appearance of multidrug-resistant tuberculosis and extremely drug-resistant tuberculosis (Sepkowitz and Eisenberg, 2005). SARS is another example of the inhalation route of transmission. In the 2002 to 2003 SARS epidemic in China, Singapore, and Toronto, Canada, there were 8,098 cases, of whom 1,707 (21 %) were healthcare workers. The overall death rate was 9.6 % (Sepkowitz and Eisenberg, 2005).

Appendix D

Physics and Dosimetry

Table D.1 summarizes the principal radiation quantities used in this Report and Table D.2 provides conversion factors from some common presentations in other literature and regulatory documents to those used in this Report. Appendix D.1 discusses current dosimetry approaches for FGI procedures and Appendix D.2 discusses the nature of the x-ray beams used in FGI procedures.

D.1 Dosimetry Technology

Peak skin dose ($D_{skin,max}$) is the highest absorbed dose in any portion of a patient's skin accumulated during a procedure. $D_{skin,max}$ (expressed in gray) includes contributions from the primary x-ray beam, leakage radiation, and scattered radiation. As of 2010, no commercially-available fluoroscopy unit is capable of calculating or displaying $D_{skin,max}$. It is possible to determine $D_{skin,max}$ using arrays of thermoluminescent dosimeters or optically-stimulated luminescence dosimeters, or with radiographic film or Gafchromic® film (International Specialty Products, Wayne, New Jersey) placed between the patient and the table (Figure D.1), but these methods do not provide real-time measurements during a procedure (Mantovani *et al.*, 2006; Struelens *et al.*, 2005; Vano *et al.*, 1997).

$K_{a,r}$ (gray) is the cumulative air kerma at a specific point in space (*i.e.*, the air-kerma reference point) relative to the fluoroscopy gantry. It does not include backscatter (Figure D.2). The concept of $K_{a,r}$ for x-ray equipment for interventional procedures first appeared in IEC (2000). It was subsequently adopted as part of the FDA performance standard for medical x-ray fluoroscopy systems (FDA, 2009a). All interventional fluoroscopes conforming to the IEC standard and all fluoroscopes sold in the United States after June 2006 are capable of measuring or calculating $K_{a,r}$ and displaying $K_{a,r}$ at the working position of the interventionalist.

TABLE D.1—*Radiation quantities used in this Report.*

Quantity	Symbol Used in this Report	Special SI Name	Comment	Reference (for quantity or style of notation)
Absorbed dose at a point	D	gray (Gy)	Absorbed dose at a specified point in a medium (*e.g.*, tissue, air) (includes backscattered radiation).	ICRU (2005)
Mean absorbed dose in an organ or tissue (organ dose)	D_T	gray (Gy)	Obtained by integrating or averaging absorbed doses over the whole organ or tissue.	ICRP (1991a)
Peak skin dose	$D_{skin,max}$	gray (Gy)	Absorbed dose to the most highly-irradiated local area of skin (includes backscattered radiation).	ICRU (2005)
Entrance-surface absorbed dose	$D_{skin,e}$	gray (Gy)	Absorbed dose on the central x-ray beam axis at the point where the x-ray beam enters the patient or phantom (includes backscattered radiation).	ICRU (2005)
Incident air kerma	$K_{a,i}$	gray (Gy)	Air kerma from the incident beam on the central x-ray beam axis at the focal-spot-to-surface distance (does not include backscattered radiation).[a]	ICRU (2005)
Entrance-surface air kerma	$K_{a,e}$	gray (Gy)	Air kerma on the central x-ray beam axis at the point where the x-ray beam enters the patient or phantom (includes backscattered radiation).	ICRU (2005)

TABLE D.1—*(continued)*

Quantity	Symbol Used in this Report	Special SI Name	Comment	Reference (for quantity or style of notation)
Air kerma at the reference point	$K_{a,r}$	gray (Gy)	Application of the quantity air kerma (K_a) (ICRU, 2005) for FGI procedures (measured in air at the $K_{a,r}$ reference point in the absence of backscattered radiation).[a]	FDA (2009a), IEC (2000; 2004; 2010)
Air kerma-area product	P_{KA}	Gy cm^2	The integral of the air kerma (in the absence of backscattered radiation)[a] over the area of the x-ray beam in a plane perpendicular to the beam axis.	ICRU (2005)
Equivalent dose	H_T	sievert (Sv)	The radiation weighting factor (w_R) is unity for the x rays used in the FGI procedures relevant to this Report.	ICRP (1991a; 2007a); NCRP (1993)
Effective dose[b]	E	sievert (Sv)	The values of w_R and tissue weighting factors (w_T) used in ICRP (1991a) and NCRP (1993) have recently been updated in ICRP (2007a).	ICRP (1991a; 2007a) NCRP (1993)
FDA compliance air-kerma rate	$\dot{K}_{FDA}$	mGy min^{-1}	Measured under FDA compliance conditions.	FDA (1999; 2009a)
Personal dose equivalent	$H_p(10)$ $H_p(0.07)$	gray (Gy)	The dose equivalent in soft tissue at a depth of 10 mm (or 0.07 mm) below a specified point on the body (used in personal monitoring of staff participating in FGI procedures).	ICRU (1993)

[a]That is, in the absence of a patient or phantom (free-in-air).

[b]A previous formulation of E, named effective dose equivalent (H_E) (ICRP, 1977; NCRP, 1987), is currently reported in various regulatory programs in the United States. Some values of w_R and w_T for H_E are different from those for E.

TABLE D.2—*Conversion factors from other literature presentations to those used in this Report.*

Convert From	Convert To	Conversion	Comments
Exposure (R)	Air kerma (K_a) (mGy)	1 R = 8.77 mGy 1 mGy = 0.114 R	The radiation field may or may not have a backscattered radiation component
Dose-area product (Gy cm^2)	Air kerma-area product (P_{KA}) (Gy cm^2)	Unity	Newer terminology emphasizes that this quantity is measured in the absence of backscattered radiation (*i.e.*, free-in-air) (ICRU, 2005)
Air kerma-area product (cGy cm^2 or μGy m^2)	Air kerma-area product (P_{KA}) (Gy cm^2)	1 Gy cm^2 = 100 cGy cm^2 1 Gy cm^2 = 100 μGy m^2	Simple scaling
Air kerma-area product (mGy cm^2)	Air kerma-area product (P_{KA}) (Gy cm^2)	1 Gy cm^2 = 1,000 mGy cm^2	Simple scaling

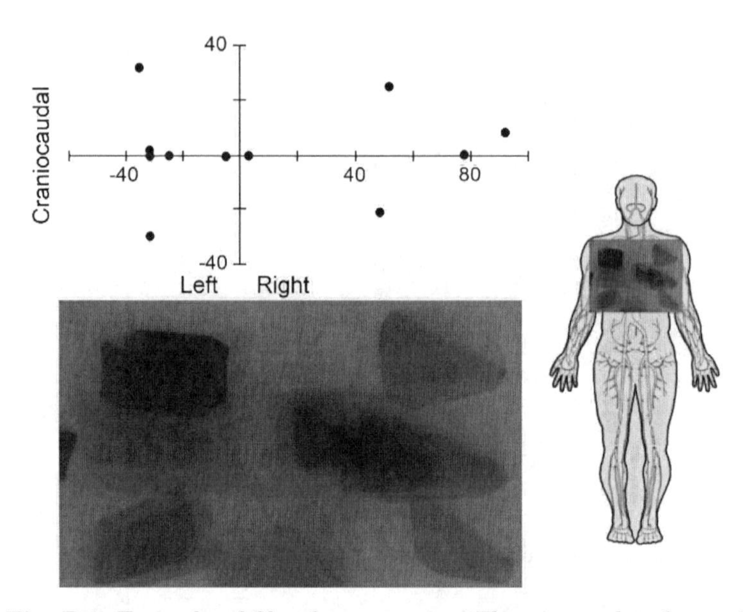

Fig. D.1. Example of film dose mapping. The upper plot shows the angular coordinates of the cinefluorography series recorded. The digitized picture of the slow film shows both cinefluorography and fluoroscopy fields. The film image has been adapted to approximately match the graph. Angular data for every projection are referred to the image intensifier entrance position, whereas radiation enters the patient in the reverse direction (Vano *et al.*, 2003).

For isocentric fluoroscopes (C-arms), the reference point for measuring $K_{a,r}$ lies on the central axis of the x-ray beam, 15 cm on the x-ray tube side of isocenter (IEC, 2000; 2010). This approximates the location of the patient's entrance skin. The location of the reference point for $K_{a,r}$ relative to the x-ray gantry does not change when SSD is changed or the gantry is rotated. Changing table height can have a profound influence on the relationship between $K_{a,r}$ and $K_{a,i}$ as demonstrated in Figures D.3 and D.4. The location of the FDA reference point for other geometries is the same as the fluoroscopic dose-rate compliance point described in the FDA compliance manual (FDA, 1999).

The location of the reference point does not change when SID is changed. However, FDA compliance measurements for interventional C-arms are made 30 cm in front of the image receptor at any SID. Therefore, the location of the FDA air-kerma-rate compliance point will change relative to the x-ray gantry when SID is changed (FDA, 1999). The reference point used to report $K_{a,r}$ is seldom identical to the FDA compliance point. This is shown in Figure D.5.

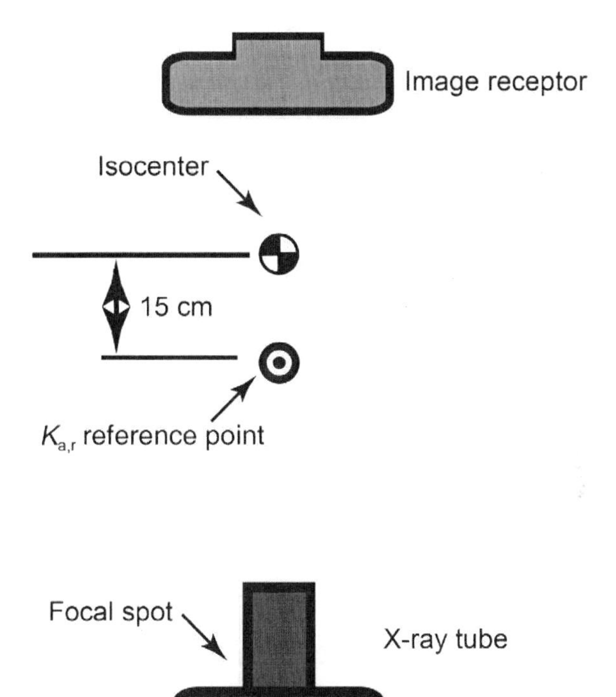

Fig. D.2. The quantity $K_{a,r}$ represents the air kerma at a defined point with all optional attenuators (such as the table top) and all removable sources of scattered radiation (such as the patient) removed from the x-ray beam. The locations of the $K_{a,r}$ reference point, focal spot, and isocenter are marked. For isocentric systems the reference point is 15 cm away from the isocenter toward the x-ray tube.

Since the location of $K_{a,r}$ is defined relative to the x-ray equipment, rather than the patient, and fixed with respect to the gantry, it moves relative to the patient when the gantry moves. The quantity $K_{a,r}$ is an approximation of the total absorbed dose delivered to skin, summed over all the areas of the skin irradiated during the procedure. While the position of the reference point is usually close to the patient's skin, it is rarely on the entrance-skin surface. Also, during the course of most but not all interventional radiology procedures, the x-ray beam is moved periodically with respect to the patient, and is directed at different areas of the patient's skin. It is the interventionalist's responsibility to be aware of actual beam motion during every potentially-high radiation dose procedure. In those individual cases where the x-ray beam is frequently moved,

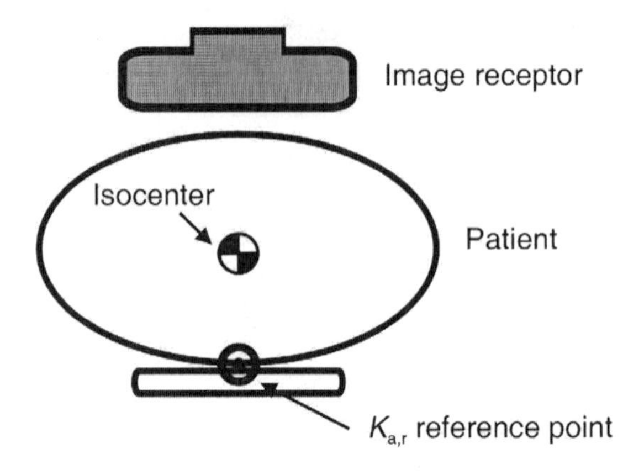

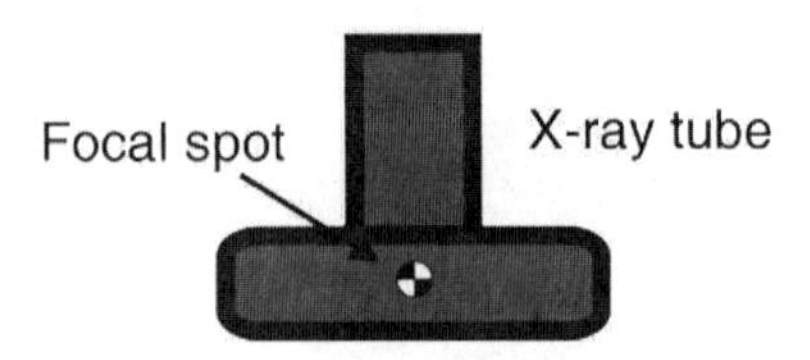

Fig. D.3. The $K_{a,r}$ reference point is nominally at the patient's entrance-skin surface as shown here. In this exact geometry and in the absence of beam motion, the incident air kerma ($K_{a,i}$) and entrance-skin absorbed dose ($D_{skin,e}$) are: $K_{a,i} = K_{a,r} A\ BSF$; and $D_{skin,e} = K_{a,i} f$; where A is the attenuation of the table top, f is the ratio of mass attenuation coefficients of skin to air, and BSF is the backscatter factor (dependent on x-ray spectrum and field size at the surface). Under these conditions: $D_{skin,e} \simeq 1.3\ K_{a,r}$ (approximate range 1.2 to 1.5).

estimates of the likelihood of deterministic effects in the skin that are based on $K_{a,r}$ tend to overestimate this risk (Miller *et al.*, 2003b). This is shown in Figure D.6. However, the risk of skin injury increases as beam motion decreases. Most of the reported skin reactions have sharp borders, evidence of limited beam motion during major parts of the procedure.

P_{KA} (Gy cm^2), is a measure of the total x-ray energy leaving the x-ray tube. It is typically measured with an ionization chamber located near the collimator. Because dose decreases proportionately to the square of the distance from the focal spot and the area of the irradiated field increases proportionally in the same way, P_{KA} is independent of SID. If the radiation field is confined to the patient, P_{KA} is a good measure of the total x-ray energy entering

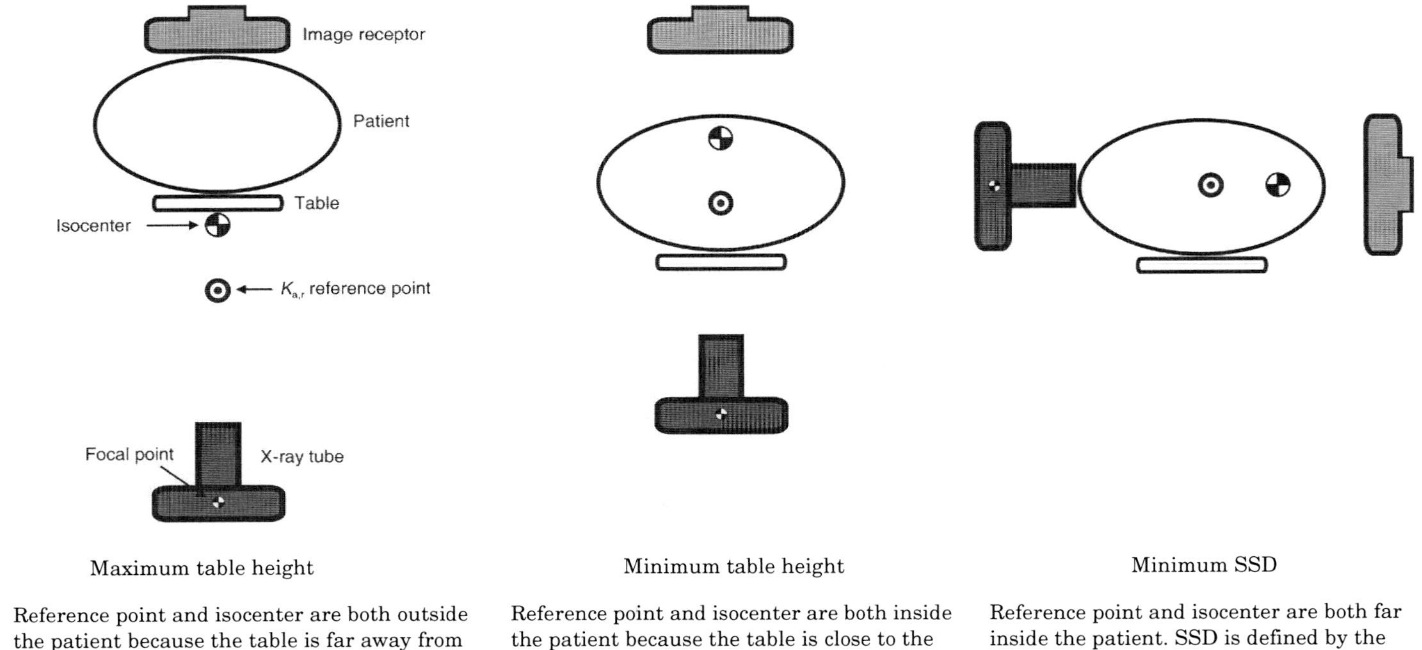

Maximum table height	Minimum table height	Minimum SSD
Reference point and isocenter are both outside the patient because the table is far away from the x-ray tube	Reference point and isocenter are both inside the patient because the table is close to the x-ray tube	Reference point and isocenter are both far inside the patient. SSD is defined by the spacer on the x-ray tube assembly
$K_{a,r} > K_{a,i.}$	$K_{a,r} < K_{a,i.}$	$K_{a,r} \ll K_{a,i.}$

Fig. D.4. Effect of SSD changes.

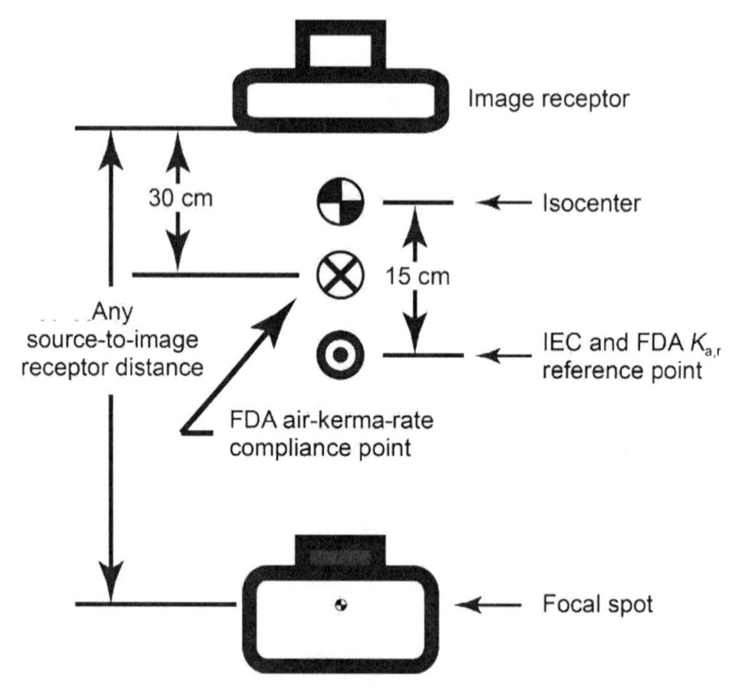

Fig. D.5. Location of the IEC and FDA $K_{a,r}$ reference point and the FDA air-kerma-rate compliance point for isocentric fluoroscopes. For these systems, the FDA air-kerma-rate compliance point is 30 cm from the image receptor at any SID.

the patient. Since fluoroscopic x-ray beams are almost totally absorbed by the patient's tissues, P_{KA} also provides a good estimate of the total x-ray energy imparted to the tissues of the patient, which relates to the stochastic effects (*e.g.*, cancer) of radiation. P_{KA} is widely used in Europe to monitor patient dose during interventional procedures. P_{KA} is less useful as a metric for estimating, $D_{skin,max}$ because a large radiation dose delivered to a small skin area can yield the same P_{KA} as a small radiation dose delivered to a large skin area. Estimation of absorbed skin dose from P_{KA} data has a potential error of at least 30 to 40 % (McParland, 1998b).

D.2 Fluoroscopic Radiation Beams

FGI procedures are performed using x-ray beams generated by peak voltages ranging from 50 to 125 kVp, and with filtrations ranging from the minimum permitted up to 1 mm copper. Photon spectral distributions under specified conditions can be calculated using standard methods (Poludniowski *et al.*, 2009).

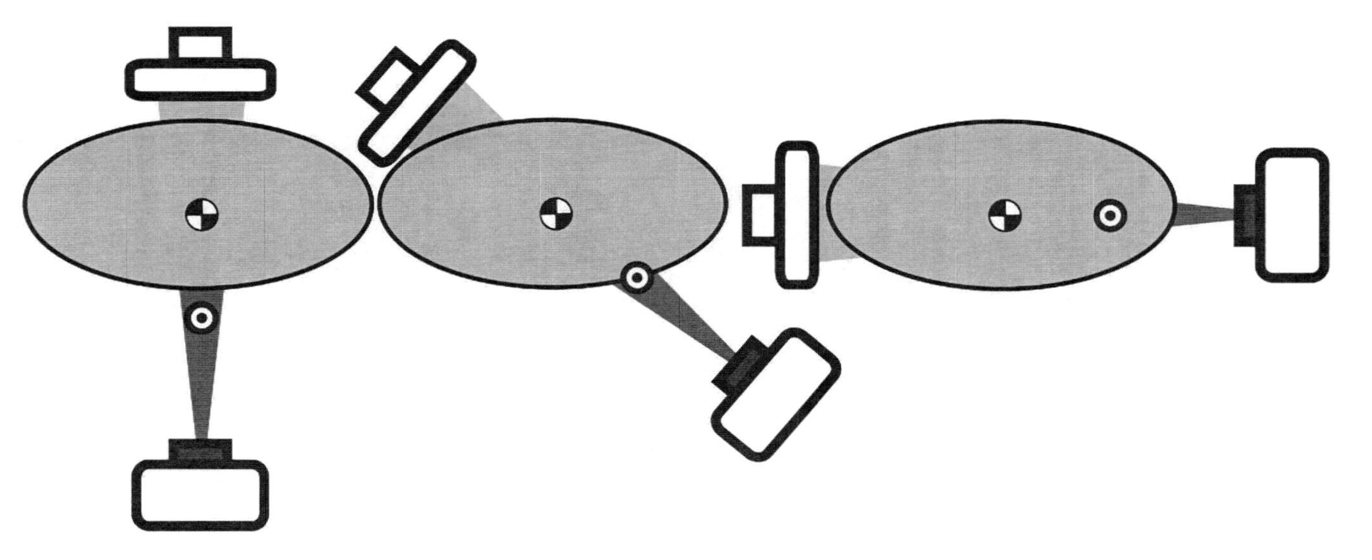

Fig. D.6. The $K_{a,r}$ reference point moves with the gantry. The reference point may be outside the patient (left), on the patient's skin (center), or inside the patient (right), depending on gantry angulation (shown) and patient table height (not shown). Peak incident air kerma (peak $K_{a,i}$) will be lower than, equal to, or higher than $K_{a,r}$ for these three conditions (assuming no motion) (Balter, 2008). Different portions of the patient's skin are irradiated when the beam moves during the procedure. With sufficient beam motion, any portion of the patient's skin receives a value of $K_{a,i}$ that is a fraction of the peak $K_{a,i}$ associated with zero motion.

Dose rates at the skin surface range from <1 mGy min^{-1} (very low dose-rate fluoroscopy) to several Gy min^{-1} (DSA or cineangiography in a large patient). The operating parameters (kilovolt peak, milliampere, pulse duration, pulse rate, and often beam filtration) are driven by the system's automatic dose rate control. These values also vary from machine to machine and in most cases with the exact clinical mode selected to perform a particular procedure. Values of $D_{\text{skin,max}}$ for single procedures on the order of several tens of gray have occurred (Koenig *et al.*, 2001a). The same region of skin on a patient may be irradiated during a subsequent procedure, occurring hours to years after the preceding procedure. These irradiation patterns differ from those associated with most therapeutic and accidental exposures.

Figure D.7 indicates the general effect spectral shaping of an FGI-procedure beam. The spectrum is narrowed to the region just above the K-absorption edge of iodine by a combination of decreasing the kilovolt peak and adding a copper filter to the beam. Total output is adjusted by increasing the x-ray tube current to compensate for losses due to the reduced kilovolts and increased filtration. The net effect is an increase in the visibility of iodine-containing contrast media using the same total patient radiation level or maintenance of iodine visibility with a lower radiation level.

Radiation therapy with 250 kVp x-ray beams (orthovoltage radiotherapy) provides a great deal of insight into the effects of radiation on skin (Cohen *et al.*, 1972; Krizek, 1979; MacKee and Cipollaro, 1946; Moss, 1959). The physical dose distribution near the skin surface produced by 250 kVp x-ray beams is not too dissimilar to modern fluoroscopy beams. Dose rates to skin were usually in the range of several hundred mGy min^{-1}. Orthovoltage radiotherapy was for the most part replaced by radiation therapy with higher energy x-ray beams by the late 1970s. The maximum dose delivered by beams in the million-volt range occurs at a depth of millimeters to centimeters below the skin. The build-up of dose below the surface can minimize injury of skin. This is commonly called "skin sparing."

Because orthovoltage radiotherapy does not exhibit skin sparing, orthovoltage radiotherapy equipment was prescribed in a manner that minimized skin injury by allowing repair of sublethal radiation damage and some tissue regeneration between treatment fractions (Ellis, 1942; Paterson, 1948). A typical orthovoltage radiotherapy prescription was 2 to 3 Gy per fraction to the skin, usually given in five fractions per week, for a treatment course of approximately six weeks. Single-field skin doses were always higher than the tumor doses. Because of this, multiple field treatment plans

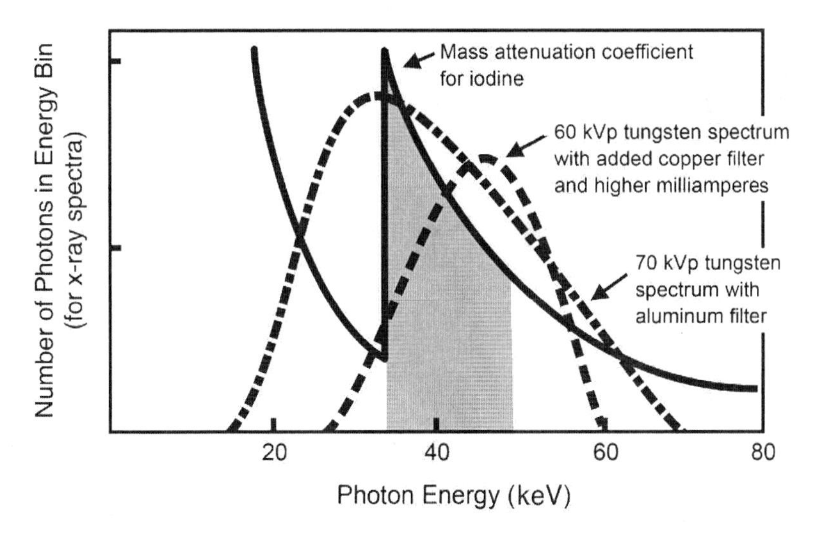

Fig. D.7. Spectral modification of an FGI-procedure x-ray beam. The goal is to concentrate the photons (x rays) in the beam to the shaded region just above the K-absorption edge of iodine (37 keV). This increases the visibility of contrast media containing iodine while reducing the photon energy bands incident on the patient that make small contributions to the useful image. In this schematic example, the 60 kVp copper-filtered beam results in a lower value of P_{KA} than that for the conventional 70 kVp aluminum-filtered beam. The mass absorption coefficient of iodine as a function of energy is also shown schematically.

were used as a means of delivering a higher dose to the tumor than to the skin. The transfer of experience from orthovoltage radiotherapy to fluoroscopy is relevant. However, some individual interventional procedures result in skin doses of a few tens of gray. These high single-fraction doses are outside of most radiation-therapy experiences.

Skin dose is influenced by x-ray field size. Backscatter from the patient increases with increasing field size. Appendix A in ICRU Report 74 presents detailed information (ICRU, 2005). For typical x-ray fields used in FGI procedures, the backscatter factor is ~1.3. Calculation of skin dose also requires application of a correction factor accounting for the mass energy absorption coefficient for tissue relative to air. The value of this coefficient is 1.06 for typical x-ray beam energies used during FGI procedures (McParland, 1998b). An additional physical factor related to field size is the increased likelihood of overlap between two or more larger fields in comparison to smaller fields.

Appendix E

Lens of the Eye

E.1 Lens of the Eye Anatomy and Pathology

The lens of the eye is an optically-clear, avascular tissue, which continues to grow in size and cell number throughout life and whose primary pathology is cataract (Kleiman and Worgul, 1994; van Heyningen, 1975). Lens of the eye transparency depends on the proper function of lens fiber cells, anuclear, amitochondrial, and differentiated progeny of the epithelial cell monolayer on the lens anterior surface.

The predominant type of opacity following radiation exposure is termed "posterior subcapsular cataract" (Figure E.1). It develops due to aberrant differentiation and migration of lens epithelial cells and results in opacity at the posterior pole (Kleiman, 2007).

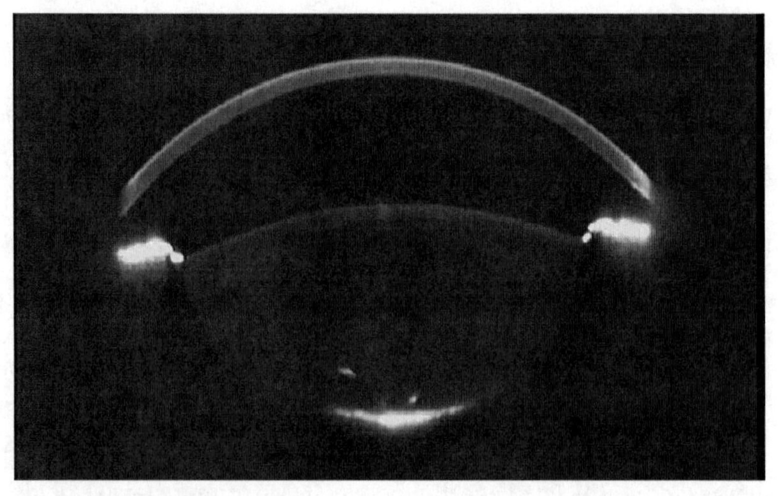

Fig. E.1. Scheimpflug slit lamp biomicroscopic image of a human posterior subcapsular cataract.

Radiation cataracts characteristically develop in a sequential fashion (Merriam and Focht, 1962) with the earliest lens changes consisting of development of an opalescent sheen to the posterior capsule followed by the appearance of dots, vacuoles, or diffuse opacities centered around the posterior lens suture. Continued cataract development leads to progression of these posterior changes, the involvement of the anterior subcapsular region and, ultimately, total lens opacification (Merriam and Worgul, 1983).

E.2 Epidemiology of Radiation Effects on the Lens of the Eye

Findings from a variety of recent human epidemiological studies following acute or chronic low-dose radiation exposure (*e.g.*, Chodick *et al.*, 2008; Ciraj-Bjelac *et al.*, 2010; Neriishi *et al.*, 2007; Vano *et al.*, 2010; Worgul *et al.*, 2007) are in agreement with those from experimental low-dose radiation cataract animal studies (Merriam and Worgul, 1983). The types of opacities observed in both human and animal studies are compatible with the generally accepted genotoxic pathomechanism of radiation cataract development (Worgul *et al.*, 1989). In fact, recent findings demonstrate dose-related significant lens opacification within a reasonable fraction of the lifespan of the mouse or rat after exposure to as little as 100 mGy (x-ray dose) (Worgul *et al.*, 2005a) and 325 mGy (^{56}Fe) (Worgul *et al.*, 2005b). These animal studies utilized doses far lower than the presumptive threshold human dose for cataractogenesis. That 100 mGy (x-ray dose) is cataractogenic within a third of the lifespan of the rat is of particular relevance to human regulatory guidelines and risk estimates and suggests that significantly-lower doses might also be cataractogenic.

Some epidemiological studies of humans exposed to low doses of radiation suggest an absence of a dose threshold for cataractogenesis (Shore and Worgul, 1999). More recent studies (Chodick *et al.*, 2008; Neriishi *et al.*, 2007; Worgul *et al.*, 2007) support this hypothesis. It is critical to determine whether radiation-induced lens changes are, in and of themselves, sufficient for the purposes of setting regulatory standards and establishing risk estimates for cataractogenesis. This approach assumes that, given sufficient latency, these lens changes will progress to clinical cataract and eventual loss of visual acuity requiring surgical removal of the lens.

In this context, a recent report that examined dose response and threshold in atomic-bomb survivors who had cataract surgery (Neriishi *et al.*, 2007) is the first to document that clinically-relevant and visually-disabling cataract occurs after low-dose exposure.

A statistically-significant dose-response increase in the prevalence of cataract surgery was reported in this study, with an odds ratio at 1 Gy of 1.39 and an absorbed dose threshold of 0.1 Gy (95 % confidence interval, <0.8 Gy), significantly lower than the current exposure thresholds (equivalent dose) for visual disability of 5 Sv fractionated (ICRP, 1991a) and 2 Sv acute (NCRP, 2000b). The authors concluded that their data were incompatible with an absorbed dose threshold greater than 0.8 Gy. At the time of the study the youngest atomic-bomb survivors were 55 y of age while the average age for cataract surgery was 73 y, suggesting that additional surgical cases may occur in this population in the future.

E.3 Protocol for a Slit-Lamp Examination

The pupils of the eyes should be fully dilated before examination. A comprehensive ocular and medical history should be taken.

Both slit-imaging and retro-illumination are effective for detecting radiation cataracts and both approaches should be used. Early lens changes at the posterior pole are best seen with a narrow slit beam and oblique illumination. The examination technique depends upon the fact that lens opacities associated with ionizing radiation exposure develop in a relatively unique, characteristic and sequential fashion, usually beginning as nascent posterior subcapsular opacities.

Thus, radiation cataract is primarily associated first with the posterior lens pole. Less frequently, anterior subcapsular opacities may be noted. In general, posterior opacities precede any anterior lens changes. Cortical opacities, most commonly those associated with age-related lens changes, or nuclear opacities, including nuclear sclerosis, are not associated with ionizing radiation exposure. While they may be noted by the ophthalmologist, no conclusions regarding ocular ionizing radiation exposure should be made based on lens cortical or nuclear morphology.

Systemic steroid use should be noted as steroid-induced opacities typically present as a posterior subcapsular cataract. It is important that the ophthalmologist note all lens changes associated with ionizing radiation exposure, both to establish a baseline for future examinations and to permit distinguishing such early lens changes from the more common forms of lens opacification that accompany aging.

A variety of methods have been developed to quantify the severity of ionizing radiation associated lens changes. One such approach utilizes a so called "modified Merriam-Focht scoring methodology" (Merriam and Focht, 1962; Worgul *et al.*, 2007). This scoring scale

(Stages 0.5 to 4.0) is shown in Table E.1. It should be noted that Stages 2.0 and higher are generally associated with visual disability. Lesser degrees of opacification are not usually perceived by the subject as a change in vision. The earliest radiation-induced changes (Stage 0.5) consist of a polychromatic sheen associated with the posterior lens capsule and small numbers of vacuoles and/or discrete dots centered around the posterior lens sutures.

TABLE E.1—*Stages of the modified Merriam-Focht methodology for scoring lens of the eye changes (Merriam and Focht, 1962; Worgul et al., 2007).*

Stage	Notes
0.5	Polychromatic sheen and/or five or fewer dots and vacuoles noted in the posterior subcapsular region.
1.0	More than 10 posterior dots or five vacuoles are noted but the anterior region remains transparent.
	May be further defined as discrete, superficial subcapsular opacities including small spots (visible on retro-illumination), aggregates of dots or vacuoles, cortical spokes, waterclefts, and granulated opacities .
1.5	Stage 1.0 plus; opacities roughly between Stage 1.0 and 2.0.
	Characterized by less than five anterior dots or vacuoles.
2.0	Both posterior and anterior regions contain significant opacities with at least 10 dots or vacuoles posteriorly and five anteriorly.
	Stage 2.0 opacities generally include extensive cortical changes collectively occupying ~25 % of the posterior lens surface.
2.5	The vitreous is still visible through scattered anterior opacification.
3.0	No visualization of the posterior vitreous.
3.5	The opacity has not become severe enough to prevent passage of the slit beam into the posterior region, but detailed visualization is impossible.
4.0	Complete anterior opacification prevents visualization of the remainder of the lens.

Appendix F

Tissue Reactions: Skin and Hair

F.1 Biological Factors Influencing Skin Reactions

The pathophysiology of radiation-induced skin injury has been reviewed in detail (Hymes *et al.*, 2006). Tissues at risk also include the skin, hair, subcutaneous fat, and muscle. The expression of this injury varies, and is dependent on a number of factors that affect the dose-response relationship and the kinetics of healing (Geleijns and Wondergem, 2005; Hymes *et al.*, 2006). Total dose and the interval between dose fractions can affect the expression and severity of radiation injury. Physical and patient-related factors that affect the expression of the injury include smoking, poor nutritional status, compromised skin integrity, obesity, overlapping skin folds, and the location of the irradiated skin (Hymes *et al.*, 2006). The anterior aspect of the neck is the most sensitive site. The flexor surfaces of the extremities, the trunk, the back and extensor surfaces of the extremities, the nape of the neck, the scalp, and the palms of the hands and soles of the feet are less sensitive, in that order. The scalp is relatively resistant to the development of skin damage, but scalp hair epilation occurs at lower doses in comparison to hair loss elsewhere on the body (Geleijns and Wondergem, 2005). Hair loss is illustrated in Figure F.1, as is the differential sensitivity of the skin of the neck and scalp. Ethnic differences in skin coloration are also associated with differences in radiation sensitivity; individuals with light-colored hair and skin are most sensitive.

A separate form of radiation-related drug toxicity is termed "radiation recall." This is an inflammatory skin reaction of unknown etiology (occasionally seen following radiation therapy) that occurs in a previously-irradiated body part after drug administration (Azria *et al.*, 2005; Hird *et al.*, 2008). Radiation recall may occur minutes to days after drug exposure and weeks to years after

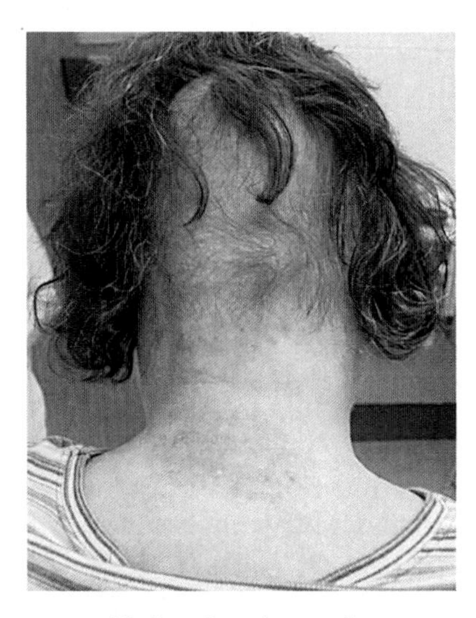

Fig. F.1. Sixty year old female who underwent a successful neurointerventional procedure for the treatment of acute stroke and subsequently developed a radiation injury. Estimated fluoroscopy time was >70 min; 43 imaging series were performed during the course of the procedure. The head was not shaved. Note focal epilation on the scalp and skin injury on the neck but not on the scalp. There are no dose estimates available for this case.

radiation exposure (Hymes *et al.*, 2006). It occurs with chemotherapeutic agents (doxorubicin, etoposide, paclitaxel, bleomycin, epirubicin and gemcitabine, among others), antibiotics (cefotetan), statins (simvastatin) and herbal preparations [hypericin (St. John's wort)] (Ayoola and Lee, 2006; Hird *et al.*, 2008; Putnik *et al.*, 2006).

Radiation recall has also been anecdotally attributed to ultraviolet exposure from sunlight with no apparent contributing factors. Clinical practice after radiation therapy is to caution patients about sun exposure. However, there is no definitive scientific literature on this point.

F.2 Biology of Radiation Injuries

Avoidance of critical damage to normal tissues is one of the basic principles of radiation therapy. Clinical time-dose fractionation prescriptions (classically, 30 fractions spread over six weeks) are largely based on this consideration. The response of normal tissues and structures to the effects of radiation therapy depends to a large extent on the radiation dose, the fraction size, and the integrity of the tissue (affected by prior surgery, chemotherapy, or radiation therapy), among other factors. Until relatively-recent times, conventional fractionation regimens were routinely utilized. Hypofractionation treatment schedules (a smaller number of fractions and a larger dose per fraction) were more often reserved for a palliative treatment setting. With x-ray image-guided radiation-therapy techniques, more precise targeting can be achieved with enhanced accuracy. As a result, there has been increasing interest in the use of treatment regimens employing several large dose fractions of 6 to 12 Gy. Stereotactic radiosurgery and image-guided radiation-therapy schedules have also delivered single doses of 20 to 30 Gy to finite tumor volumes, with excellent tumor control rates. Clinicians should be aware that, depending upon the location of the treatment, skin reactions are possible with the administration of these ultra-high dose fractions that result in relatively-high doses to the skin surface.

Interventional procedures are typically performed in one to a few sessions over a time period ranging from days to months. The number of sessions and their timing are dependent on the underlying disease process and on procedural factors such as contrast material use. Different interventional sessions may or may not irradiate the same portion of the patient's skin.

The time interval between sessions is important because of cellular DNA repair processes and repopulation. The effects of radiation on tissue are mitigated by both the repair of sublethal damage by viable cells and the replacement of killed cells by repopulation. Repair processes are usually considered to be complete within 1 to 2 d of exposure. Repopulation, on the other hand, can take up to several months to complete.

Experimental models demonstrate these effects. In the pig, moist desquamation occurs 50 % of the time after a single dose of 28 Gy (van den Aardweg *et al.*, 1988). When the total dose is delivered in two equal fractions separated by 24 h, it requires a higher total dose of 36 Gy (2 × 18 Gy) to achieve the same effect. Further extension of the time interval, up to 14 d, does not permit a larger total dose to be delivered for the same skin effect. This indicates that for this tissue reaction, repair of sublethal damage to viable cells is completed within 24 h. The kinetics repair for cells associated with other clinical endpoints for the skin (*e.g.*, late dermal necrosis) are similar.

The timing of repopulation in irradiated skin depends on the type of skin effect. For the early reaction of moist desquamation, repopulation is not seen in the first two weeks because a certain level of cell depletion is required before repopulation by surviving cells is initiated. In the epidermis, this is a 50 % reduction of the density of cells in the basal layer (Hopewell, 1990). Following skin doses <15 Gy, repopulation of depleted cells in the basal layer is essentially complete within two months. Repopulation occurs more slowly as the dose increases, because of the dose-related decrease in the number of surviving cells in the basal layer. Repopulation of depleted cells occurs over a much longer time scale in dermal tissue (van den Aardweg *et al.*, 1988).

Repeated interventional procedures are typically separated by days to months. Skin recovery between sessions is therefore governed by the kinetics of both the repair of sublethal damage in viable cells and the repopulation of skin from surviving, reproductively viable cells.

F.3 Initiating Dose and Subsequent Time Course of Radiation Injury

Damage may be expressed in the epidermis, dermis and subcutaneous tissues. When this damage becomes evident depends on the dose of radiation and on biological factors. Clinically apparent damage can be classified loosely as being prompt, early, mid-term, or late in terms of the time of expression. A summary of skin and hair effects, as a function of dose and time, is given in Table 2.5. Due to dosimetric uncertainty and biological variability, the dose and time boundaries between the rows and columns in the table are not rigid. There is overlap between events in any one time-dose zone and all adjacent zones.

The characteristics and timing of the signs and symptoms of radiation-induced injuries are also influenced by a variety of aggravating and mitigating factors. Specifically, anything that damages irradiated skin (*e.g.*, sunburn, abrasion, biopsy) is likely to aggravate the tissue response and may increase the probability of infection. Areas of skin that are thin and lack redundant dermal tissue, such as the anterior tibia and portions of the sole of the foot, may be more prone to radiation injury. Caution is exercised when exposing such areas to radiation (Fitzgerald *et al.*, 2008).

The clinical concept that field size influences the radiation response of the skin first came from the publication in the 1940s of tables of skin tolerance doses for different areas of human skin exposed to orthovoltage irradiation (Ellis, 1942; Jolles and Mitchell, 1947; Paterson, 1948). The same skin dose delivered to different-sized fields will produce the same initial reaction. For a small field, the dose prescribed was intended to produce moist desquamation. This was considered a tolerable skin reaction, because when small skin areas are irradiated, moist desquamation heals quickly, largely by cell migration from the field edges. The same reaction (moist desquamation) from the same dose to a large field does not heal quickly and thus was clinically unacceptable. As a result, doses had to be reduced for large fields, to prevent the development of moist desquamation. This phenomenon led to the term tolerance doses (doses that are tolerated). Tolerance doses are not iso-effective. Tables of skin tolerance dose were based on experience with orthovoltage x-ray beams and radium. These data were then inappropriately used to establish mathematical relationships between dose and treatment area with the premise that the quoted tolerance doses represented biologically iso-effective doses.

In subsequent studies of human skin, no influence of field size on the development of moist desquamation was found for fields ≥(40 mm × 40 mm) (Joyet and Hohl, 1955). Comparable results have been obtained for pig skin. For fields ≥22.5 mm in diameter, field size had no influence on the development of moist desquamation (Hopewell, 1991). The difference between tolerance and iso-effect was first recognized by von Essen and was subsequently discussed in greater depth (Hopewell, 1997; von Essen, 1969). Since most interventional fields are larger than 40 mm × 40 mm at the patients' skin, the effect of field size can be ignored.

Because of clinical variability, it is appropriate to assume that any skin changes observed following an interventional procedure are radiogenic in origin unless an alternative diagnosis is established.

The National Cancer Institute has defined five grades of skin toxicity for radiation dermatitis (NCI, 2006). This grading scale and the appearance of the skin reaction resulting from experimental (animal) or interventional (human) irradiation for the first four grades are shown in Appendix G. The same scale can be used to define the skin toxicity associated with FGI procedures.

F.3.1 *Prompt Reactions (less than two weeks)*

The most frequently reported prompt reaction is the so-called early or 24 h erythematous reaction. This can occur from a few hours up to 24 h after an absorbed dose >2 Gy. Once this reaction has resolved, there is no conclusive evidence suggesting that it has any influence on subsequent responses. It is believed to represent an acute inflammatory reaction with an associated increase in vascular permeability. In animal studies, this increase in permeability is dose related, but only up to 8 Gy (Jolles, 1972).

There are very few reports of early erythematous reactions following FGI procedures. Two cases are described here. In the first case, the patient, shown in Figure F.2, had undergone a PTCA of the right coronary artery six months prior to a subsequent procedure which involved two PTCAs, performed an hour apart, in a branch of the left anterior descending coronary artery. The latter two PTCAs entailed nearly 1.5 h of fluoroscopy, with concomitant cinefluorography. The physician noted in the procedure report that some prompt erythema was apparent on the back of the patient at the time the patient was removed from the table (no picture available). Several weeks afterward the patient noticed an erythema that progressed over time into a large area of necrosis. The necrosis possibly represents a severe mid-term reaction, as discussed below.

In the second case, the patient reported stabbing pain in the right thorax 24 h following an unsuccessful PTCA of the right coronary artery. Three days after this prompt event, an erythema developed which then progressed into what appeared to be a superficial acute ulceration as a result of exposure to a very-high radiation dose (see next paragraph). However, the timing of events was not recorded clearly in this case.

Very serious prompt reactions occur at very-high absorbed doses (>80 Gy) (Barabanova and Osanov, 1990; Hopewell, 1986; NCRP, 1999). Lesions, varyingly described as ulceration or total necrosis, develop between 14 to 25 d after exposure. The appearance of this pattern of response after an interventional procedure should be viewed with extreme concern. Exposure to such high doses in an interventional setting should be carefully investigated, to prevent the likelihood of reoccurrence. The patient is also followed for possible late effects in other irradiated organs.

F.3.2 *Early Reactions (two to eight weeks)*

These effects occur in the basal cells of the epidermis and the germinal region of the hair follicles (collectively called stem cells). These are the more rapidly proliferating cells in the skin. The underlying mechanism is comparable in both systems: cell differentiation and cell loss continue at the normal rate, but radiation inhibits cell proliferation and new cell production. Thus, the timing of the responses is dependent on the turnover time of the system and independent of the radiation dose. While the timing is independent of the radiation dose, the severity of the radiation response in the epidermis and the hair is dose dependent, since the reproductive survival of stem cells is dose related.

The main erythematous reaction is a secondary inflammatory reaction to effects occurring in the epidermis (Hopewell, 1983). Animal studies, including pig studies with exposure to single absorbed doses of orthovoltage radiation ≥15 Gy or daily fractionated absorbed doses of ~2 Gy, demonstrate a very marked fall in the number of cells synthesizing DNA in preparation for cell division and in the number of cells seen in mitosis. This represents a marked reduction in cell production (Archambeau *et al.*, 1988; Morris and

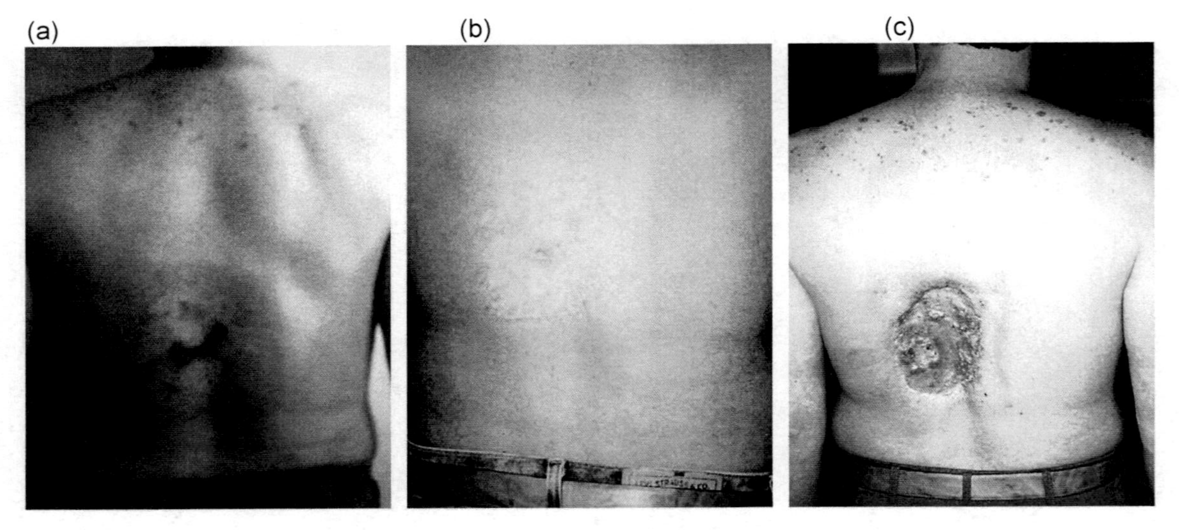

Fig. F.2. Forty year old male who underwent multiple coronary angiography and angioplasty procedures. The photographs show the time sequence of a major radiation injury (Shope, 1996): (a) 6 to 8 weeks post-exposure (prolonged erythema with mauve central area, suggestive of ischemia); (b) 16 to 21 weeks post-exposure (de-pigmented skin with central area of necrosis); and (c) 18 to 21 months post-exposure (deep necrosis with atrophic borders). This sequence is available on the FDA website, and is in the public domain (FDA, 1995b). This sequence of photographs that appeared on the FDA website has provided the first hint to other individual patients that their injury was related to a previous FGI procedure.

Hopewell, 1988). Irradiated cells in the basal layer of the epidermis that are not reproductively viable continue to differentiate and migrate into the upper layers of the epidermis. This causes a steady decline in both the density of cells in the basal layer and in the number of viable cell layers in the epidermis. The rate of decline in basal-cell density depends on the rate of epidermal turnover at the site of irradiation (usually four to six weeks). It is independent of the radiation dose and the dose-fractionation schedule used to administer that dose (Hopewell, 1990; Morris and Hopewell, 1986). A decline in the basal-cell density to ~50 % of its normal value appears to provide a stimulus to the remaining viable clonogenic stem cells in the basal layer to proliferate rapidly; the number of such cells will depend on the radiation dose (Hopewell, 1990). These proliferating cells form colonies of viable cells within the otherwise degenerating epidermis.

The number of developing colonies, which is dose related, dictates the subsequent outcome. If a large number of colonies develop (lower dose), they will coalesce, moist desquamation will be avoided, and the main erythema reaction will resolve. The epidermis may display hyperplasia prior to full recovery. This is evident clinically as dry desquamation. If a limited number of colonies develop (higher doses), the areas between them will continue to lose cells from the basal layer at the same rate. All cells and viable cell layers will be lost from the areas between colonies at a time associated with the normal turnover time of the epidermis (typically four to six weeks). Clinically, this is apparent as moist desquamation. Colonies of viable cells may be seen within areas of moist desquamation (Withers, 1967). Regeneration of the epidermis takes place due to the continued proliferation of cells in these cell colonies and cell proliferation from around the field edges. In this situation, bright red erythema persists until repopulation nears completion.

With high radiation doses, few or no viable cells remain in the irradiated area. Repopulation progresses slowly, primarily from the edges of the field. Regeneration after a high radiation dose to the lateral aspect of the hip during radiographic localization of a brachytherapy applicator is illustrated in Figure F.3 (Thomadsen *et al.*, 2000). Regeneration is seen both from the edges of the irradiated area and from surviving cells that were partially shielded by the lead cross wires.

In areas where moist desquamation develops, infection and dehydration of subepithelial tissues after high radiation doses can lead to the development of secondary ulceration, the loss of dermal tissue over a mid-term time frame. These skin lesions are likely to

(a)

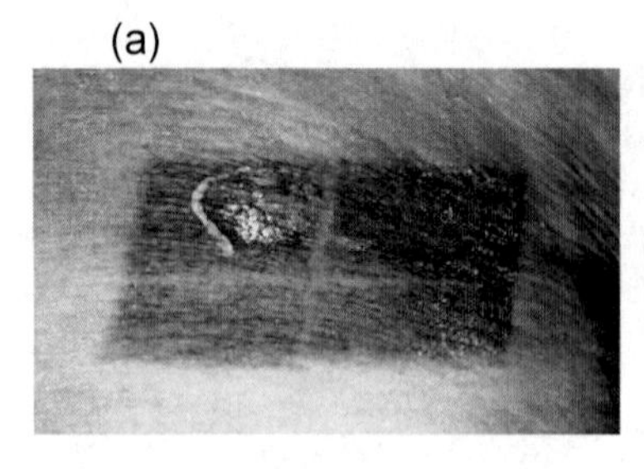

(b)

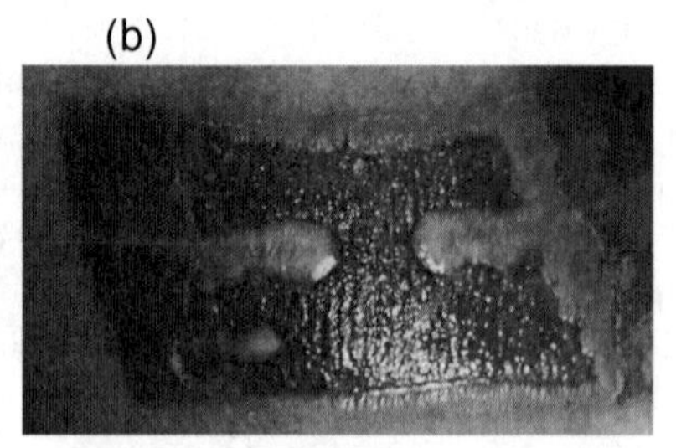

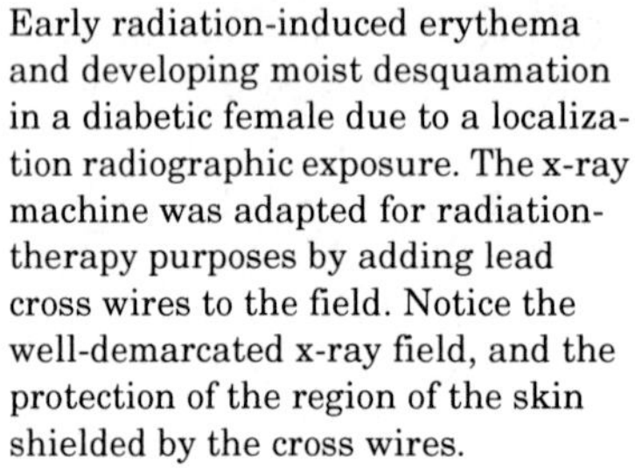

Early radiation-induced erythema and developing moist desquamation in a diabetic female due to a localization radiographic exposure. The x-ray machine was adapted for radiation-therapy purposes by adding lead cross wires to the field. Notice the well-demarcated x-ray field, and the protection of the region of the skin shielded by the cross wires.

Healing of moist desquamation by epithelial regeneration, both from epithelial stem cells extending inward from the margin of the irradiated area and from the shadow of the lead cross wires (Thomadsen *et al.*, 2000).

Fig. F.3. Skin injuries induced by diagnostic radiography (images supplied by Bruce Thomadsen).

require surgical intervention to remove irradiated dermal and subcutaneous tissue. Extensive scar tissue formation will result if these lesions are left to heal slowly.

When hair is irradiated, the cells at the base of the hair follicle are affected. The degree of radiation response depends on the number of cells that remain reproductively viable. The timing of the appearance of these radiation effects depends on the normal growth rate of the hair. After low absorbed doses (1 to 8 Gy in the pig and 5 to 14 Gy in man), only a relatively small, transient loss of stem cells occurs. This is associated with a transient reduction in the diameter of the hairs. Subsequent recovery to a normal diameter will occur (Sieber *et al.*, 1986; 1992). The maximum reduction in hair diameter in both species appears to be 30 %. Past this limit, the terminal part of the hair distal to the thinned region tends to snap off, giving the appearance of temporary (and usually partial) epilation prior to regrowth of the remaining hair to a normal diameter. In the pig, the probabilities of both detectable hair loss and >50 % hair loss after six weeks are clearly dose related. The absorbed dose to produce these effects in 50 % of irradiated porcine and human sites is 9.8 ± 0.6 Gy and 13.8 ± 0.9 Gy, respectively (Sieber and Hopewell, 1990). Total permanent epilation occurs at the radiation dose where all stem cells in the follicle are reproductively sterilized and the hair follicle is lost.

F.3.3 *Mid-Term Reactions (6 to 52 weeks)*

Mid-term reactions are associated with the development of delayed lesions in the walls of blood vessels in the dermis and subcutaneous fat (Hopewell and Young, 1982). Skin with a "dusky" or "mauve" appearance provides evidence for the presence of ischemia (Figure F.2a) with a measurable reduction in blood flow and a reduction in vascular density 12 weeks after irradiation (Archambeau *et al.*, 1984; Moustafa and Hopewell, 1979). The probability of developing either dusky-mauve erythema or full- or partial-thickness dermal necrosis depends on the radiation dose (Hopewell, 1990). Slices of skin taken over this time period demonstrate that necrosis may affect only part of the thickness of the dermis, while at the same time there is evidence of vascular changes in the subcutaneous fat (Figure F.4a). In some instances these vascular changes, with associated necrosis, may involve the fat and not the dermis (Figure F.4b). As areas of partial-thickness necrosis heal, the epidermis migrates under the dead tissue to form a new covering. Small areas of full-thickness necrosis may heal or may require surgical intervention to remove necrotic dermal and subcutaneous tissue.

At irradiated sites that do not develop dermal necrosis, the blood flow per unit volume of dermis returns to normal concomitant with the development of dermal thinning and contraction in size of the originally irradiated area over the period from 12 to 16 weeks after exposure (Moustafa and Hopewell, 1979). This has been demonstrated in pig skin by serial measurements of relative dermal thickness after irradiation with single doses of $^{90}Sr/^{90}Y$ beta particles (Hopewell, 1990).

F.3.4 *Long-Term Reactions (>40 weeks)*

Primary long-term reactions include the further development of dermal thinning. Dermal thinning is one measure of dermal atrophy; the surface dimensions are also reduced and because of the lines of tension in the skin, irradiated sites frequently take on a slightly stellate shape. Therefore, dermal thinning can be assessed as a reduction in the dimensions of the original irradiated site, or as clinically-detectable induration due to the atrophy of both dermal and subcutaneous fat (Hopewell, 1980; 1990). Serial measurements of relative dermal thickness after irradiation with single doses of $^{90}Sr/^{90}Y$ beta particles show that this phase of dermal thinning develops between 52 and 78 weeks after irradiation. For single absorbed doses in the range of 10 to 33 Gy, the timing is independent of the radiation dose, although the severity of the reaction is dose related. The severity of dermal thinning, as assessed by reduction in

(a)

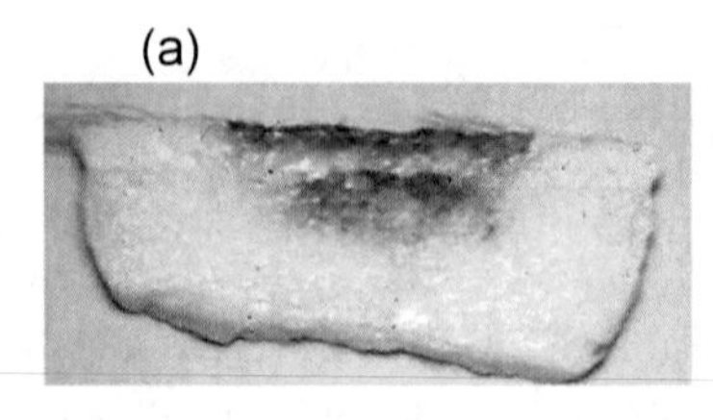

Note necrosis of a partial thickness (dark grey area) of the dermis and evidence vascular changes in the underlying subcutaneous fatty layer.

(b)

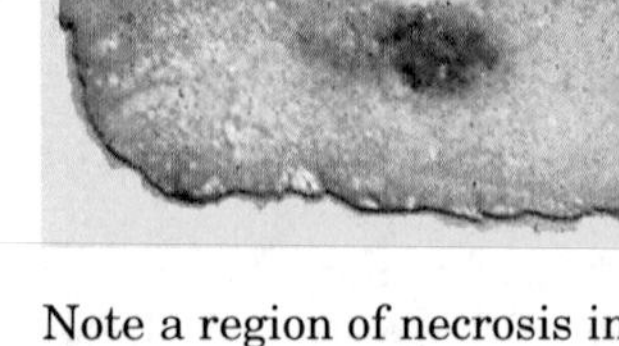

Note a region of necrosis in the subcutaneous fatty layer; the overlying dermis shows no evidence of necrosis.

Fig. F.4. Slices though a region of pig skin, 16 weeks after irradiation with a single absorbed dose of 20.7 Gy of 250 kVp x rays. (These circa 1960 images have been cleaned and enhanced for this publication.)

the dimensions of the original irradiated site, may be influenced by the initial development of transient moist desquamation.

In addition to these late deterministic effects, late stochastic effects are possible. Experience from radiation therapy demonstrates that there is a small but real increased risk for the development of malignancies such as carcinoma, melanoma and sarcoma, with latency periods that can extend beyond 20 y (Fitzgerald *et al.*, 2008; ICRP, 1991c).

Radiation-induced telangiectasia is different from the clinical telangiectasia that results from wound healing. It is a well-recognized, long-term reaction of human skin, representing the dilation of capillaries in the superficial papillary dermis. These vessels are often visible though the epidermis. They are rarely seen prior to 52 weeks after the completion of radiation therapy, but then increase in both incidence and severity for at least 10 y. The rate of progression is dose related (Turesson and Notter, 1986). The origin of this type of lesion is unclear. It has been suggested that the development of telangiectasia is secondary to smooth muscle degeneration in end arterioles in the dermis. In pig skin, there is histological evidence for the presence of telangiectasia after >52 weeks, with associated hyaline change in the walls of end arterioles, petechial hemorrhages, and focal dermal necrosis. In humans, areas of skin that show transient moist desquamation tend to show more pronounced telangiectasia in the long term (Bentzen and Overgaard, 1991).

Appendix G

Graded Examples of Skin Reactions to Radiation

TABLE G.1—*NCI (2006) skin reaction grades adapted to FGI procedures.*

NCI Skin Toxicity Grade (NCI, 2006)	Description	Representative Images
1	Faint to moderate erythema	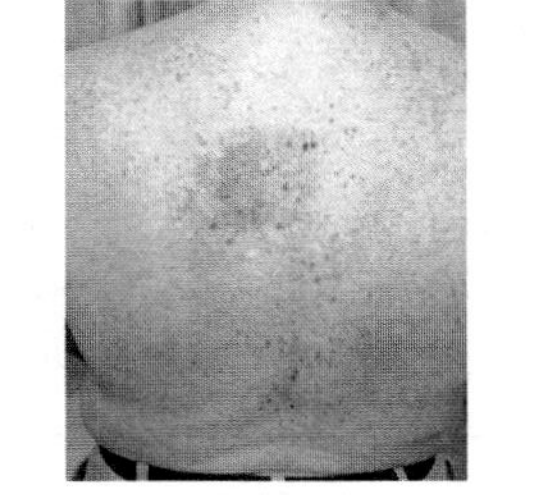 **Fig. G.1.** Sixty-five year old male who had two FGI procedures performed through overlapping skin ports. Note the enhanced reaction in the overlap zone. The first procedure was performed six weeks before this photograph was obtained. The second procedure was performed two weeks before the photograph was obtained.

TABLE G.1—(*continued*)

NCI Skin Toxicity Grade (NCI, 2006)	Description	Representative Images	
2	Moderate to brisk erythema; patchy moist desquamation, mostly confined to skin folds and creases; moderate edema	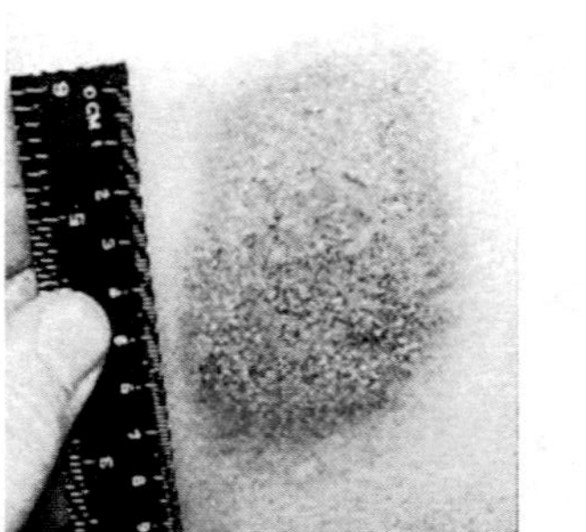 **Fig. G.2a.** Subacute radiation dermatitis from fluoroscopy during coronary artery stenting. Photograph obtained two months after fluoroscopy (Stone *et al.*, 1998).	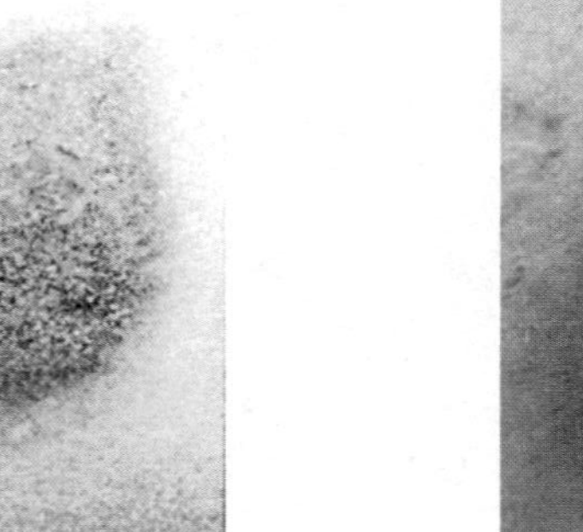 **Fig. G.2b.** The lesion progressed to hyperpigmentation, sclerosis, and ulceration five months after fluoroscopy (Stone *et al.*, 1998).

2 | Moderate to brisk erythema; patchy moist desquamation, mostly confined to skin folds and creases; moderate edema

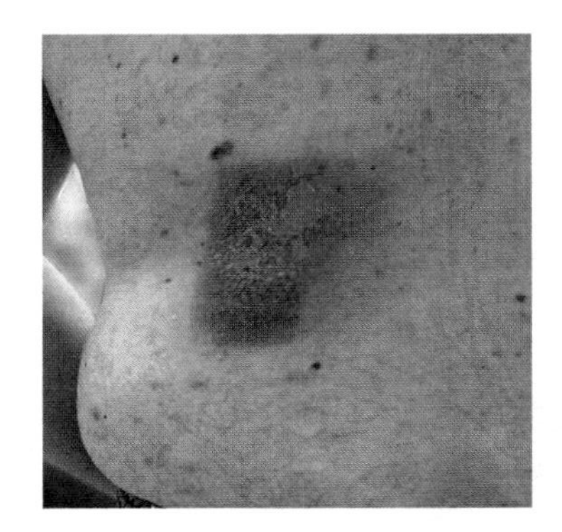

Fig. G.3a. Fifty year old male with radiation injury. Photograph taken two months after ~10 Gy ($D_{skin,max}$).

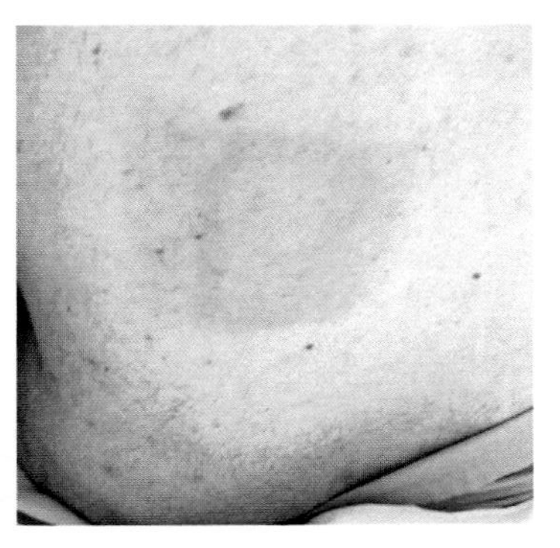

Fig. G.3b. Photograph of the same patient taken six months after ~10 Gy ($D_{skin,max}$).

TABLE G.1—(*continued*)

NCI Skin Toxicity Grade (NCI, 2006)	Description	Representative Images
2	Moderate to brisk erythema; patchy moist desquamation, mostly confined to skin folds and creases; moderate edema	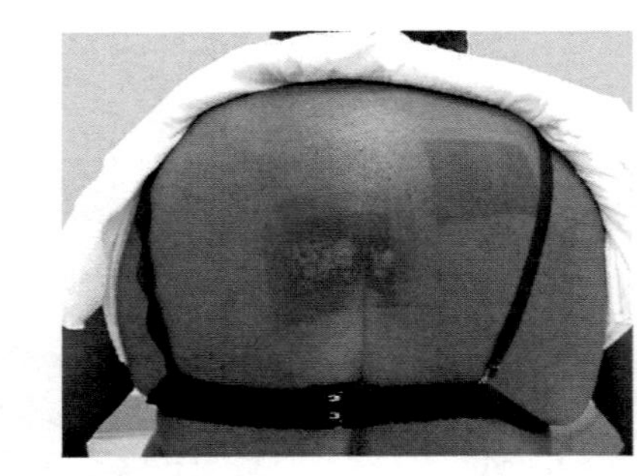 **Fig. G.4.** Eighty year old female with radiation injury due to overlapping radiation fields. Superior region of injury 12 weeks after ~10 Gy ($D_{skin,max}$). Midline region of injury (with overlap) 10 weeks after ~8 Gy ($D_{skin,max}$) in overlap area. Reactions had faded at six months (telephone interview).

3 Moist desquamation other than skin folds and creases

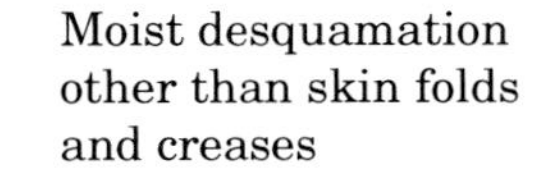

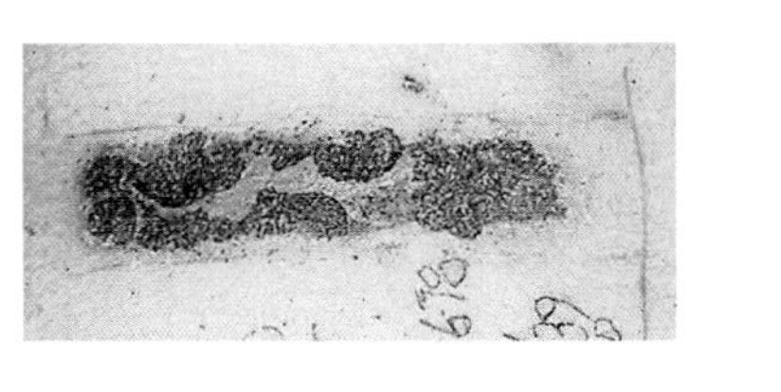

Fig. G.5. Pig model demonstrating early radiation reaction; moist desquamation with dried serum exudates on the final day of irradiation with a schedule involving 18 dose fractions over 39 d (three times per week; total dose 72 Gy). Reaction developed in the final week of irradiation. Dried exudate on surface due to serosanguinous fluid leakage. (This circa 1960 image has been cleaned and enhanced for this Report.)

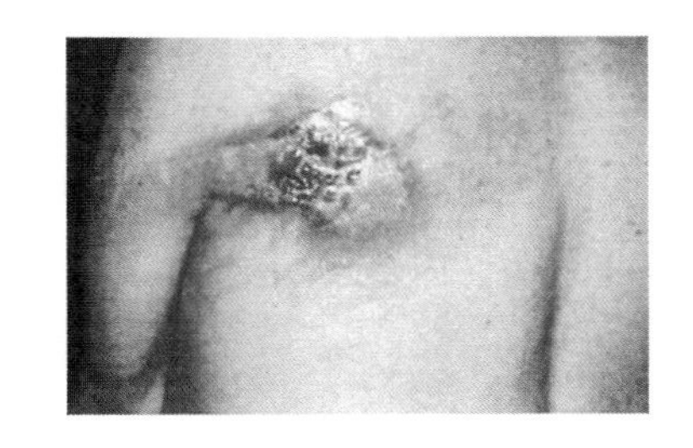

Fig. G.6. Increased severity of reaction in an area of radiation field overlap (Granel *et al.*, 1998).

TABLE G.1—(*continued*)

NCI Skin Toxicity Grade (NCI, 2006)	Description	Representative Images	
4	Skin necrosis or ulceration of full-thickness dermis; spontaneous bleeding from involved site	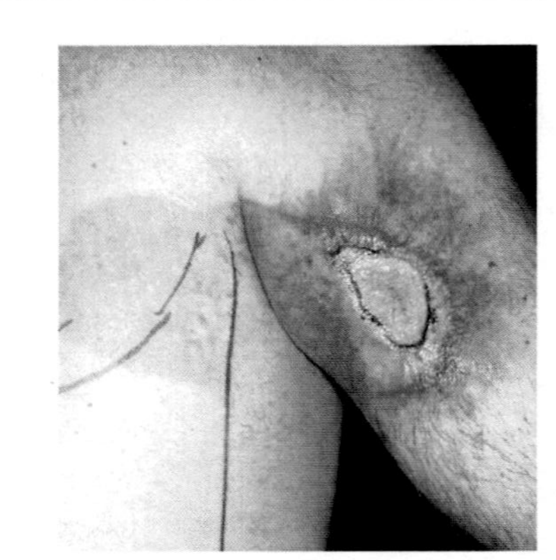 **Fig. G.7.** Four months after an EP ablative procedure. The patient's arm had accidentally been positioned within the radiation field during the 10 h procedure. Estimated 15 to 20 Gy ($D_{skin,max}$). After plastic surgery, the patient's ulceration healed, and his pain resolved (Wong and Rehm, 2004).	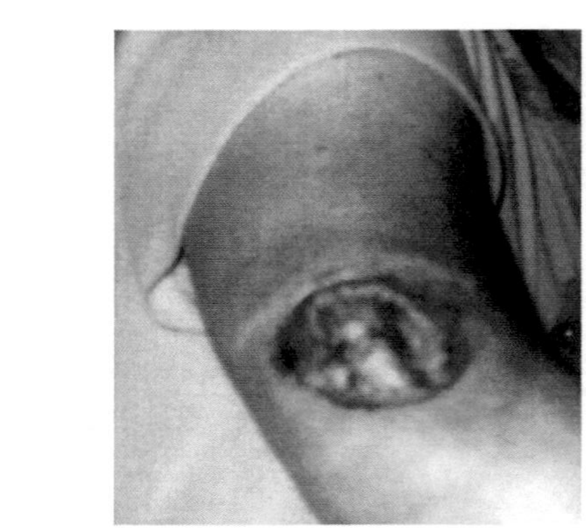 **Fig. G.8.** Deep necrosis in the arm at the elbow months after a procedure. Under the cover of the sterile drapes, the patient had unknowingly rested an arm over the port of the x-ray tube during an EP and ablation procedure (Wagner and Archer, 2004).

4	Skin necrosis or ulceration of full-thickness dermis; spontaneous bleeding from involved site 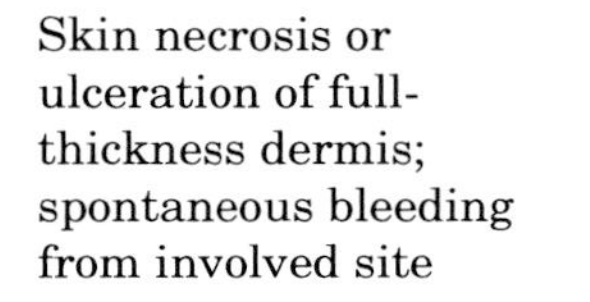	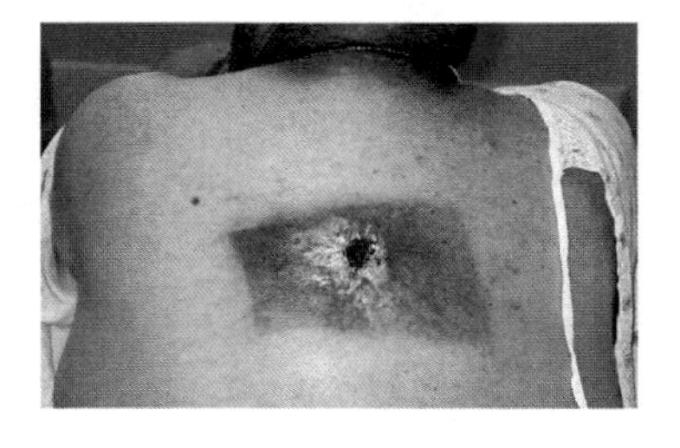 **Fig. G.9a.** Sixty year old male, 30 weeks after a coronary angioplasty. Central area of deep necrosis surrounded by indurated and depigmented skin within an area of prolonged erythema. Necrosis probably induced by unhealed punch biopsy.	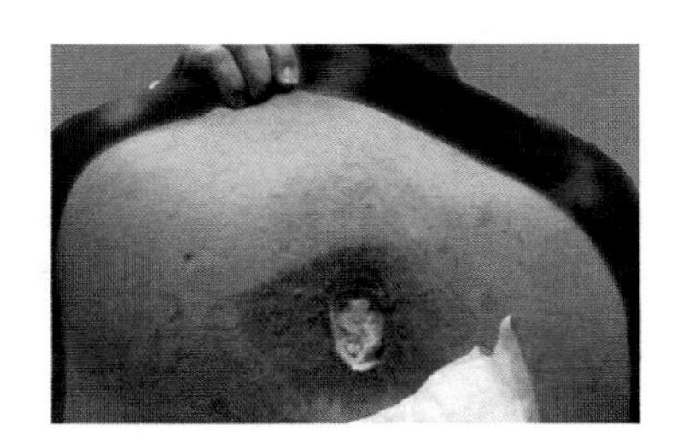 **Fig. G.9b.** Same patient, 38 weeks after the procedure.

TABLE G.1—(*continued*)

NCI Skin Toxicity Grade (NCI, 2006)	Description	Representative Images	
4	Skin necrosis or ulceration of full-thickness dermis; spontaneous bleeding from involved site	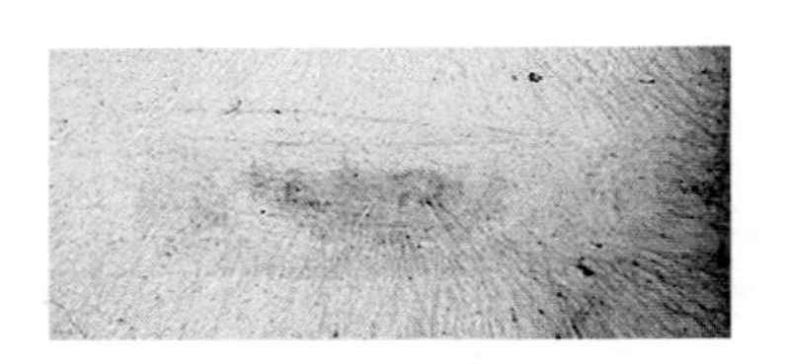 **Fig. G.10a.** Mid-term radiation reaction in a pig model; dusky mauve reaction indicating dermal ischemia 12 to 14 weeks after a single absorbed dose of 23 Gy. (This circa 1960 image has been cleaned and enhanced for this Report.)	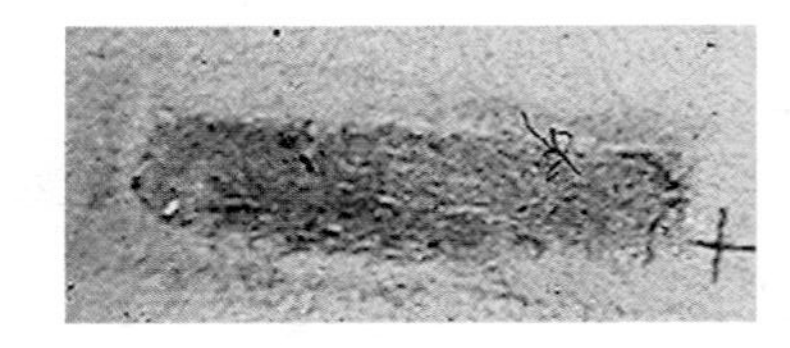 **Fig. G.10b.** Subsequent dermal necrosis in a pig model, 14 to 16 weeks after a single absorbed dose of 23 Gy. (This circa 1960 image has been cleaned and enhanced for this Report.)

Appendix H

Theory of Operation of Equipment

H.1 Introduction

This appendix describes the theory of operation of FGI-procedure equipment. The performance of FGI-procedure equipment has significantly improved over the past 30 y (Strauss, 2002). Better stability of electronic components and improvements in real-time monitoring of the equipment's performance contribute to more accurate production and control of x rays.

The technical evolution of FGI-procedure equipment continues, as further developments occur in many areas. These developments are based on both the evolving needs of interventional examinations and on investments made in research and development of imaging technologies. Evolving needs include more complex and longer interventions; use of new, more complex interventional tools that should be integrated in the examination workflow; and other needs described in the main part of this document. Evolving imaging technologies include large-area thin-film transistor array image receptors, refined image processing based on the availability of greater computational power, anti-collision devices based on capacitive sensors, and the integration of technologies developed in other areas (*e.g.*, touch-screen controls at tableside, liquid-crystal displays, wireless communication).

Some of these evolutions enable a reduction in radiation dose, either by permitting a lower radiation dose rate (*e.g.*, pulsed fluoroscopy, copper filtration) or by eliminating the need for repeating a part of the examination (*e.g.*, storage and replay of fluoroscopy loops). It is expected that further technical improvements will occur, making FGI-procedure equipment more efficient, safer and easier to use in the future. However, some available equipment settings will increase dose even when used appropriately and should be used with caution.

H.2 Workflow

Interventional equipment is composed of several devices that are used together as a system during patient treatment. This section presents typical devices that may be used during an interventional procedure. The actual equipment available for clinical use depends on the configuration that was purchased.

Usually, the equipment is located in a dedicated room. Any device (*e.g.*, monitor suspension, radiation shield) that is attached to the ceiling is moved out of the way. The patient typically enters the room on a stretcher and is placed on the imaging table with a mattress and drapes. The table is moved up and down to facilitate patient transfer. Equipment motion may be disabled by the actuation of a switch to avoid unintentional movement of the patient table or imaging gantry while the patient is being moved, since the patient's body typically crosses over the tableside controls. Either the patient or operator may inadvertently interact with these controls. Devices that are required to be attached to the patient (*e.g.*, electrocardiogram leads, oxygen) are located near the table. The fluoroscopy monitor suspension and radiation shields are placed at their normal working locations after the patient is positioned.

The patient typically lies supine on the tabletop. When a biplane system is used, the tabletop should be elevated to place the anatomy of interest at the height of the isocenter. This can be checked with fluoroscopy in the lateral plane to verify that the patient anatomy is in the vertical center of the image. Next, the tabletop is moved during fluoroscopy in the posterior-anterior projection to position the anatomy of interest in the center of the image. Placing the anatomy of interest at the isocenter of the imaging equipment prevents patient anatomy from shifting across the image as different projections through the anatomy are selected. This is less often necessary when a single-plane system is used.

In conjunction with patient placement, the patient's identification and status are entered into the imaging system, either manually with a keyboard or by accessing the appropriate patient information from the hospital information system. The correct anatomical program is selected to provide the proper fluoroscopic and acquisition modes. This sets the necessary acquisition and image-processing parameters for the unique imaging challenges presented by the patient.

The operator initiates x-ray production with a foot switch that is depressed continuously when x rays are desired. The foot switch contains a switch for fluoroscopy that produces low-dose-rate live images. This mode is used intermittently over extended periods of time to allow the operator to position guide wires, catheters, and other devices during the interventional procedure. Fluoroscopic images are not usually saved. However, some equipment permits retrospective storage of the last few seconds of fluoroscopy when requested by the operator. The operator depresses a second switch on the foot switch when higher-dose-rate images are created and recorded (fluorography). These images of higher quality are obtained for review at a later time or to document the results of the examination. Foot pedals are designed to provide clear differentiation between fluoroscopy and fluorography modes.

During the procedure, the x-ray beam orientation relative to the patient can be changed to minimize the superimposition of anatomical structures on the image. The x-ray beam is produced by an x-ray tube that is typically attached to one end of a C-arm. The image receptor (x-ray detector) is attached to the other end of the C-arm. It creates a visible image from the x-ray pattern in space emitted from the patient. The C-arm is attached to mechanical supports that allow the necessary degrees of freedom for positioning during the procedure. Tableside controls are used to rotate the x-ray tube and image receptor around the patient in two independent directions. One of these directions is left/right angulation, resulting in image projections that are typically labeled RAO or LAO for a supine patient. Oblique projections are named for the side (right or left) and the body surface (anterior or posterior) closest to the image receptor. Therefore, the entrance-skin region is opposite (*i.e.*, the entrance-skin area for an RAO projection is the left, posterior region of the patient). The other rotation is towards the patient's head or feet in a plane parallel to the long axis of the patient. This rotation results in image projections typically labeled cranial or caudal. The distance between the x-ray tube and detector can usually be adjusted to optimize C-arm angulation and management of scattered radiation.

The operator may review the recorded acquisitions. With some equipment, reference images may be created and selected for display on a second monitor. Additional image processing may be applied (*e.g.*, to quantify stenosis, to assess the volume of the left ventricle at various points in the cardiac cycle, to enhance a part of the image). Other imaging modalities (*e.g.*, ultrasound) and other medical devices (*e.g.*, for navigation or ablations) may be used during interventions.

The recorded acquisitions are typically stored on the system during the examination and transferred during or after the examination to a long-term storage archive using the DICOM® (National Electrical Manufacturers Association, Rosslyn, Virginia) protocol (DICOM, 2005). Image archives may include all recorded acquisitions or a selected subset. Images viewed during fluoroscopy are not typically archived, though some systems allow for the recording and storage of short fluoroscopy sequences. A review workstation is generally integrated in the system to allow post-processing of the images, use of other imaging modality sequences during the examination, or review of the images during preparation of the patient report.

At the end of the procedure, the patient is transferred from the imaging table to a stretcher. The patient examination is then closed on the system. This prevents further emission of x rays or acquisition of additional image sequences in the patient record. The imaging room and equipment are cleaned before the beginning of the next examination.

H.3 Main Acquisition Modes

Interventional equipment is provided with a number of acquisition modes. Some of these are standard, and others are extra-cost options. This section presents the two important FGI-procedure acquisition modes.

The first mode is unrecorded image acquisition, called *fluoroscopy.* This mode generates x-ray images and presents them instantaneously and continuously as visible images that capture motion. FDA limits normal-dose fluoroscopy to a maximum air-kerma rate (30 cm from the image receptor) of 88 mGy min^{-1} (FDA, 2009a). This mode is used to guide devices to specific locations within the patient and to observe moving structures. High-dose-rate fluoroscopy operates at a maximum air-kerma rate (30 cm from image receptor) of 176 mGy min^{-1}. The latter mode should be used sparingly, and only when the image quality from normal-dose fluoroscopy is inadequate. Two features that the operator should exploit to reduce fluoroscopy time are last-image hold and the ability to record fluoroscopy image sequences. All fluoroscopy units currently sold in the United States have last-image hold capability; the ability to record fluoroscopy image sequences is typically an extra-cost option.

The second acquisition mode is the recording of high-dose-rate images called *fluorography.* This mode records one or a series of images for interpretation after termination of the exposure. It is used to create permanently stored images for diagnosis or documentation. A number of different fluorography modes exist for specific imaging tasks. Some of these are described below. They are typically extra-cost options, and it is uncommon for all of them to be available on the same fluoroscope.

Cardiac angiography records images at a high frame rate, 10 to 60 images per second, which allows capture of the rapid movement of the beating heart. Fifteen images per second are common for most digital systems. Higher heart rates require higher frame rates.

DSA mode is usually performed at rates between 0.5 and 4 images per second; it is intended to image stationary anatomy. Only the contrast material bolus is visible on the subtracted image. It improves visualization of blood vessels. The grayscale in an image from among the first images obtained prior to the injection of contrast material is electronically inverted and the resultant image is combined with subsequent images.

Three-dimensional or cone-beam CT acquires angiographic images at a rapid rate as the gantry rotates the x-ray beam through at least 180 degrees of rotation. The angiographic images may be viewed sequentially to view the vasculature at different angles. Alternatively, the angiographic images may be reconstructed to create an image of the injected vessels in a three-dimensional view or to create cross-sectional images of the patient. The use of a fluoroscopic flat-panel detector provides good spatial resolution of high contrast objects with the cone-beam CT. However, the ability to visualize low-contrast objects is decreased relative to standard CT images due to more scattered radiation with a cone beam.

Bolus chasing mode moves either the patient and table or the C-arm relative to each other after an injection of contrast material. This permits the interventionalist to follow the progression of the contrast material through the arteries of the abdomen, pelvis and legs when arteriography of the lower extremities is performed. This longitudinal motion enables coverage of a large region of anatomy with a relatively small x-ray detector.

H.4 Interventional Suite Layout

The imaging suite should be designed to provide adequate space for all required equipment as well as additional workspace sufficient to allow all staff members, with their required equipment and supplies, to work effectively together as a team. This additional workspace is necessary to ensure quality patient care and to minimize hazards to the patient and staff members (Roeck, 1994; Rossi and Hendee, 1980; Strauss *et al.*, 1996). Other areas needed to support the imaging suite include sterile supply storage, sedation/recovery rooms, changing rooms, patient lockers, patient toilets, soiled/clean utility rooms, patient examination rooms, reception/registration areas, and waiting areas. These areas are not discussed here, but are equally important.

Equipment capable of performing high-risk interventions safely should be permanently installed. In such cases, an equipment room houses part of the fluoroscopy equipment (*e.g.*, x-ray generator, controls, x-ray tube chiller, noninterruptible power supply, and other components). Some of the fluoroscopy equipment (*e.g.*, controls for the imaging equipment, image monitors, physiological display monitors, and controls for nonimaging equipment) is housed in an adjacent control room as illustrated in Figure H.1.

The image chain, gantry, and patient support are installed in the middle of the procedure room. Adequate space should be provided in the procedure room to allow effective separation of sterile and nonsterile workspaces. Any special requirements for the imaging equipment placed in the procedure room should be identified during the planning phase of the installation, in consultation with the equipment supplier, to ensure that facility deficiencies do not restrict the designed function of the imaging equipment. Workflow patterns of staff should be considered carefully to ensure that nonsterile staff, devices and equipment do not transgress established sterile zones. Considerable space at the periphery of the procedure room is required for storage of sterile devices that may be needed during the procedure.

The procedure room should provide appropriate medical gases, adjustable lighting, cleanable surfaces in support of infection control, adequate ventilation and temperature control, a sufficient number of appropriate electrical outlets, a telephone, and an appropriate intercom or other communication system. Support structures that secure the image equipment and eliminate mechanical hazards are necessary (Strauss *et al.*, 1996). Facility systems that safely supply large amounts of electrical power to the imaging equipment, at high voltages, are necessary. Since the imaging equipment produces considerable heat, a heating, ventilation, and air conditioning (HVAC) system for the facility is required that eliminates this heat gain by providing frequent, clean-air exchanges to the room.

The equipment room is adjacent to or near the procedure room to minimize cable/connection lengths. This room contains imaging equipment components that emit noise, heat, and exhaust air (from fans moving potentially-contaminated dust/particles). The air circulation of the HVAC system provided to eliminate heat is carefully controlled by a thermostat in the equipment room and routed to prevent entry of exhaust air into the procedure room.

The adjacent control room (Figure H.1) houses the controls of the equipment (*e.g.*, on/off, reset, patient data acquisition). The common wall of the procedure and control rooms has a large, shielded window so that staff within the control room can see the ongoing examination. An intercom between these two rooms facilitates verbal communication between staff in the control and procedure rooms.

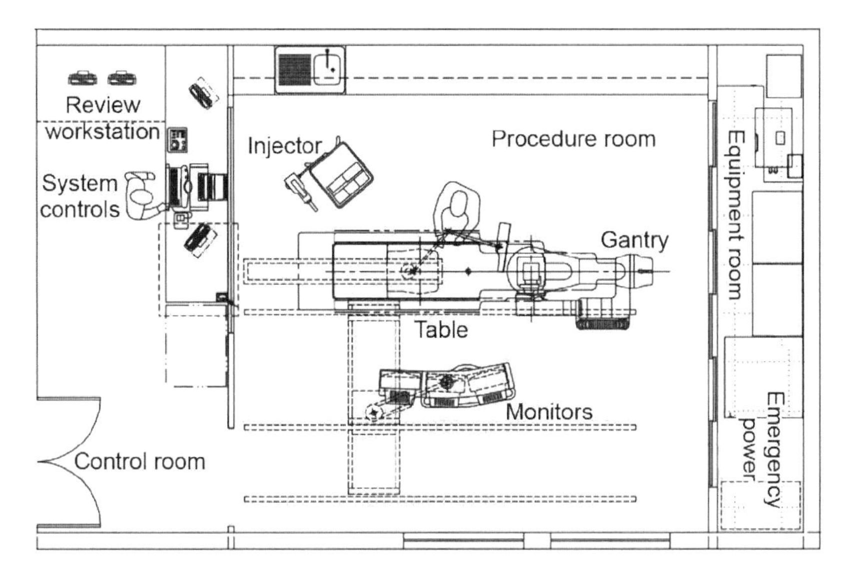

Fig. H.1. Example of an interventional equipment layout.

Image monitors are generally installed in both rooms so that staff can see the acquisition images from the control room. A review workstation, suitable for additional post-processing of acquired images and capable of managing images from other imaging modalities, may be provided in the control room; sometimes this workstation has its own additional remote monitor. Since operators wearing sterile garb may occupy the control room during portions of the examination, the control room is large enough to allow their movement without risk of contacting nonsterile surfaces.

Fluoroscopy equipment typically requires 5 to 10 min from the time that power is initiated until its computers have rebooted. An electrical power supply anomaly, such as a millisecond interruption or a voltage sag, during an interventional procedure could trigger the full 10 min reboot process of the fluoroscopy equipment computers and have devastating consequences for the patient. An uninterruptible power supply capable of maintaining the fluoroscopy equipment computer systems during power loss eliminates this risk, provided that the imaging equipment will receive sufficient power from the hospital's emergency generators before the uninterruptible power supply is exhausted. This is necessary because fluoroscopy is required to be available continuously when fluoroscopically-guided devices are being inserted, placed or withdrawn from the patient's body.

H.5 Interventional Image Chain

This section describes the image chain used in FGI-procedure equipment (Figure H.2). This chain consists of devices for: x-ray generation, x-ray detection, x-ray control, x-ray image-processing, and x-ray image display. The enhanced capabilities required of fluoroscopy equipment used for FGI procedures, compared to standard radiography/fluoroscopy equipment, are briefly discussed.

The x-ray generation device is composed of an x-ray generator that provides electrical energy to the x-ray tube. The x-ray tube converts electrical energy into x-ray energy. Pulsing is accomplished either by switches in the generator or by use of a grid-controlled x-ray tube. Typical FGI-procedure x-ray tubes contain two focal spots. The small focal spot is used for fluoroscopy of all body parts and for image acquisition on smaller body parts. The large focal spot is used for image acquisition on thick body parts where attenuation of the x rays is more pronounced. To provide pulsed fluoroscopy, the x-ray generator either rapidly switches the high voltage on and off or uses a grid-switched x-ray tube.

A collimator that contains appropriate spectral and wedge filters is mounted to the exit port of the x-ray tube. The collimator controls the geometric shape and the effective energy of the x-ray beam. The train of energy pulses during fluoroscopy is required to be at high power (many photons per unit time) to properly penetrate the spectral and wedge filters and provide adequate imaging of thick body parts at a reduced radiation dose. Since x-ray generation is at high power, a means of cooling the x-ray tube is necessary. Typically, a liquid circulating through a radiator improves the rate of heat removal from the x-ray tube assembly.

The x-ray detection device is an x-ray image receptor with an anti-scatter grid mounted at its face. The image receptor is designed to capture the majority of x rays exiting the patient, converting this x-ray pattern-in-space into a high-quality visible image. Traditionally, this image receptor has been an image intensifier. Image intensifiers are manufactured in a variety of sizes ranging from 15 to 35 cm in diameter. The type of clinical studies determines the required size of the image intensifier. The largest diameter image intensifiers are found on angiographic equipment used to image the entire chest or abdomen. A 30 cm diameter image intensifier is required for adult neuroangiographic studies. A 23 cm diameter image intensifier is typically found on adult cardiac angiographic equipment.

Flat-panel image receptors have become the image receptor of choice replacing the traditional image intensifier (Seibert, 2006).

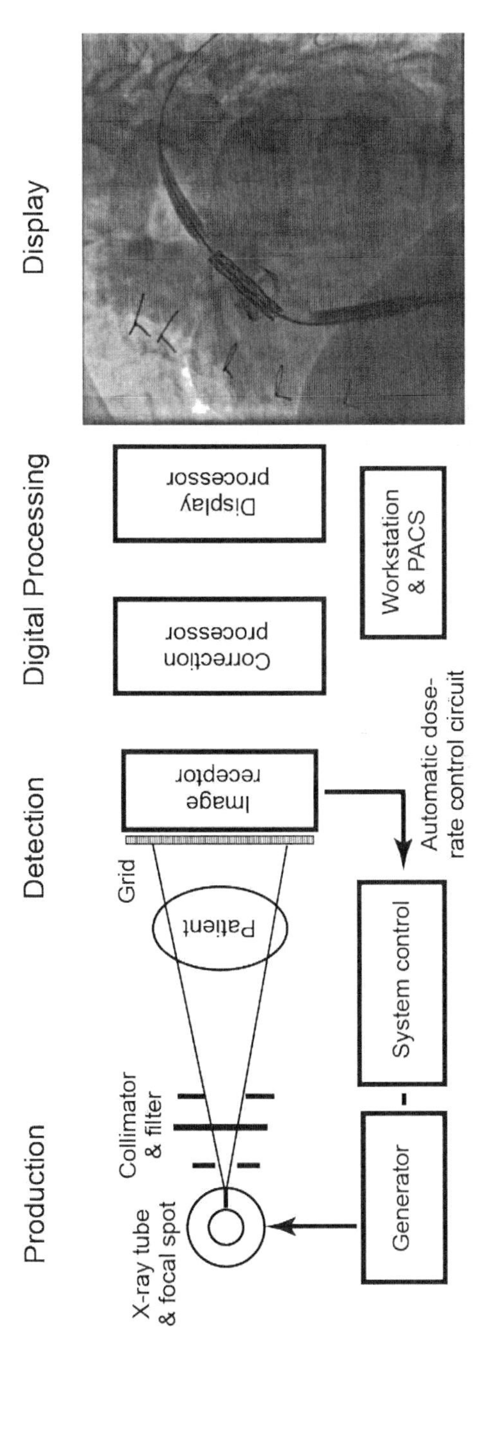

Fig. H.2. Example of an image chain and its typical components. PACS is picture archiving and communication system.

Flat-panel detectors provide better image uniformity and do not suffer from changes in image brightness seen with image intensifiers. The flat-panel detector can image a larger dynamic range of radiation attenuation in the patient (*e.g.*, heart/spine versus lung tissue) than an image intensifier. At the higher exposure values per image frame associated with recorded images of interventional cases, the flat-panel detector provides better image quality compared to the image intensifier. However, currently, the image intensifier still provides better image quality at the lower exposure rate used for fluoroscopic images (Seibert, 2004; 2006). In typical operation for interventional cardiology procedures, flat-panel detectors do not yield significantly-lower entrance doses (Chida *et al.*, 2009c).

Similar to image intensifiers, flat-panel image receptors are manufactured in a variety of sizes with the type of clinical studies determining the size selection. A large format, with 40 cm of coverage, is typically found on units intended for interventional radiology applications. A small format, with ~20 cm of coverage, is typically found on cardiac units. One manufacturer currently offers a medium format flat-panel detector with 30 cm of coverage for neuroangiographic applications.

The x-ray image processing may be performed by a single processor or may be distributed across multiple processors. Image corrections are applied initially to the x-ray detector output (*e.g.*, gain, offset for flat-panel image receptors). The initial corrections are followed by basic image enhancement processing (*e.g.*, edge enhancement, smoothing). After this second step, application processing (*e.g.*, subtraction, volume reconstruction, enhanced visualization) may be applied.

The fully-processed images are displayed in the procedure room and typically also in the adjacent control room on liquid-crystal display flat-panel monitors. These monitors are calibrated by properly adjusting their maximum brightness and by setting the appropriate gray scale calibration curve (*e.g.*, the DICOM® Grayscale Standard Display Function). The Grayscale Standard Display Function (DICOM, 2008) provides a predictable display of image contrast on video monitors, causing images to have a similar appearance, even on monitors of different maximal luminances. It also attempts to provide a display of image contrast that is optimally matched to the human visual system. The number of monitors needed in the procedure and control rooms depends on considerations such as the number of imaging planes and whether there is a need to display images from previous cases or other modalities.

H.6 Patient Support and Image-Chain Gantry

H.6.1 *Patient Support*

A typical table used to support the patient provides a flat-top mounted to a pedestal that is secured to the floor. The operator manipulates a tableside control to adjust the height of the tabletop to provide a reasonable working height. A different tableside control allows the operator to rotate the tabletop about the pedestal in the horizontal direction to assist with positioning of the patient with respect to the imaging equipment or to assist with loading and unloading the patient on the tabletop. The tableside controls can also be used to release the brakes that immobilize the tabletop, so that the operator can "free float" the tabletop horizontally in longitudinal or transverse directions to center the anatomy of interest in the displayed image.

The patient table may be designed to tilt the tabletop in addition to the basic motions discussed above. This optional feature lowers or raises the patient's head relative to their feet. When the tabletop is tilted, the ability to free float the tabletop in the longitudinal or horizontal direction is suspended as a safety precaution. The operator uses a different tableside control to allow power driven longitudinal or transverse positioning of the patient when the tabletop is not horizontal.

Some patient tables are designed to rotate the tabletop with respect to the patient's longitudinal axis. When this feature is provided, the patient is secured to the tabletop with straps. This optional motion is typically found on fluoroscopy equipment installed in operating rooms. The rotation of the tabletop and patient may be necessary to allow comfortable access to the surgical site for the operator.

The National Electrical Manufacturers Association has produced a standard for the design and function of the primary tableside controls (NEMA, 2008). The standard addresses the controls for C-arm and table movement, as well as the controls for the collimator and foot switch.

Patient tables for FGI-procedure systems require adequate strength to support large patients, uniform x-ray attenuation to reduce the appearance of image artifacts, and minimal x-ray attenuation for reduced tube loading. To satisfy these requirements, carbon-fiber composite material is used. Generally, pads are also placed on the table support to improve patient comfort. When the x-ray tube is positioned under the table and pad, the x-ray beam attenuation factor is typically between 1.2 and 1.4 (Geiser *et al.*, 1997). Thick rubber-based pads might be even more attenuating.

H.6.2 *Image-Chain Gantry*

The gantry of the interventional fluoroscope supports the x-ray tube relative to the image receptor. The frontal gantry may be mounted to the floor or suspended from ceiling rails. A ceiling-suspended gantry typically may be moved farther away from the patient table than a floor-mounted gantry, allowing easier positioning of a transport stretcher near the tabletop. When two planes of imaging are provided in the same room (a biplane setup), the frontal plane is typically floor mounted and the lateral plane is mounted on rails from the ceiling. Biplane units offer the ability to acquire two simultaneous images in different projections in both fluoroscopy and record modes of operation. Biplane units are most commonly used for interventional procedures in the brain and heart.

A C-arm is typically constructed so that the image receptor and x-ray tube are mounted at opposite ends of a C-shaped arm. This curved arm is mounted inside a curved support system, on rollers. This design allows the x-ray tube and image receptor to be rotated about the patient, in the plane of the C-arm, with no motion of the x-ray tube relative to the image receptor (see curved arrow labeled "C-arm roll" in Figure H.3). The range of motion is typically at least 180 degrees. The image receptor is mounted to the C-arm in such a way that it can be moved closer to or farther away from the x-ray tube focal spot (see straight arrow labeled "Image receptor adjustment" in Figure H.3). The distance between the x-ray tube focal spot and the image receptor is called the source-to-image-receptor distance (SID) and is typically 85 to 115 cm. Independent tableside controls allow the operator to rotate the C-arm or adjust SID. Figure H.3 presents a typical floor-mounted C-arm fluoroscope.

The C-arm, mounted within its curved support, is typically mounted on a pivot shaft which allows the x-ray tube and image receptor to be rotated in a plane at right angles to the plane of the C-arm's rotation (Rauch and Strauss, 1998). The C-arm assembly and its pivot shaft are designed so that both rotations are about the same point in space called the isocenter of the gantry. This position is typically 75 to 80 cm from the source of x rays.

In Figure H.3, the plane of the C-arm is parallel to the long axis of the patient table. Rotation in the plane of the C-arm is referred to as craniocaudal rotation. L-arm rotation in Figure H.3 (left to right with respect to the patient) is referred to as oblique rotation. The floor-mounted C-arm is oriented as shown in Figure H.3 when combined with a ceiling suspended lateral imaging plane as illustrated in Figure H.4. In Figure H.4, the frontal plane floor-mounted C-arm fluoroscope has right cranial angulation while the lateral plane has left caudal angulation.

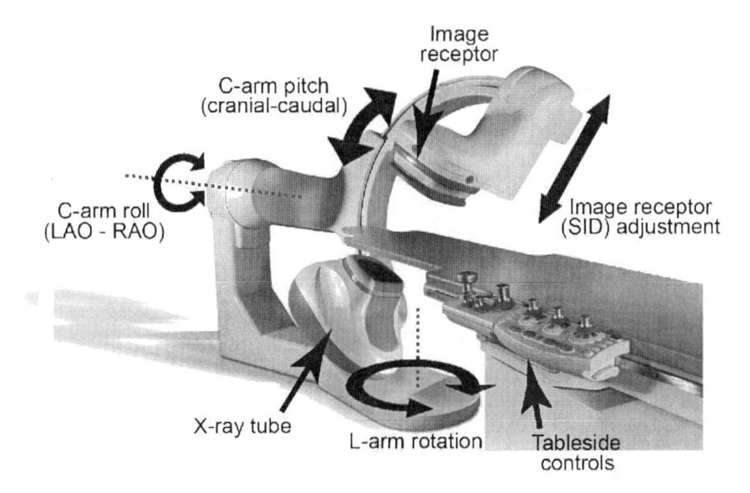

Fig. H.3. Picture of a floor-mounted C-arm fluoroscope. The unit is shown with ~30 degrees of craniocaudal and 15 degrees of oblique angulation.

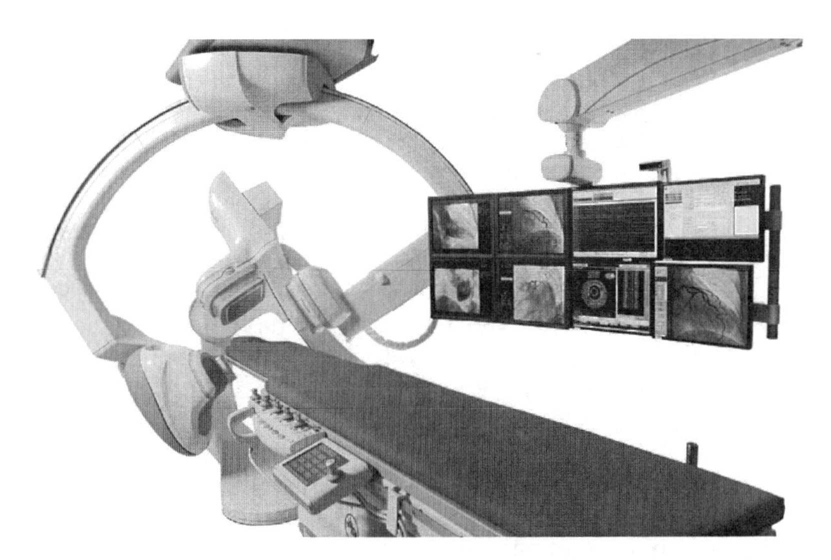

Fig. H.4. Picture of a C-arm fluoroscope with right cranial angulation for the frontal plane and left caudal angulation for the lateral plane. The lateral plane is attached to the ceiling while the frontal plane is mounted to the floor.

Most frontal plane gantries allow the C-arm to approach the patient table from either side, or from the head end of the table. In Figure H.3, this is achieved by rotating the L-arm ±90 degrees about the base where it is attached to the floor. If the C-arm approaches from the head of the table, rotation in the plane of the C-arm provides craniocaudal angulation, and rotation at the pivot shaft, at right angles to the C-arm, provides oblique angulation. Most permanently installed fluoroscopy units allow operators to program standard craniocaudal and oblique gantry angles into the anatomical programs. More complete descriptions of angulations of the gantry and the standardized notation which describes these angles may be found in Rauch and Strauss (1998).

H.7 Automatic Exposure-Rate Control Devices

The automatic exposure-rate control device monitors changes in the radiation exposure rate at the entrance plane of the image receptor due to variations in attenuation of the x-ray beam and adjusts one or more of the available image capture controls to maintain the appropriate exposure rate at the image receptor. The exposure rate at the image receptor is adjusted to fall within a range of allowable values. The allowable range is a function of the clinical imaging task and other image capture control settings that are selected in response to the patient size.

Image production controls include setting the machine parameters kilovolt peak, milliampere, and pulse width. Other image production controls include focal-spot size, anode rotation speed, and x-ray collimator parameters including spectral filters (type and thickness), and wedge filter position(s). The detection parameters of the image receptor, which include the analog to digital selection range, pixel reading method (binning factor), and the active area of the detector are controlled. Monitoring of the x-ray tube's thermal status is essential. Some of these parameters may be defined in advance by the user (*e.g.*, by the selection of an anatomical protocol or a specific field-of-view) or preset in a limited-access service menu.

In summary, there are a large number of parameters that are controlled for a given acquisition, either statically or dynamically. They are controlled based on a programmed goal that may not be selectable by the user, may be selected inherently with other controls, or may be explicitly selected by the user. The programmed goal is influenced by the fluoroscopy unit's unique design characteristics. However, it should attempt to comply with the ALARA principle of using the least amount of radiation dose necessary to obtain the desired image quality.

The following are two scenarios of specific programmed goals. The first scenario is pulsed fluoroscopy. It is necessary to control the air-kerma rate at the entrance plane of the image receptor relative to the frame rate. This can be done in several ways. One method is a constant air-kerma rate per x-ray pulse regardless of frame rate. This results in a change in air-kerma rate proportional to the change in frame rate. Lower frame rates result in lower dose to the patient (Figure H.5). Another method is an increased air-kerma rate (compared to the first method) per x-ray pulse as the frame rate decreases (Figure H.5) (Aufrichtig *et al.*, 1994). This method prevents an increase in perceived image noise at reduced pulse rates, but sacrifices some of the patient dose savings achieved with the first method. On the other hand, the second method makes it less likely that operators will refuse to use lower pulse rates because of objectionable noise levels. Deterioration of image quality (jerkiness or strobing) due to the reduced motion sampling associated with lower frame rates is typically better tolerated by operators than is an increase in perceived noise levels.

The second scenario of a specific programmed goal is the change of dose rate as a function of the selected field-of-view. With flat-panel technology, a reduction of field-of-view can be accompanied by either a constant dose rate or an increased dose rate (to maintain or improve the image-quality level) (Srinivas and Wilson, 2002). The actual implementation of changes in dose rate as a function of field-of-view change varies among manufacturers for both flat-panel and image intensifier detectors. With image intensifiers,

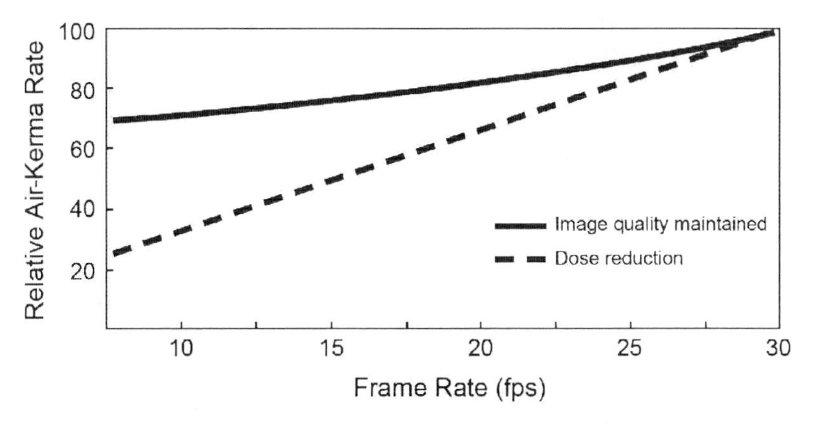

Fig. H.5. Examples of the variation of relative air-kerma rate as a function of frame rate [frames per second (fps)]. The peak kilovoltage is approximately constant across frame rate in this example.

the dose rate is scaled using physical means (*e.g.*, an iris) to compensate for either the programmed target or for the light flux reduction in the image intensifier. Examples of dose rate variation with field-of-view for flat-panel systems can be found in Mackenzie *et al.* (2006).

H.8 Interventional Image Processing

Three fundamental types of image-processing controls are edge enhancement, smoothing, and gray-level dynamic range adjustment. The degree of each image-processing parameter is carefully controlled for different imaging tasks. Proper control of image-processing parameters during fluoroscopy is analogous to properly developing the photographer's exposed film prior to making the photographic print.

X-ray image processing may be a single or a distributed function for basic or application processing. The distinction between basic and application processing is arbitrary and simply a convenient means to present various types of processing in this section.

Basic processing (Figure H.6) transforms pixel values and may combine neighboring pixels in order to present the x-ray information in an appropriate visual format for the human eye. Examples

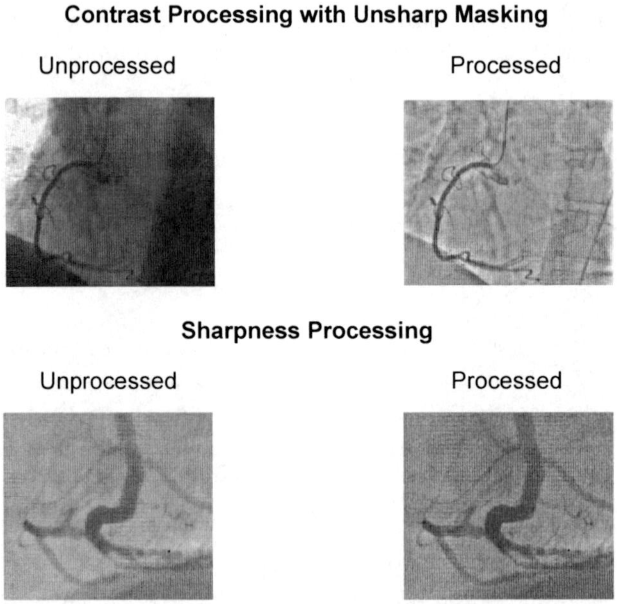

Contrast Processing with Unsharp Masking

Unprocessed

Processed

Sharpness Processing

Unprocessed

Processed

Fig. H.6. Examples of basic processing.

of basic processing are edge enhancement (spatial filtering) and dynamic range optimization (compression of the less useful part of the dynamic range to enhance the contrast of objects). Tradeoffs between different image-quality parameters are necessary when applying basic image processing. For example, a spatial filter has to find a balance between increased image sharpness and increased image noise. Dynamic range processing has to find a balance between improved contrast and increased image noise. While basic processing is not unique to fluoroscopy equipment, it plays an important role in providing adequate imaging.

Application processing (Figure H.7) transforms and combines pixel values using some sort of *a priori* knowledge of the image content. DSA processing assumes that an injection will be done after

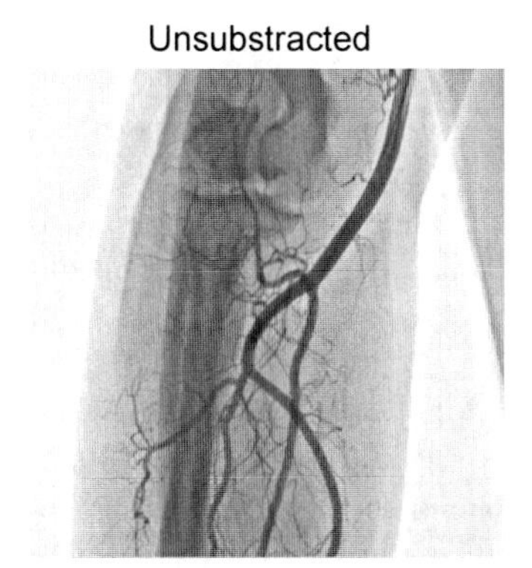

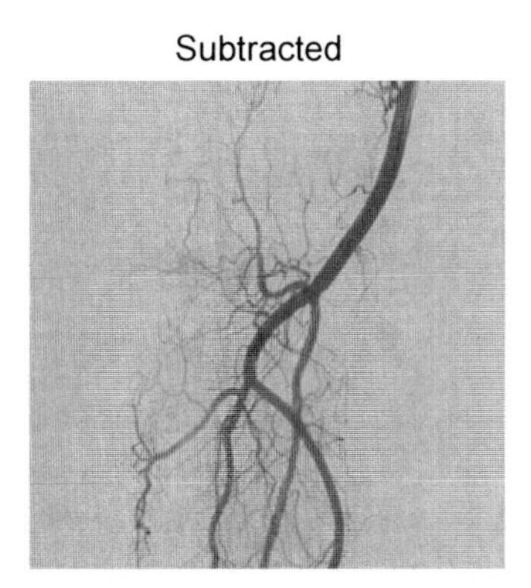

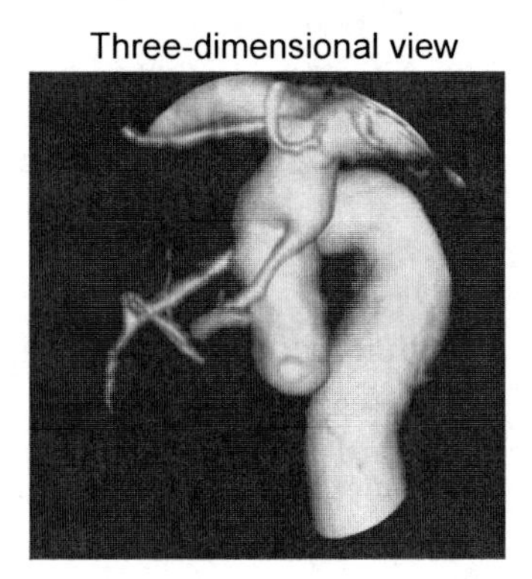

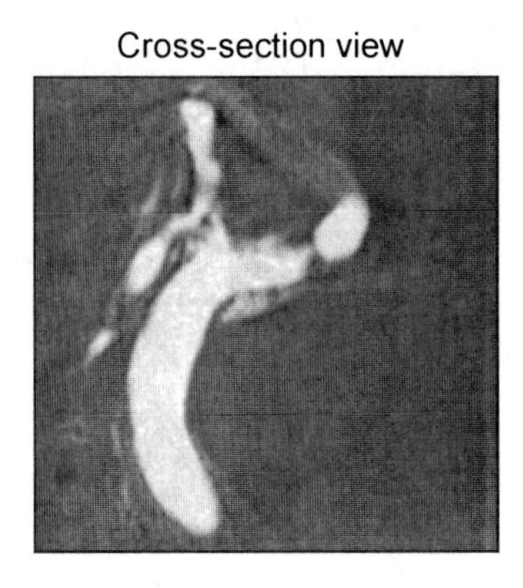

Fig. H.7. Examples of application processing.

the start of an acquisition. Three-dimensional or CT-like volume reconstruction assumes that the images represent a series of views rotating around the patient's body. While application processing capabilities are most typically found on fluoroscopy equipment used for FGI procedures, the most general forms of application processing could be found on noninterventional equipment as well.

This section does not provide an exhaustive description of all types of processing; manufacturers have developed many types of specific processing with methods that are beyond the scope of this presentation. Many of these are options. The level of image processing available depends on the configuration that was selected and purchased. Appropriate image-processing options are, in fact, essential, and are mandatory for fluoroscopy equipment used for potentially-high radiation dose procedures. However, no room needs every available processing capability, only the capabilities necessary for the tasks that are anticipated.

H.9 Anatomical Programming

All current imaging equipment is controlled by software. This feature allows the manufacturer to change the function of the imaging device in response to evolving clinical needs by modifying software instead of redesigning and changing the hardware components of the imaging equipment. Software control also allows proper management of the large number of parameters that are controlled in response to the variety of patient conditions and sizes encountered in the clinical setting. Image production controls contribute to better optimized production of x rays, which helps to reduce patient dose. Image-processing controls affect the appearance of the displayed image.

Software-controlled imaging equipment permits automated control of the wide variety of image production and processing controls that affect the final image. Settings for these parameters for a specific anatomical region or specific examination are stored under a software icon, hence the name "anatomical programming." Anatomical programming is typically configured at the factory to optimize imaging of adults. After the installation of the unit, the manufacturer's application specialist may alter these software settings in response to unique imaging challenges identified by the physicians or qualified physicists of the department. Any changes should be carefully documented to avoid confusion in the future.

While anatomical programming can be a great asset, it creates challenges. One should not assume that the "out-of-the-box" default settings provided by the manufacturer are properly optimized for their facility's patients. The vendor's default settings are based on the vendor's understanding of average adult imaging needs. These settings may work reasonably well for routine interventional studies on adults, but additional settings should be developed for specialized examinations.

Anatomical programming tends to hide the actual machine features that are being controlled, since multiple parameter changes are affected by the selection of a single setting. Anatomical programming also makes it difficult to determine the actual settings of the various image production and processing controls. The physician operator and the qualified physicist should take an active role in setting up the anatomical programming, but they are dependent on cooperation and technical information from the vendor during acceptance testing of the fluoroscopy equipment.

H.9.1 *Anatomical Program Selection: Fluorography (image recording mode)*

The operator's anatomical program selection at the beginning of the case sets both image capture and processing controls for the record mode of the imaging equipment. Included in these controls are automatic brightness curve (ABC) algorithms that describe how the settings of these controls will vary as a function of patient thickness. During the actual recording of images, the brightness of the image on the image display is monitored; the generator is programmed to adjust the acquisition control parameters as described by the ABC algorithms in response to changes in the thickness or density of patient anatomy.

Good image quality at a properly managed patient dose is highly dependent on proper selection of the ABC algorithm. The control scheme of the ABC algorithm is properly matched to patient size, the body part being imaged, and the type of clinical study. For example, the duration of each radiation pulse should be <5 ms for small body parts (especially pediatric studies) to freeze patient motion. The duration of each radiation pulse should be as high as 8 to 10 ms for large body parts to balance the need for additional radiation to penetrate the larger patient thickness against the need to freeze patient motion. Also, focal-spot size should be appropriate for the patient's size. A large focal spot is needed for large body parts, to allow increased x-ray tube current and a greater number of x rays produced per unit time. This increase also allows a reduction in x-ray tube voltage, which improves image contrast. Conversely, for small body parts a small focal spot should be selected automatically to improve image sharpness.

One important image capture parameter, added filtration in the x-ray beam, may or may not be automatically controlled by an ABC algorithm within each anatomical program selection. Thicker filters significantly reduce the patient dose without a significant decrease in image quality. Filter thickness is reduced as the body part increases in thickness, due to increased attenuation by the thicker body part. Failure to appropriately reduce the filter thickness for thicker body parts increases motion unsharpness and reduces contrast in the image. If the ABC algorithms control the filter thickness, no adjustment by the operator is required. If the filter thickness is fixed within an anatomical program selection, the operator changes the anatomical program selection as patient thickness changes. Patient dose and image quality will not be optimal unless the physician operator understands the capabilities and limitations of each control.

Two important image capture parameters, the exposure per pulse at the image receptor and the pulse rate, are typically controlled by the operator, either by making specific selections at the tableside control or by selecting a different anatomical program. The exposure per pulse at the image receptor is a primary determinant of both patient radiation dose rate and image quality. Most fluoroscopy equipment provides two to three different dose-rate levels. Typically, each dose-rate setting to the image receptor differs by a factor of two, which changes the dose-rate incident on the patient's skin by approximately a factor of two. The lowest level may be appropriate when reduced image quality is adequate. The highest level provides the best image quality, but should be used judiciously since this level may deliver four times the dose rate of the lowest setting.

The type of clinical examination should determine the selected pulse rate in the record mode. Pediatric cardiology examinations tend to require 30 pulses per second to adequately image the rapidly beating heart of a child. Adult cardiologists typically find 15 pulses per second to be adequate. Peripheral angiography is typically recorded at one to six pulses per second; the operator should tailor DSA frame rates to the region of the vasculature being imaged.

The selected anatomical program also automatically adjusts the image processing applied to the captured image prior to its display on the television monitors (see Appendix H.8 for more details). These controls affect the degree of edge enhancement, the degree of edge smoothing, the range of anatomy thicknesses displayed in a single image, and the brightness and contrast of the image on the monitor. The type of examination and body thickness determines the proper mix of these interdependent settings. The operator may choose more edge enhancement of arteries for cardiac angiography, but this comes at the cost of increased noise in the image. For arteriography of the legs the operator may elect to smooth the image to reduce noise levels in digitally subtracted images, but this reduces the edge sharpness in the image. Cardiac angiography requires imaging the large range of soft-tissue thicknesses in the thorax. This mandates a wide grayscale range in the image and results in reduced image contrast. The range of thicknesses is quite small in arteriography of the legs when DSA images are recorded, which may permit different contrast settings.

H.9.2 *Anatomical Program Selection: Fluoroscopy*

In addition to controlling the recorded images, the anatomical program selection for an examination also controls the fluoroscopy mode of operation of the equipment. Since the vast majority of FGI procedures are performed with a pulsed, as opposed to a continuous x-ray beam, the majority of the discussion in Appendix H.9.1 also applies to fluoroscopy images. The primary differences between the two different modes of operation are the dose per pulse at the image receptor and the range of useful pulse rates.

Appendix I

Pediatric Equipment Considerations

Since children are not just small adults (Strauss, 2006a), their disease states differ from adults. Children, as a function of their age, are three to five times more sensitive to the stochastic effects of radiation exposure than a middle-aged adult (Hall, 2002). Since many pediatric diseases are managed by multiple FGI procedures over time as opposed to a single procedure (Chung and Kirks, 1998), in addition to the frequent use of CT for diagnosis and observation, each pediatric radiation dose should be carefully managed to minimize stochastic risk.

Relative to an adult, a neonate or infant requires dramatically reduced radiologic technique factors. A neonate with a posterior-anterior thickness of 6 cm presents a short path length compared to a large adult with a posterior-anterior thickness of 30 cm (Jones, 1997). The half-value layer of soft tissue is ~3 cm at 70 kVp for imaging equipment with standard total filtration of ~3 mm aluminum. This difference of patient size (24 cm) requires a reduction of milliampere value per pulse of radiation by a factor of 256 (2^8 or eight half-value layers) to maintain an optimum 70 kVp as the patient path length decreases from 30 to 6 cm (Strauss, 2006a).

Image quality should be carefully controlled in infants and small children. First, excellent high contrast resolution (ability to distinguish two small objects that are close together) is required to image the significantly smaller anatomy of an infant and of smaller devices used during the intervention. This image-quality parameter is improved by a smaller focal-spot size, which still provides enough power to penetrate the small pediatric body. The smallest focal spot that allows adequate penetration of the body part should be used to improve image quality. When the smallest focal spot is used for pediatric imaging, an x-ray tube with three focal spots (0.3, 0.6, and 1 mm), typically found in neuroangiographic suites, provides better high contrast resolution than the standard dual focal-spot tube with a typical 0.5 mm small focal spot.

Second, the low subject contrast of pediatric soft tissue requires more enhancement than similar adult tissues. Iodine, as a contrast medium, should be used with great care in these small patients. Contrast within a vessel is a function of both the concentration of iodine in the vessel and the diameter of the vessel (Kruger and Riederer, 1984). Smaller diameter vessels require higher concentrations of contrast material to obtain adequate subject contrast. The allowable iodine burden is limited by the toxicity of the contrast agent, and equates to 4 to 6 mL kg^{-1} of 320 to 350 mg mL^{-1} iodinated contrast material (Chung and Kirks, 1998). This limitation in contrast material volume limits the number of contrast material injections per procedure to approximately half a dozen. Biplane units offer the advantage of obtaining simultaneous images from two different projections in both fluoroscopy and acquisition modes of operation. The primary disadvantages of biplane fluoroscopy equipment are the additional purchase and maintenance costs and additional bulky hardware surrounding the patient.

Third, increased motion blur in children should be controlled by pulsed radiation. While 8 ms is a reasonable maximum setting for pulse width in adults, small children require pulse widths that are ideally <5 ms (Strauss, 2006a).

Total radiation exposure to the patient can be reduced by either reducing the radiation exposure used to create each image or by modifying the protocol of the examination to reduce the number of images. Both approaches should be exploited to minimize pediatric radiation dose during FGI procedures (Strauss, 2006a).

The type of clinical study determines the required size of the image receptor for pediatric patients. Ironically, a 30 cm diameter image receptor (as opposed to 17 cm) may be installed in the pediatric cardiac interventional suite to allow simultaneous imaging of both lung fields. While large-format image receptors provide the greatest coverage of anatomy, image quality diminishes as the field of view increases. A larger format image receptor is also bulkier, which may limit the beam angulations that can be attained with the C-arm. All these factors should be carefully considered when specifying image-receptor size for a clinical unit.

Anatomical programming can be a great asset, but it creates challenges for pediatric imaging. One should not assume that the "out-of-the-box" anatomical programming settings provided by the manufacturer are optimized for the facility's patients. The vendor's default settings are typically based on the vendor's understanding of average adult imaging needs. These settings may work reasonably well for routine interventional studies on adults, but adult settings will not perform well for pediatric imaging.

Appendix J

Example of Information for Informed Consent for Radiation Risk[7]

You have been scheduled for an interventional procedure. This involves the use of x rays for imaging during the procedure and documenting the results. Because of the nature of the planned procedure, it is possible that we will have to use substantial amounts of radiation. What does this mean?

Potential radiation risks to you may include:

- There may be a slightly elevated risk of cancer several years later in life. This risk is typically less than 0.5 %. This risk is low in comparison to the natural risk of developing cancer. According to the American Cancer Society, the natural risk of developing cancer over a lifetime is 33 % for women and 50 % for men.[8]
- Skin rashes occur infrequently; on very rare occasions they may result in tissue breakdown and possibly severe ulcers. Hair loss may occur which can be temporary or permanent. The likelihood of either of these occurring depends on the difficulty of the procedure and whether you are sensitive to radiation due to previous procedures, disease, or genetic conditions.

You or your family will be advised if we actually used substantial amounts of radiation during the case. If this happens, you will be given written instructions stating that you are expected to have a family member check you for any of the above mentioned signs of skin damage or hair loss.

[7]Adapted from Stecker *et al.* (2009).

[8]Source for the natural lifetime risk of cancer (women and men) is the American Cancer Society (ACS, 2010). The reference citation does not have to be included in the informed consent document given to the patient.

Appendix K

Example of Post-Procedure Patient Discharge Instructions for a Substantial Radiation Dose Level[9]

Your procedure required the use of substantial amounts of x rays. Radiation side-effects are unlikely but possible.

Please have a family member inspect the skin on your _______ for signs of hair loss, redness or rash two and four weeks from today.

Please call us at (xxx) xxx-xxxx and tell us whether or not anything is seen.

[9]Adapted from Stecker *et al.* (2009).

Appendix L

Equipment Acquisition and Testing

During the planning process for acquisition of new equipment, identification of the clinical requirements is essential, both for appropriate specification of the imaging equipment and suite and for development of a budget to ensure that adequate funding is secured for the project. An appropriate request for proposals, with specific requirements, can be used to obtain matching competitive bids. Once the vendor is selected, a purchase contract is developed and signed. It should be designed to protect the interests of the buyer, not the seller. A detailed description of the entire image equipment acquisition process is provided elsewhere (Strauss, 2002; 2006b).

Effective communication among all stakeholders during planning of the imaging suite is required to ensure successful installation and commissioning of FGI-procedure equipment (Strauss, 2006b). To properly design the facility, the architect and design engineer should understand the clinical needs of the operators and staff and the environmental requirements of the imaging equipment. The effort required to communicate needs effectively will depend in part on the extent previous experience of the architect and engineer in designing FGI-procedure suites. The length and tediousness of facility planning meetings may tempt physicians and their staff to assume that "experienced" architects and engineers should be familiar with clinical needs. This assumption almost invariably leads to an unsatisfactory final design. The physicians and their staff should participate in the planning process.

After the imaging equipment is installed, the equipment and the facility systems (including radiation shielding) should be acceptance tested (Bushong, 1994; Rauch, 1982; Rossi, 1982; 1994; Rossi *et. al.*, 1985; Strauss, 1982; 1996) to verify proper performance of the imaging equipment in its environment prior to first clinical use.

Twenty-five years ago, acceptance testing was important to identify and eliminate substandard components (Bhatnagar and Rao, 1970; Randall and Horn, 1977; Rauch, 1982; Rauch and Block, 1976; 1977; Rossi, 1982; Starchman *et al.*, 1976). Current software-controlled and calibrated imaging equipment exhibits fewer hardware problems and eliminates the human errors associated with manual calibration of older style equipment (Strauss, 2006b). Today, acceptance testing should be completed to identify the infrequent performance issue. More important, acceptance testing is an opportunity for important functional testing of the equipment, to answer the following questions (Strauss, 2006b):

- Are all the purchased features of the equipment activated by the current software configuration?
- Do the necessary facility systems (*e.g.*, HVAC) properly support the imaging equipment during actual use?
- Does the facility design allow all features of the imaging equipment to be utilized?
- Is the imaging equipment properly configured on the facility network to allow typical clinical image sets to arrive at their appropriate destinations?
- Does each selected anatomical program setting properly adjust machine function for the unique imaging requirements associated with that anatomical program?

The qualified physicist conducting the acceptance testing should understand the clinical needs of the operators and staff and the design capabilities and limitations of the fluoroscopy unit with respect to image capture and processing parameters. This should allow the physicist to assist the vendor's application specialists, design engineers, and others in developing additional anatomical program settings. Proper management of basic image quality and patient radiation dose rate is partly determined prior to acceptance testing, by properly setting those image production controls that determine the radiation dose delivered to the image receptor. After the proper dose rate at the image receptor is set up, the image-processing controls are set to further enhance image quality as displayed on the video monitors. Acceptance testing provides the opportunity to test default and onsite-developed anatomical programs with appropriate phantoms prior to first clinical use.

TQC testing may use different test methods and procedures than those used for acceptance/commissioning tests. A full set of TQC tests should be performed as part of the acceptance/commissioning testing. The full TQC evaluation should be performed periodically. A subset should be performed after any modifications (*e.g.*, hardware, software, clinical configuration) or repair that might affect radiation output or x-ray beam confinement.

TQC tests, tools and/or procedures may be fully prescribed by local regulations. Where the regulations are not explicit, the qualified physicist should be allowed to perform alternative or supplementary testing to ensure compliance with regulatory performance requirements and local safety standards. The choice of test tools and procedures can be guided by professional documents (AAPM, 1998; 2001; 2002; 2005) and the literature. An example of alternative and supplementary testing (*i.e.*, minimum half-value layer and table-top output) would be:

- *Minimum half-value layer (HVL)*: FDA performance standards specify the minimum acceptable HVL as a function of kilovolt peak (FDA, 2009d). Minimum HVL is verified using appropriate attenuation measurements during acceptance testing or after repairs. Since minimum HVL at a given kilovolt peak is mechanically determined by the material installed between the x-ray tube port and collimator exit, there is no need to make an attenuation measurement provided that there has been no mechanical work on the x-ray tube or collimator and that the radiation output is similar to the value measured during acceptance testing.
- *Table-top output (air-kerma rate)*: FDA performance standards limit fluoroscopic output to a maximum of 88 mGy min^{-1} for normal operation (FDA, 2009e). Additional guidance (FDA, 1999) defines the measuring point for several fluoroscopic system geometries. The specified point for C-arm fluoroscopes is 30 cm in front of the image receptor. SID is not specified by FDA (1999). There are no limits on acquisition mode outputs. Acceptance or TQC testing that does not measure and report maximum acquisition outputs can often result in a gross underestimation of the potential incident air kerma for the patient.

Appendix M

List of Practical Actions to Reduce Unnecessary Radiation Exposure to Patients and Staff During FGI Procedures[a,b]

Practical Action	Sections in This Report
Keep the beam-on time and the dose accumulation in a single area of the skin to the lowest level commensurate with the benefits of the procedure.	4.2.4.2
Set the dose and dose-rate controls for the best compromise in image quality and in radiation dose accumulation.	4.2.4.2
Remember, dose rates are greater and dose accumulates more rapidly as patient size and as tissue penetration thickness increases.	2.6.1
Don't overuse geometric or electronic magnification.	4.2.4.2
Collimate to the area of interest.	4.2.4.3
Keep the image receptor as close to the patient as practicable.	4.2.4.2
Keep the patient at the maximum practicable distance from the x-ray tube.	4.2.4.2
If image quality is not compromised, remove the grid during procedures on small patients or when the image receptor cannot be placed close to the patient.	4.2.4.2
Monitor radiation utilization and maintain a quality-control program to ensure that radiation is managed properly.	4.2.4.1 and 4.2.4.2
Commensurate with their duties, ensure that personnel have mastered radiation safety and management.	5

[a]Adapted from Wagner and Archer (2004).
[b]See also Section 4.3.3.1.

Glossary

absorbed dose: The energy imparted to matter by ionizing radiation per unit mass of irradiated material at the point of interest. In the Systeme Internationale (SI), the unit is J kg^{-1} with the special name gray (Gy).

advisory data set (ADS): In this Report, the ADS is the full data set for a measured dose-related radiation field quantity (*i.e.*, $K_{a,r}$ or P_{KA}) that is used to evaluate the local dose distribution [*i.e.*, the facility data set (FDS)] for a group of patients for a particular type of FGI procedure, in order to better manage patient doses at the local facility. An ADS for an FGI procedure is obtained by collecting data for a large number of procedures from each of many facilities. The entire data set (typically a lognormal distribution) is used and includes sufficient data to characterize the entire distribution, rather than just the 10th and 75th percentile values typically used for DRLs. The percentages of cases in the ADS and FDS that exceed the substantial radiation dose level (SRDL) selected for the FDS, and the values of $K_{a,r}$ or P_{KA} that define the potentially-high radiation dose procedure category are also used in the evaluation (see *facility data set*, *potentially-high radiation dose procedure,* and *substantial radiation dose level*).

air kerma: (see *FDA compliance air-kerma rate, kerma, total air kerma at the reference point*).

air kerma-area product (P_{KA}): The integral of the air-kerma free-in-air (*i.e.*, in the absence of backscatter) over the area of the x-ray beam in a plane perpendicular to the beam axis.

air kerma at the reference point ($K_{a,r}$): The air kerma at a point in space located at a fixed distance from the focal spot expressed in gray. For isocentric fluoroscopes (C-arms), the reference point lies on the central axis of the x-ray beam, 15 cm on the x-ray tube side of isocenter (IEC, 2000; 2010). The location of the reference point relative to the x-ray gantry does not change when the source-to-image-receptor distance is changed. Referred to as reference air kerma by IEC (2000; 2004; 2010).

angiography: The radiographic visualization of blood vessels following introduction of contrast material.

as low as reasonably achievable (ALARA): A principle of radiation protection philosophy that requires that exposures to ionizing radiation be kept as low as reasonably achievable, economic and societal factors being taken into account. The ALARA principle is satisfied when the expenditure of further resources would be unwarranted by the reduction in exposure that would be achieved.

attenuation: The reduction of radiation intensity upon passage of radiation through matter.

cancer: A general term for more than 100 diseases characterized by abnormal cells and altered control of proliferation of malignant cells.

cardiac catheterization: Passage of a small catheter through a vessel in an arm, leg or neck and into the heart, permitting the securing of blood samples, determination of intracardiac pressure, detection of cardiac anomalies, and injection of contrast media for imaging of vessels.

C-arm: A fluoroscopic system where the image receptor and x-ray tube are mounted at the opposite ends of a C-shaped arm. This design allows the x-ray tube and image receptor to be rotated about the patient through at least 90 degrees relative to the patient with no motion of the x-ray tube relative to the image receptor.

cinefluorography: The production of motion picture photographic records of the image formed on the output phosphor of an image intensifier by the action of x rays transmitted through the patient.

clinical privileges: The authorities granted to a physician by a hospital governing board to provide patient care in the hospital. Clinical privileges are limited to the individual's license, experience and competence.

computed tomography (CT)**:** An imaging procedure that uses multiple x-ray transmission measurements and a computer program to generate tomographic images of the patient.

computed-tomography-guided interventional (CTGI) **procedure:** An interventional diagnostic or therapeutic procedure which uses external ionizing radiation in the form of computed tomography.

controlled area: A limited-access area in which the occupational exposure of personnel to radiation is under the supervision of an individual in charge of radiation protection. This implies that access, occupancy and working conditions are controlled for radiation protection purposes.

conventional fluoroscopy: An imaging technique using x rays to visualize the dynamics of bodily functioning. For example, a material with high x-ray absorption is injected or ingested and fluoroscopy is used to monitor the progress of the material through the blood vessels or gastrointestinal tract. The image receptor can be either an image intensifier and video-camera tube, or a large-area solid-state detector.

conventional radiography: An imaging technique where the image receptor consists of a combination of (usually two) intensifying(s) screens in intimate contact with a photographic film (usually a dual-emulsion film). After exposure to the x-ray image, the photographic film is then processed in chemical solutions. Photographic film is relatively insensitive to x rays; the light from the intensifying screens produces most of the film optical density. Also referred to as screen-film radiography.

deterministic effects: Effects that occur in all individuals who receive greater than a threshold dose; the severity of the effect varies with the dose above the threshold. Examples are radiation-induced cataracts (lens of the eye) and radiation-induced erythema (skin).

detriment: (see *radiation detriment*).

diagnostic reference level (DRL)**:** A measured dose-related radiation field quantity (*e.g.*, $K_{a,i}$ or P_{KA}) used to evaluate whether the patient dose (with regard to stochastic effects) for a standardized noninvasive imaging task performed on a standard-size patient or standard phantom is unusually high or low. The upper DRL is typically selected as the 75th percentile of the distribution of the dose-related radiation field quantity collected from a large number of facilities performing the noninvasive imaging task.

dose (radiation dose): A general term used when the context is not specific to a particular dose quantity. When the context is specific, the name or symbol for the quantity is used [*e.g.*, mean absorbed dose in a tissue or organ (D_T), effective dose (E)].

dose limit: A limit on dose that is applied for exposure to individuals in order to prevent the occurrence of radiation-induced deterministic effects or to limit the probability of radiation-related stochastic effects.

dose rate (radiation dose rate): Dose delivered per unit time. Dose rate can refer to any dose quantity (*e.g.*, absorbed dose, dose equivalent).

dosimeter: (see *personal monitoring*).

dosimetry: The science or technique of determining radiation dose.

effective dose (E)**:** The sum over specified organs and tissues of the products of the equivalent dose in an organ or tissue (H_T) and the tissue weighting factor for that organ or tissue (w_T):

$$E = \sum_T w_T H_T . \tag{G.1}$$

The tissue weighting factors have been developed from a reference population of equal numbers of both males and females and a wide range of ages (ICRP, 1991a; 2007a). Effective dose (E) applies only to stochastic effects. The unit is joule per kilogram (J kg^{-1}) with the special name sievert (Sv).

embryo: In the human, the developing individual from one week after conception to the end of the second month.

entrance-surface absorbed dose ($D_{skin,e}$)**:** Absorbed dose on the central x-ray beam axis at the point where the x-ray beam enters the patient or phantom (includes backscattered radiation).

entrance-surface air kerma ($K_{a,e}$)**:** Air kerma on the central x-ray beam axis at the point where the x-ray beam enters the patient or phantom (includes backscattered radiation).

equivalent dose (H_T)**:** Mean absorbed dose in a tissue or organ ($D_{T,R}$) weighted by the radiation weighting factor (w_R) for the type and energy of radiation incident on the body:

$$H_T = \sum_R w_R D_{T,R} . \tag{G.2}$$

The SI unit of equivalent dose is joule per kilogram (J kg^{-1}) with the special name sievert (Sv). 1 Sv = 1 J kg^{-1}.

exposure: A general term used to express the act of being exposed to ionizing radiation (also called irradiation). Exposure is also a defined ionizing radiation quantity. It is a measure of the ionization produced in air by x or gamma rays. The unit of exposure is coulomb per kilogram ($C\ kg^{-1}$). The special name for exposure is roentgen (R), where 1 R = $2.58 \times 10^{-4}\ C\ kg^{-1}$.

facility data set (FDS): In this Report, the FDS consists of all of the data for a measured dose-related radiation field quantity (*i.e.*, $K_{a,r}$ or P_{KA}) for all cases of the specified procedure performed at an individual facility. A FDS is compared against the advisory data set (ADS) for the FGI procedure in order to better manage patient doses at the local facility. The percentages of cases in the FDS and ADS that exceed the substantial radiation dose level (SRDL) selected for the FDS, and the values of $K_{a,r}$ or P_{KA} that define the potentially-high radiation dose procedure category are also used in the evaluation (see *advisory data set*, *potentially-high radiation dose procedure,* and *substantial radiation dose level*).

FDA compliance air-kerma rate: An FDA compliance measurement for air-kerma rate. For C-arm systems the measurement is made 30 cm in front of the image receptor at any source-to-image-receptor distance. Therefore, the location of the reference point for this measurement will change relative to the x-ray gantry when the source-to-image-receptor distance is changed.

fetus: In the human, the developing young in the uterus after the second month.

film: A thin, transparent sheet of polyester or similar material coated on one or both sides with an emulsion sensitive to radiation and light.

fluorography: Another mode of operation of a fluoroscope. Fluorography is intended to record images of moving objects for a few seconds at a time (*e.g.*, cinefluorography of the heart).

fluoroscope (medical): An instrument used in medical procedures for observing the internal structure of the body by means of the image of the anatomy examined formed on an image receptor when the patient is placed between the image receptor and an x-ray beam.

fluoroscopically-guided interventional (FGI) procedure: An interventional diagnostic or therapeutic procedure performed *via* percutaneous or other access routes, usually with local anesthesia or intravenous sedation, which uses external ionizing radiation in the form of fluoroscopy to: localize or characterize a lesion, diagnostic site, or treatment site; monitor the procedure; and control and document therapy.

fluoroscopy: A medical x-ray procedure used for observation of the internal features of the body by means of the fluorescence produced on a screen by a continuous field of x rays transmitted through the body. Fluoroscopy is intended to observe moving objects for relatively long periods of time (seconds to minutes) without the intent of preserving the images.

genetic effects: Changes in reproductive cells that may result in detriment to offspring.

gray (Gy)**:** The special name for the SI unit J kg $^{-1}$ (*i.e.*, energy imparted per unit mass of a material). 1 Gy = 1 J kg $^{-1}$.

heritable (effects): Effects expressed in offspring due to alteration of reproductive cells in the parent(s).

image intensifier: An x-ray image receptor which increases the brightness of a fluoroscopic image by electronic amplification and image minification.

image quality: The overall clarity of a radiographic image. Image sharpness, contrast and noise are three common measures of image quality.

image receptor: A system for deriving a diagnostically-usable image from the x rays transmitted through the patient. Examples: screen-film system, photostimulable storage phosphor, solid-state detector.

incident air kerma ($K_{a,i}$**):** Air kerma from the incident beam on the central x-ray beam axis at the focal-spot-to-surface distance (does not include backscattered radiation).

intended use: Use of a product, process or service in accordance with the specifications, instructions and information provided by the manufacturer.

interventionalist: In this Report, an individual who has been granted clinical privileges to perform or supervise FGI procedures in a facility, and who is personally responsible for the use of radiation during a specific FGI procedure in that facility (see *clinical privileges*).

ionizing radiation: Particulate or electromagnetic radiation that is capable of removing electrons from a neutral atom or molecule either directly or indirectly, resulting in an excess charge.

irradiation: In this Report, the process of being exposed to ionizing radiation.

justification (in radiation protection): The principle of radiation protection that any decision that alters the existing radiation exposure situation should do more good than harm. For a discussion of the application of this principle to medical exposure of patients see ICRP (2007a; 2007b).

kerma (kinetic energy released per unit mass) (K**):** The sum of the initial kinetic energies of all the charged particles liberated by uncharged particles in a mass of material. The unit for kerma is J kg^{-1}, with the special name gray (Gy). Kerma can be quoted for any specified material at a point in free space or in an absorbing medium (*e.g.*, air kerma).

lognormal distribution: A probability density function in which the logarithms of a set of values are normally distributed.

magnification (in medical x-ray imaging): (1) geometric magnification is an imaging procedure carried out with magnification usually produced by purposeful introduction of distance between the subject and the image receptor; and (2) electronic magnification is selection of an imaging field-of-view with a smaller size using an image intensifier or flat-panel detector image receptor.

***may*:** The term *may* (or *may not*) (in italics) indicates a reasonable practice that is permissible. When the term "may" appears in this Report in the context of its general usage, it is not italicized.

mean: Sum of the measured values divided by the number of measurements. The mean value is also often called the (arithmetic) average value. The mean of a distribution is the weighted average of the possible values of the random variable.

mean absorbed dose (D_T): The mean absorbed dose in an organ or tissue (organ dose), obtained by integrating or averaging absorbed doses at points in the organ or tissue.

median: Of a set of n values, the median is the value that is as frequently exceeded (by other values in the set) as not. The median value of a distribution is the 50th percentile.

minimum detectable level: A general term for the lowest value of a quantity that a measurement device is capable of detecting.

monitoring (dose): In this Report, the actions taken to monitor radiation dose to patients and staff during FGI procedures.

normal distribution: The normal distribution is an unbounded symmetric distribution, characterized by its mean and standard deviation. In a normal distribution, the median is equal to the mean.

occupational exposures: Radiation exposures to individuals that are incurred in the workplace as a result of situations that can reasonably be regarded as being the responsibility of management (radiation exposures associated with medical diagnosis of or treatment for the individual are excluded).

optimization (of radiation protection): The principle of radiation protection that the likelihood of incurring exposures, the number of people exposed, and the magnitude of their individual doses should be kept as low as reasonably achievable, taking into account economic and societal factors (see *as low as reasonably achievable)*. For a discussion of the application of this principle to medical exposure of patients see ICRP (2007a; 2007b).

organ dose: (see *mean absorbed dose*).

peak skin dose ($D_{skin,max}$): The maximum absorbed dose to the most heavily irradiated localized region of skin (*i.e.*, the localized region of skin that lies within the primary x-ray beam for the longest period of time during an FGI procedure).

personal dose equivalent at 0.07 mm [$H_p(0.07)$]: {see *personal dose equivalent at 10 mm [$H_p(10)$]*}.

personal dose equivalent at 10 mm [$H_p(10)$]: The dose equivalent in soft tissue at a depth of 10 mm below a specified point on the body. The SI unit of $H_p(10)$ is J kg^{-1} and its special name is sievert (Sv). There are also personal dose equivalent quantities measured at other depths [*e.g.*, $H_p(0.07)$ at a depth 0.07 mm; $H_p(3)$ at a depth 3 mm].

personal dosimeter: (see *personal monitoring*).

personal monitoring: The use of a small radiation detector (dosimeter) that is worn by an individual. Common personal dosimeters contain film, thermoluminescent, or optically-stimulated luminescent materials as the radiation-detection device.

photon: Quantum of electromagnetic radiation, having no charge or mass, that exhibits both particle and wave behavior, such as a gamma or x ray.

potentially-high radiation dose procedure: An FGI procedure for which more than 5 % of cases of that procedure result in $K_{a,r}$ exceeding 3 Gy or P_{KA} exceeding 300 Gy cm^2.

practicable: Likely to meet a need, but not yet tested in practice or proved in service or use (implies an expectation).

practical: Proven effective in use (implies an actual established usefulness).

projection: The direction of the central ray (*e.g.*, mediolateral, craniocaudal) in an x-ray exam.

protective devices (medical exposure): Devices such as gloves, aprons and gowns made of radiation absorbing materials, used to reduce occupational radiation exposure in medical procedures.

qualified physicist: A medical physicist or medical health physicist who is competent to conduct the radiation protection and patient safety functions described in this Report for FGI-procedure equipment and facilities. The qualified physicist is a person who is certified by the American Board of Radiology, American Board of Medical Physics, American Board of Health Physics, or Canadian College of Physicists in Medicine.

quality assurance: The planned and systematic activities necessary to provide adequate confidence that a product or service will meet the given requirements.

quality control: (see *technical quality control*).

radiation (ionizing): Electromagnetic radiation (x or gamma rays) or particulate radiation (alpha particles, beta particles, electrons, positrons, protons, neutrons, and heavy charged particles) capable of producing ions by direct or secondary processes in passage through matter.

radiation detriment: Measure of stochastic effects from exposure to ionizing radiation that takes into account the probability of fatal cancers, probability of severe heritable effects in future generations, probability of nonfatal cancers weighted by the lethality fraction, and relative years of life lost per fatal health effect.

radiation safety credentials: Documentation of successful completion of training in the practical radiation protection aspects of the use of FGI procedures (*e.g.*, knowledge of equipment operation, optimal imaging techniques, patient and staff radiation dose management, benefit-risk tradeoffs, and the potential for early or late detrimental radiation effects). Radiation safety credentials include initial and refresher training.

radiation weighting factor (w_R): A factor used to allow for differences in the biological effectiveness between different radiations when calculating equivalent dose (H_T) (see *equivalent dose*). These factors are independent of the tissue or organ irradiated.

radiation worker: An employee who works in a controlled area (see *controlled area*). A radiation worker has significant potential for exposure to radiation in the course of the employee's assignments or is directly responsible for or involved with the use and control of radiation. A radiation worker generally has training in radiation management and is subject to routine personal monitoring.

radiograph: A film or other record produced by the action of x rays on a sensitized surface.

radiography: The production of images on film or other media by the action of x rays transmitted through a patient.

radiology: That branch of healing arts and sciences that deals with the use of images in the diagnosis and treatment of disease.

roentgen (R)**:** The special name for the unit of exposure. Exposure is a specific quantity of ionization (charge) produced by the absorption of x- or gamma-radiation energy in a specified mass of air under standard conditions. 1 R = 2.58×10^{-4} coulomb per kilogram (C kg^{-1}).

***shall*:** The term *shall* (or *shall not*) (in italics) indicates a recommendation from NCRP that is necessary to meet the currently accepted standards of radiation protection.

***should*:** The term *should* (or *should not*) (in italics) indicates an advisory recommendation from NCRP that is to be applied when practicable or practical (*e.g.*, cost-effective). When the term "should" appears in this Report in the context of its general usage, it is not italicized.

sievert (Sv)**:** The special name (in the SI system) for the unit of equivalent dose and effective dose (or effective dose equivalent); 1 Sv = 1 J kg^{-1}.

source (or radiation source): Radiation-producing equipment or an aggregate of radioactive nuclei.

stochastic: Of, pertaining to, or arising from chance; involving probability; random.

stochastic effects: Effects, the probability of which, rather than their severity, is assumed to be a function of dose without a threshold. For example, cancer and heritable effects of radiation are regarded as being stochastic.

substantial radiation dose level (SRDL)**:** An appropriately-selected reference value used to trigger additional dose-management actions during a procedure and medical follow-up for a radiation level that might produce a clinically-relevant injury in an average patient. There is no implication that radiation levels above an SRDL will always cause an injury or that radiation levels below an SRDL will never cause an injury. The quantities and their SRDLs recommended in this Report are provided in Table 4.7.

Systeme Internationale (SI)**:** The International System of Quantities and Units as defined by the General Conference of Weights and Measures in 1960 and periodically revised since. These units are generally based on the meter/kilogram/second units, with special quantities for radiation including the becquerel, gray and sievert.

technical quality control (TQC)**:** The routine performance of equipment function tests and tasks, the interpretation of data from the tests, and the corrective actions taken.

thermoluminescent dosimeter: A dosimeter containing a phosphor for measuring dose. When heated or otherwise stimulated (*e.g.*, laser), a thermoluminescent dosimeter that has been exposed to ionizing radiation gives off light proportional to the energy absorbed. When used for personal monitoring, filters (absorbers) are included to help characterize the types of radiation.

threshold dose: (see *deterministic effects*).

tissue weighting factor (w_T)**:** A factor that indicates the ratio of the risk of stochastic effects attributable to irradiation of a given organ or tissue (T) to the total risk when the whole body is uniformly irradiated. When calculating effective dose equivalent, tissue weighting factor represents the risk of fatal cancers or severe heritable effects. When calculating effective dose, tissue weighting factor represents total detriment.

tomography: A special technique to show in detail images of structures lying in a predetermined plane of tissue, while blurring or eliminating detail in images of structures in other planes.

uncertainty: Lack of sureness or confidence in predictions of models or results of measurements. Uncertainties may be categorized as those due to stochastic variation, or as those due to lack of knowledge founded on an incomplete characterization, understanding or measurement of a system.

uncontrolled area: Any space not meeting the definition of controlled area (see *controlled area*).

view (radiographic): The image on film resulting from projection of the x-ray beam through a patient, usually named according to the direction of the x-ray beam relative to the body (*e.g.*, antero-posterior).

x rays: Penetrating electromagnetic radiation having a range of wavelengths (energies). X rays are usually produced by interaction of the electron field around certain nuclei or by the slowing down of energetic electrons.

Acronyms and Symbols

AAA	abdominal aortic aneurysm
ABC	automatic brightness curve
ACR	American College of Radiology
ADS	advisory data set
AF	atrial fibrillation
AHA	American Heart Association
ALARA	as low as reasonably achievable
ARSPI	Alliance for Radiation Safety in Pediatric Imaging
ATM	ataxia telangiectasia mutated (gene)
AVM	arteriovenous malformation
CABG	coronary artery bypass grafting
CPR	cardiopulmonary resuscitation
CT	computed tomography
CTGI	computed-tomography-guided interventional (procedures)
DICOM®	Digital Imaging and Communications in Medicine
DNA	deoxyribonucleic acid
DRL	diagnostic reference level
DSA	digital-subtraction angiography
$D_{\mathrm{skin,e}}$	entrance-skin absorbed dose
$D_{\mathrm{skin,max}}$	peak skin dose
D_{T}	mean absorbed dose in an organ or tissue
E	effective dose
EP	electrophysiology
EVAR	endovascular aneurysm repair
FDS	facility data set
FGI	fluoroscopically-guided interventional (procedure)
fps	frames per second
HBV	hepatitis B
HCV	hepatitis C
H_{E}	effective dose equivalent
HIV	human immunodeficiency virus
$H_{\mathrm{p}}(0.07)$	personal dose equivalent at 0.07 mm
$H_{\mathrm{p}}(10)$	personal dose equivalent at 10 mm
H_{T}	equivalent dose
HVAC	heating, ventilation, and air conditioning
HVL	half-value layer
IEC	International Electrotechnical Commission
K_{a}	air kerma

$K_{a,e}$	entrance-surface air kerma
$K_{a,i}$	incident air kerma
$K_{a,r}$	air kerma at the reference point
$\dot{K}_{FDA}$	U.S. Food and Drug Administration compliance air-kerma rate
LAO	left anterior oblique
PACS	picture archiving and communication system
PCI	percutaneous coronary intervention
P_{KA}	air kerma-area product
PTCA	percutaneous transluminal coronary angioplasty
RAO	right anterior oblique
RDSR	Radiation Dose Structured Report
RF	radiofrequency
SARS	severe acute respiratory syndrome
SCD	sudden cardiac death
SI	Systeme Internationale (International System of Quantities and Units)
SID	source-to-image-receptor distance
SRDL	substantial radiation dose level
SSD	source-to-skin distance
TAA	thoracic aortic aneurysm
TIPS	transjugular intrahepatic portosystemic shunt
TQC	technical quality control
UFE	uterine fibroid embolization
w_R	radiation weighting factor
w_T	tissue weighting factor

References

AAPM (1998). American Association of Physicists in Medicine. *Instrumentation Requirements of Diagnostic Radiological Physicists*, AAPM Report No. 60, http://www.aapm.org/pubs/reports/rpt_60.pdf (accessed February 1, 2011) (American Association of Physicists in Medicine, College Park, Maryland).

AAPM (2001). American Association of Physicists in Medicine. *Cardiac Catheterization Equipment Performance*, AAPM Report No. 70, http://www.aapm.org/pubs/reports/rpt_70.pdf (accessed February 1, 2011) (American Association of Physicists in Medicine, College Park, Maryland).

AAPM (2002). American Association of Physicists in Medicine. *Quality Control in Diagnostic Radiology*, AAPM Report No. 74, http://www.aapm.org/pubs/reports/rpt_74.pdf (accessed February 1, 2011) (American Association of Physicists in Medicine, College Park, Maryland).

AAPM (2005). American Association of Physicists in Medicine. *Assessment of Display Performance for Medical Imaging Systems*, AAPM Report No. 03, http://deckard.mc.duke.edu/~samei/tg18_files/tg18.pdf (accessed February 1, 2011) (American Association of Physicists in Medicine, College Park, Maryland).

ACR (2008a). American College of Radiology. *ACR Practice Guideline for Imaging Pregnant or Potentially Pregnant Adolescents and Women with Ionizing Radiation*, http://www.acr.org/SecondaryMainMenuCategories/quality_safety/guidelines/dx/Pregnancy.aspx (accessed February 1, 2011) (American College of Radiology, Reston, Virginia).

ACR (2008b). American College of Radiology. *ACR Technical Standard for Management of the Use of Radiation in Fluoroscopic Procedures*, http://www.acr.org/SecondaryMainMenuCategories/quality_safety/RadSafety/RadiationSafety/standard-manage-radiation.aspx (accessed February 1, 2011) (American College of Radiology, Reston, Virginia).

ACR (2009). American College of Radiology. *ACR Appropriateness Criteria*, http://www.acr.org/SecondaryMainMenuCategories/quality_safety/app_criteria.aspx (accessed February 1, 2011) (American College of Radiology, Reston, Virginia).

ACS (2010). American Cancer Society. *Cancer Facts and Figures 2010*, http://www.cancer.org/acs/groups/content/@nho/documents/document/acspc-024113.pdf (accessed February 1, 2011) (American Cancer Society, Atlanta, Georgia).

ADLER, O.B., ROSENBERGER, A. and PELEG, H. (1983). "Fine-needle aspiration biopsy of mediastinal masses: Evaluation of 136 experiences," Am. J. Roentgenol. **140**(5), 893–896.

AERTS, A., DECRAENE, T., VAN DEN OORD, J.J., DENS, J., JANSSENS, S., GUELINCKX, P., FLOURl, M., DEGREEF, H. and GARMYN, M. (2003). "Chronic radiodermatitis following percutaneous coronary interventions: A report of two cases," J. Eur. Acad. Dermatol. Venereol. **17**(3), 340–343.

AHA (2008). American Heart Association. *Heart Disease and Stroke Statistics – 2008 Update*, http://circ.ahajournals.org/cgi/content/short/CIRCULATIONAHA.107.187998v1 (accessed February 1, 2011) (American Heart Association, Dallas, Texas).

AMA (2004). American Medical Association. "Physician obligation in disaster preparedness and response," page 601 in *Health and Ethics Policies of the AMA House of Delegates*, http://www.ama-assn.org/ad-com/polfind/Hlth-Ethics.pdf (accessed February 1, 2011) (American Medical Association, Chicago).

ANDERSON, J.L., ADAMS, C.D., ANTMAN, E.M., BRIDGES, C.R., CALIFF, R.M., CASEY, D.E., JR., CHAVEY, W.E., II, FESMIRE, F.M., HOCHMAN, J.S., LEVIN, T.N., LINCOFF, A.M., PETERSON, E.D., THEROUX, P., WENGER, N.K., WRIGHT, R.S., SMITH, S.C., JR., JACOBS, A.K., HALPERIN, J.L., HUNT, S.A., KRUMHOLZ, H.M., KUSHNER, F.G., LYTLE, B.W., NISHIMURA, R., ORNATO, J.P., PAGE, R.L. and RIEGEL, B. (2007). "ACC/AHA 2007 guidelines for the management of patients with unstable angina/non-ST-elevation myocardial infarction – executive summary. A report of the American College of Cardiology/American Heart Association Task Force on Practice Guidelines," J. Am. Coll. Cardiol. **50**(7), 652–726.

ANDREWS, R.T. and BROWN, P.H. (2000). "Uterine arterial embolization: Factors influencing patient radiation exposure," Radiology **217**(3), 713–722.

ANGLE, J.F., NEMCEK, A.A., JR., COHEN, A.M., MILLER, D.L., GRASSI, C.J., D'AGOSTINO, H.R., KHAN, A.A., KUNDU, S., OSNIS, R.B., RAJAN, D.K., SCHWARTZBERG, M.S., SWAN, T.L., VEDANTHAM, S., WALLACE, M.J. and CARDELLA, J.F. (2008). "Quality improvement guidelines for preventing wrong site, wrong procedure, and wrong person errors: Application of the Joint Commission 'Universal Protocol for Preventing Wrong Site, Wrong Procedure, Wrong Person Surgery' to the practice of interventional radiology," J. Vasc. Intervent. Radiol. **19**(8), 1145–1151.

ANTMAN, E.M., HAND, M., ARMSTRONG, P.W., BATES, E.R., GREEN, L.A., HALASYAMANI, L.K., HOCHMAN, J.S., KRUMHOLZ, H.M., LAMAS, G.A., MULLANY, C.J., PEARLE, D.L., SLOAN, M.A., SMITH, S.C., JR., ANBE, D.T., KUSHNER, F.G., ORNATO, J.P., JACOBS, A.K., ADAMS, C.D., ANDERSON, J.L., BULLER, C.E., CREAGER, M.A., ETTINGER, S.M., HALPERIN, J.L., HUNT, S.A., LYTLE, B.W., NISHIMURA, R., PAGE, R.L., RIEGEL, B., TARKINGTON, L.G. and YANCY, C.W. (2008). "2007 focused update of the ACC/AHA 2004 guidelines for the management of patients with ST-elevation myocardial infarction: A report of the American College of

Cardiology/American Heart Association Task Force on Practice Guidelines," J. Am. Coll. Cardiol. **51**(2), 210–247.

ARCHAMBEAU, J.O., INES, A. and FAJARDO, L.F. (1984). "Response of swine skin microvasculature to acute single exposures of x rays: Quantification of endothelial changes," Radiat. Res. **98**(1), 37–51.

ARCHAMBEAU, J.O., HAUSER, D. and SHYMKO, R.M. (1988). "Swine basal cell proliferation during a course of daily irradiation, five days a week for six weeks (6000 rad)," Int. J. Radiat. Oncol. Biol. Phys. **15**(6), 1383–1388.

ARCHER, B.R. and WAGNER, L.K. (2000). "Protecting patients by training physicians in fluoroscopic radiation management," J. Appl. Clin. Med. Phys. **1**(1), 32–37.

ARSPI (2008). Alliance for Radiation Safety in Pediatric Imaging. *Image Gently*®, http://spr.affiniscape.com/associations/5364/ig (accessed February 1, 2011) (Alliance for Radiation Safety in Pediatric Imaging, Reston, Virginia).

ARSPI (2009). Alliance for Radiation Safety in Pediatric Imaging. *Image Gently*®. *Step Lightly*, http://spr.affiniscape.com/associations/5364/ig/index.cfm?page=584 (accessed February 1, 2011) (Alliance for Radiation Safety in Pediatric Imaging, Reston, Virginia).

AUFRICHTIG, R., XUE, P., THOMAS, C.W., GILMORE, G.C. and WILSON, D.L. (1994). "Perceptual comparison of pulsed and continuous fluoroscopy," Med. Phys. **21**(2), 245–256.

AYOOLA, A. and LEE, Y.J. (2006). "Radiation recall dermatitis with cefotetan: A case study," Oncologist **11**(10), 1118–1120.

AZRIA, D., MAGNE, N., ZOUHAIR, A., CASTADOT, P., CULINE, S., YCHOU, M., STUPP, R., VAN HOUTTE, P., DUBOIS, J.B. and OZSAHIN, M. (2005). "Radiation recall: A well recognized but neglected phenomenon," Cancer Treat. Rev. **31**(7), 555–570.

BALDWIN, D.R., EATON, T., KOLBE, J., CHRISTMAS, T., MILNE, D., MERCER, J., STEELE, E., GARRETT, J., WILSHER, M.L. and WELLS, A.U. (2002). "Management of solitary pulmonary nodules: How do thoracic computed tomography and guided fine needle biopsy influence clinical decisions?," Thorax **57**(9), 817–822.

BALTER, S. (2001a). *Interventional Fluoroscopy: Physics, Technology, Safety* (Wiley-Liss, New York).

BALTER, S. (2001b). "Stray radiation in the cardiac catheterization laboratory," Radiat. Prot. Dosim. **94**(1–2), 183–188.

BALTER, S. (2008). "Capturing patient doses from fluoroscopically based diagnostic and interventional systems," Health Phys. **95**(5), 535–540.

BALTER, S. and LAMONT, J. (2002). "Radiation and the pregnant nurse," http://www.cathlabdigest.com/article/357 (accessed February 1, 2011), Cath. Lab. Digest **10**(4).

BALTER, S. and MILLER, D.L. (2007). "The new Joint Commission sentinel event pertaining to prolonged fluoroscopy," J. Am. Coll. Radiol. **4**(7), 497–500.

BALTER, S. and MOSES, J. (2007). "Managing patient dose in interventional cardiology," Catheter Cardiovasc. Interv. **70**(2), 244–249.

BALTER, S., SONES, F.M., JR. and BRANCATO, R. (1978). "Radiation exposure to the operator performing cardiac angiography with U-arm systems," Circulation **58**(5), 925–932.

BALTER, S., MILLER, D.L., VANO, E., ORTIZ LOPEZ, P., BERNARDI, G., COTELO, E., FAULKNER, K., NOWOTNY, R., PADOVANI, R. and RAMIREZ, A. (2008). "A pilot study exploring the possibility of establishing guidance levels in x-ray directed interventional procedures," Med. Phys. **35**(2), 673–680.

BALTER, S., BRENNER, D. and BALTER, R. (2009). "Informed consent: Communicating radiation risk," pages 571 to 572 in *Radiation Protection and Dosimetry, Biological Effects of Radiation*, International Federation for Medical and Biological Engineering (IFMBE) Proceedings, World Congress on Medical Physics and Biomedical Engineering, Vol. 25/3, Dossel, O. and Schlegel, W.C., Eds. (Springer, New York).

BALTER, S., HOPEWELL, J.W., MILLER, D.L., WAGNER, L.K. and ZELEFSKY, M.J. (2010). "Fluoroscopically guided interventional procedures: A review of radiation effects on patients' skin and hair," Radiology **254**(2), 326–341.

BARABANOVA, A. and OSANOV, D.P. (1990). "The dependence of skin lesions on the depth-dose distribution from β-irradiation of people in the Chernobyl nuclear power plant accident," Int. J. Radiat. Biol. **57**(4), 775–782.

BCCA (2006). British Columbia Cancer Agency. *Care of Radiation Skin Reactions*, http://www.bccancer.bc.ca/NR/rdonlyres/6065FF4A-7A08-4DA2-9C13-293890A7ABB2/16217/RadiationSkinReactions2006web.pdf (accessed January 23, 2011) (British Columbia Cancer Agency, Vancouver).

BEINFELD, M.T., BOSCH, J.L., ISAACSON, K.B. and GAZELLE, G.S. (2004). "Cost-effectiveness of uterine artery embolization and hysterectomy for uterine fibroids," Radiology **230**(1), 207–213.

BENTZEN, S.M. and OVERGAARD, M. (1991). "Relationship between early and late normal-tissue injury after post mastectomy radiotherapy," Radiother. Oncol. **20**(3), 159–165.

BERGERON, P., CARRIER, R., ROY, D., BLAIS, N. and RAYMOND, J. (1994). "Radiation doses to patients in neurointerventional procedures," Amer. J. Neuroradiol. **15**(10), 1809–1812.

BERNARDI, G., PADOVANI, R., MOROCUTTI, G., VANO, E., MALISAN, M.R., RINUNCINI, M., SPEDICATO, L. and FIORETTI, P.M. (2000). "Clinical and technical determinants of the complexity of percutaneous transluminal coronary angioplasty procedures: Analysis in relation to radiation exposure parameters," Cathet. Cardiovasc. Interv. **51**(1), 1–9.

BERRINGTON, A., DARBY, S.C., WEISS, H.A. and DOLL, R. (2001). "100 years of observation on British radiologists: Mortality from cancer and other causes 1897–1997," Br. J. Radiol. **74**(882), 507–519.

BHATNAGAR, J.P. and RAO, G.U. (1970). "Kilovoltage calibration of diagnostic roentgen ray generators," Acta Radiol. Ther. Phys. Biol. **9**(6), 555–566.

BLASZAK, M.A., MAJEWSKA, N., JUSZKAT, R. and MAJEWSKI, W. (2009) "Dose-area product to patients during stent-graft treatment of thoracic and abdominal aortic aneurysms," Health Phys. **97**(3), 206–211.

BLEESER, F., HOORNAERT, M.T., SMANS, K., STRUELENS, L., BULS, N., BERUS, D., CLERINX, P., HAMBACH, L., MALCHAIR, F. and BOSMANS, H. (2008). "Diagnostic reference levels in angiography and interventional radiology: A Belgian multi-centre study," Radiat. Prot. Dosim. **129**(1–3), 50–55.

BODEN, W.E., O'ROURKE, R.A., TEO, K.K., HARTIGAN, P.M., MARON, D.J., KOSTUK, W.J., KNUDTSON, M., DADA, M., CASPERSON, P., HARRIS, C.L., CHAITMAN, B.R., SHAW, L., GOSSELIN, G., NAWAZ, S., TITLE, L.M., GAU, G., BLAUSTEIN, A.S., BOOTH, D.C., BATES, E.R., SPERTUS, J.A., BERMAN, J.D.S., MANCINI, G.B.J. and WEINTRAUB, W.S. (2007). "Optimal medical therapy with or without PCI for stable coronary disease," N. Engl. J. Med. **356**(15), 1503–1516.

BOLCH, W., LEE, C., WATSON, M. and JOHNSON, P. (2009). "Hybrid computational phantoms for medical dose reconstruction," Radiat. Environ. Biophys. **49**(2), 155–68.

BOMMER, K.K., RAMZY, I. and MODY, D. (1997). "Fine-needle aspiration biopsy in the diagnosis and management of bone lesions: A study of 450 cases," Cancer **81**(3), 148–156.

BOOTH, C.M., MATUKAS, L.M., TOMLINSON, G.A., RACHLIS, A.R., ROSE, D.B., DWOSH, H.A., WALMSLEY, S.L., MAZZULLI, T., AVENDANO, M., DERKACH, P., EPHTIMIOS, I.E., KITAI, I., MEDERSKI, B.D., SHADOWITZ, S.B., GOLD, W.L., HAWRYLUCK, L.A., REA, E., CHENKIN, J.S., CESCON, D.W., POUTANEN, S.M. and DETSKY, A.S. (2003). "Clinical features and short-term outcomes of 144 patients with SARS in the greater Toronto area," JAMA **289**(21), 2801–2809.

BOR, D., SANCAK, T., TOKLU, T., OLGAR, T. and ENER, S. (2008). "Effects of radiologists' skill and experience on patient doses in interventional examinations," Radiat. Prot. Dosim. **129**(1–3), 32–35.

BOTWIN, K.P., THOMAS, S., GRUBER, R.D., TORRES, F.M., BOUCHLAS, C.C., RITTENBERG, J.J. and RAO, S. (2002). "Radiation exposure of the spinal interventionalist performing fluoroscopically guided lumbar transforaminal epidural steroid injections," Arch. Phys. Med. Rehabil. **83**(5), 697–701.

BRAMBILLA, M., MARANO, G., DOMINIETTO, M., COTRONEO, A.R. and CARRIERO, A. (2004). "Patient radiation doses and references levels in interventional radiology," Radiol. Med. **107**(4), 408–418.

BRAVATA, D.M., GLENGER, A.L., MCDONALD, K.M., SUNDARAM, V., PEREZ, M.V., VARGHESE, R., KAPOOR, J.R., ARDEHALI, R., OWENS, D.K. and HLATKY, M.A. (2007). "Systematic review: The comparative effectiveness of percutaneous coronary interventions and coronary artery bypass graft surgery," Ann. Int. Med. **147**(10), 703–716.

BRINKER, J. (2008). "The left main facts: Faced, spun, but alas too few," J. Am. Coll. Cardiol. **51**(9), 893–898.

BRISTOW, M.R., SAXON, L.A., BOEHMER, J., KRUEGER, S., KASS, D.A., DE MARCO, T., CARSON, P., DICARLO, L., DEMETS, D., WHITE, B.G., DEVRIES, D.W. and FELDMAN, A.M. (2004). "Cardiac-resynchronization therapy with or without an implantable defibrillator in advanced chronic heart failure," N. Engl. J. Med. **350**(21), 2140–2150.

BRYK, S.G., CENSULLO, M.L., WAGNER, L.K., ROSSMAN, L.L. and COHEN, A.M. (2006). "Endovascular and interventional procedures in obese patients: A review of procedural technique modifications and radiation management," J. Vasc. Interv. Radiol. **17**(1), 27–33.

BUCEK, R.A., PUCHNER, S. and LAMMER, J. (2006). "Mid- and long-term quality-of-life assessment in patients undergoing uterine fibroid embolization," Am. J. Roentgenol. **186**(3), 877–882.

BULS, N., PAGES, J., MANA, F. and OSTEAUX, M. (2002). "Patient and staff exposure during endoscopic retrograde cholangiopancreatography," Br. J. Radiol. **75**(893), 435–443.

BURROWS, P.E., MITRI, R.K., ALOMAI, A., PADUA, H.M., LORD, D.J., SYLVIA, M.B., FISHMAN, S.J. and MULLIKEN, J.B. (2008). "Percutaneous sclerotherapy of lymphatic malformations with doxycycline," Lymphat. Res. Biol. **6**(3–4), 209–216.

BUSHONG, S.C. (1994). "Specification/acceptance testing of radiation shielding," pages 993 to 1015 in *Specification, Acceptance Testing and Quality Control of Diagnostic X-Ray Imaging Equipment*, AAPM Monograph No. 20, Seibert, J.A., Barnes, G.T. and Gould, R.G., Eds. (American Institute of Physics, New York).

CAHILL, A.M., BASKIN, K.M., KAYE, R.D., FITZ, C.R. and TOWBIN, R.B. (2004). "CT-guided percutaneous lung biopsy in children," J. Vasc. Interv. Radiol. **15**(9), 955–960.

CAPPATO, R., CALKINS, H., CHEN, S.A., DAVIES, W., IESAKA, Y., KALMAN, J., KIM, Y.H., KLEIN, G., PACKER, D. and SKANES, A. (2005). "Worldwide survey on the methods, efficacy, and safety of catheter ablation for human atrial fibrillation," Circulation **111**(9), 1100–1105.

CARLSON, S.K., BENDER, C.E., CLASSIC, K.L., ZINK, F.E., QUAM, J.P., WARD, E.M. and OBERG, A.L. (2001). "Benefits and safety of CT fluoroscopy in interventional radiologic procedures," Radiology **219**(2), 515–520.

CDC (2001). Centers for Disease Control and Prevention. *Updated U.S. Public Health Service Guidelines for the Management of Occupational Exposures to HBV, HCV, and HIV and Recommendations for Postexposure Prophylaxis*, http://www.cdc.gov/mmwr/preview/mmwrhtml/rr5011a1.htm (accessed February 1, 2011), MMWR **50**(RR-11), 1–42.

CDC (2004). Centers for Disease Control and Prevention. *Health Care in America, Trends in Utilization*, DHHS Pub No. 2004-1031, http://www.cdc.gov/nchs/data/misc/healthcare.pdf (accessed February 1, 2011) (National Center for Health Statistics, Hyattsville, Maryland).

CHAIT, P.G., SHLOMOVITZ, E., CONNOLLY, B.L., TEMPLE, M.J., RESTREPO, R., AMARAL, J.G., MURACA, S., RICHARDS, H.F. and

EIN, S.H. (2003). "Percutaneous cecostomy: Updates in technique and patient care," Radiology **227**(1), 246–250.

CHEN, W.L., HWANG, J.S., HU, T.H., CHEN, M.S. and CHANG, W.P. (2001). "Lenticular opacities in populations exposed to chronic low-dose-rate gamma radiation from radiocontaminated buildings in Taiwan," Radiat. Res. **156**(1), 71–77.

CHIDA, K., SAITO, H., OTANI, H., KOHZUKI, M., TAKAHASHI, S., YAMADA, S., SHIRATO, K. and ZUGUCHI, M. (2006). "Relationship between fluoroscopic time, dose-area product, body weight, and maximum radiation skin dose in cardiac interventional procedures," Am. J. Roentgenol. **186**(3), 774–778.

CHIDA, K., MORISHIMA, Y., MASUYAMA, H., CHIBA, H., KATAHIRA, Y., INABA, Y., MORI, I., MARUOKA, S., TAKAHASHI, S., KOHZUKI, M. and ZUGUCHI, M. (2009a). "Effect of radiation monitoring method and formula differences on estimated physician dose during percutaneous coronary intervention," Acta Radiologica **50**(2), 170–173.

CHIDA, K., KAGAVA, Y., SAITO, H., ISHIBASHI, T., TAKAHASHI, S. and ZUGUCHI, M. (2009b). "Evaluation of patient radiation dose during cardiac interventional procedures: What is the most effective method?," Acta Radiologica **50**(5), 474–481.

CHIDA, K., INABA, Y., SAITO, H., ISHIBASHI, T., TAKAHASHI, S., KOHZUKI, M. and ZUGUCHI, M. (2009c). "Radiation dose of interventional radiology system using a flat-panel detector," Am. J. Roentgenol. **193**(6), 1680–1685.

CHODICK, G., BEKIROGLU, N., HAUPTMANN, M., ALEXANDER, B.H., FREEDMAN, D.M., DOODY, M.M., CHEUNG, L.C., SIMON, S.L., WEINSTOCK, R.M., BOUVILLE, A. and SIGURDSON, A.J. (2008). "Risk of cataract after exposure to low doses of ionizing radiation: A 20-year prospective cohort study among US radiologic technologists," Am. J. Epidemiol. **168**(6), 620–631.

CHRISTENSEN, R. (2001). "Invasive radiology for pediatric trauma," Semin. Pediatr. Surg. **10**(1), 7–11.

CHRISTODOULOU, E.G., GOODSITT, M.M., LARSON, S.C., DARNER, K.L., SATTI, J. and CHAN, H.P. (2003). "Evaluation of the transmitted exposure through lead equivalent aprons used in a radiology department, including the contribution from backscatter," Med. Phys. **30**(6), 1033–1038.

CHUGH, K., DINU, P., BEDNAREK, D.R., WOBSCHALL, D., RUDIN, S., HOFFMANN, K., PETERSON, R. and ZENG, M. (2004). "A computer-graphic display for real-time operator feedback during interventional x-ray procedures," Proc. SPIE **5367**, 464–473.

CHUNG, T. and KIRKS, D.R. (1998). "Techniques," pages 1 to 63 in *Practical Pediatric Imaging: Diagnostic Radiology of Infants and Children*, Kirks, D.R. and Giscom, N.T., Eds. (Lippincott-Raven, Philadelphia).

CHUTER, T.A.M., SCHNEIDER, D.B., REILLY, L.M., LOBO, E.P. and MESSINA, L.M. (2003). "Modular branched stent graft for endovascular repair of aortic arch aneurysm and dissection," J. Vasc. Surg. **38**(4), 859–863.

CIRAJ-BJELAC, O., REHANNI, M.M., SIM, K.H., LIEW, H.B., VANO, E. and KLEIMAN, N.J. (2010). "Risk for radiation-induced cataract for staff in interventional cardiology: Is there reason for concern?," Catheter. Cardiovasc. Interv. **76**(6), 826-834.

CLUZEL, P., MARTINEZ, F., BELLIN. M.F., MICHALIK, Y., BEAUFILS, H., JOUANNEAU, C., LUCIDARME, O., DERAY, G. and GRENIER, P.A. (2000). "Transjugular versus percutaneous renal biopsy for the diagnosis of parenchymal disease: Comparison of sampling effectiveness and complications," Radiology **215**(3), 689–693.

CMMS (2002). Centers for Medicare and Medicaid Services. "Supplementary medical insurance benefits. Medical and other health services. Diagnostic x-ray tests, diagnostic laboratory tests, and other diagnostic tests: Conditions," 42 CFR Part 410.32 (revised October 1, 2002), http://edocket.access.gpo.gov/cfr_2002/octqtr/pdf/42cfr410.32.pdf (accessed February 1, 2011) (U.S. Government Printing Office, Washington).

COGAN, D.G. and DREISLER, K.K. (1953). "Minimal amount of x-ray exposure causing lens opacities in the human eye," Arch. Ophthalmol. **50**(1), 30–34.

COGAN, D.G., DONALDSON, D.D., GOFF, J.L. and GRAVES, E. (1953). "Experimental radiation cataract. III. Further experimental studies on x-ray and neutron irradiation of the lens," Arch. Ophthalmol. **50**(5), 597–602.

COHEN, M., JONES, D.E.A. and GREENE, D., Eds. (1972). "Central axis depth dose data for use in radiotherapy," Br. J. Radiol. **11**(Suppl. 11), 8–17.

COHEN, G.N., DAUER, L.T., BALTER, S., ZELEFSKY, M. and ZAIDER, M. (2007). "SU-FF-T 371: Reducing staff exposure from fluoroscopy during ultrasound guided permanent prostate seed implantation," Med. Phys. **34**(6), 2487.

CONNOLLY, B.L. (2003). "Gastrointestinal interventions – emphasis on children," Tech. Vasc. Interv. Radiol. **6**(4), 182–191.

CORBILLON, E., BERGERON, P., POULLIE, A.I., PRIMUS, C., OJASOO, T. and GAY, J. (2008). "The French National Authority for Health reports on thoracic stent grafts," J. Vasc. Surg. **47**(5), 1099–1107.

CRCPD (2009). Conference of Radiation Control Program Directors, Inc. *Nationwide Evaluation of X-Ray Trends (NEXT). Protocol for 2008-2009 Survey of Cardiac Catheterization*, CRCPD Publication E-09-2, http://www.crcpd.org/Pubs/NEXT_Protocols/2008-2009CardiacCath/NEXT2008-2009CardiacCathProtocol.pdf (accessed February 1, 2011) (Conference of Radiation Control Program Directors, Inc., Frankfort, Kentucky).

DAMILAKIS, J., KOULOURAKIS, M., HATJIDAKIS, S., KARABEKIOS, S. and GOURTSOYIANNIS, N. (1995). "Radiation exposure to the hands of operators during angiographic procedures," Eur. J. Radiol. **21**(1), 72–75.

DAUER, L.T., THORNTON, R., ERDI, Y., CHING, H., HAMACHER, K., BOVLAN, D.C., WILLIAMSON, M.J., BALTER, S. and ST. GERMAIN,

J. (2009). "Estimating radiation doses to the skin from interventional radiology procedures for a patient population with cancer," J. Vasc. Interv. Radiol. **20**(6), 782–788.

DEHMER, G.J. (2006). "Occupational hazards for interventional cardiologists," Catheter. Cardiovasc. Interv. **68**(6), 974–976.

DEN BOER, A., DE FEIJTER, P.J., SERRUYS, P.W. and ROELANDT, J.R.T.C. (2001). "Real-time quantification and display of skin radiation during coronary angiography and intervention," Circulation **104**(15), 1779–1784.

DENIS, M.A., ECOCHARD, R., BERNADET, A., FORISSIER, M.F., PORST, J.M., ROBERT, O., VOLCKMANN, C. and BERGERET, A. (2003). "Risk of occupational blood exposure in a cohort of 24,000 hospital healthcare workers: Position and environment analysis over three years," J. Occup. Environ. Med. **45**(3), 283–288.

DHHS (2009). U.S. Department of Health and Human Services. "Criteria for IRB approval of research. Basic HHS policy for protection of human research subjects. Protection of human subjects," 45 CFR Part 46.111(a)(2) (U.S. Department of Health and Human Services, Washington).

DICOM (2005). Digital Imaging and Communications in Medicine. *Supplement 94: Diagnostic X-Ray Radiation Dose Reporting (Dose SR)*, ftp://medical.nema.org/medical/dicom/final/sup94_ft.pdf (accessed February 1, 2011) (National Electrical Manufacturers Association, Rosslyn, Virginia).

DICOM (2008). Digital Imaging and Communications in Medicine. *Part 14: Grayscale Standard Display Function*, PS 3.14-2008, http://bio.gsi.de/DOCS/Dicom/2008/08_14pu.pdf (accessed February 1, 2011) (National Electrical Manufacturers Association, Rosslyn, Virginia).

DILLAVOU, E.D. and MAKAROUN, M.S. (2008). "Predictors of morbidity and mortality with endovascular and open thoracic aneurysm repair," J. Vasc. Surg. **48**(5), 1114–1120.

DONALDSON, J.S. (2006). "Pediatric vascular access," Pediatr. Radiol. **36**(5), 386–397.

DRAGUSIN, O., WEERASOORIYA, R., JAIS, P., HOCINI, M., ECTOR, J., TAKAHASHI, Y., HAISSAGUERRE, M., BOSMANS, H. and HEIDBUCHEL, H. (2007). "Evaluation of a radiation protection cabin for invasive electrophysiological procedures," Eur. Heart J. **28**(2), 183–189.

DROMI, S., WOOD, B.J., OBEROI, J. and NEEMAN, Z. (2006). "Heavy metal pad shielding during fluoroscopic interventions," J. Vasc. Interv. Radiol. 17(7), 1201–1206.

EC (2000). European Commission. *Guidelines on Education and Training in Radiation Protection for Medical Exposures*, Radiation Protection 116, http://ec.europa.eu/energy/nuclear/radiation_protection/doc/publication/116.pdf (accessed February 1, 2011) (European Commission, Luxembourg).

EDER, H., SCHLATTI, H. and HOESCHEN, C. (2010). "X-ray protective clothing: Does DIN 6857-1 allow an objective comparison between lead-free and lead-composite materials?" Fortschr. Rontgenstr. **182**(5), 422–428.

EDWARDS, R.D., MOSS, J.G., LUMSDEN, M.A., WU, O., MURRAY, L.S., TWADDLE, S. and MURRAY, G.D. (2007). "Uterine-artery embolization versus surgery for symptomatic uterine fibroids," N. Engl. J. Med. **356**(4), 360–370.

EHRENSTEIN, B.P., HANSES, F. and SALZBERGER, B. (2006). "Influenza pandemic and professional duty: Family or patients first? A survey of hospital employees," BMC Public Health **6**(1), 311–313.

ELLIS, F. (1942). "Tolerance dosage in radiotherapy with 200 kV x rays," Br. J. Radiol. **15**(180), 348–350.

EPSTEIN, A.E., DIMARCO, J.P., ELLENBOGEN, K.A., ESTES, N.A., III, FREEDMAN, R.A., GETTES, L.S., GILLINOV, A.M., GREGORATOS, G., HAMMILL, S.C., HAYES, D.L., HLATKY, M.S., NEWBY, L.K., PAGE, R.L., SCHOENFELD, M.H., SILKA, M.J., STEVENSON, L.W., SWEENEY, M.O., SMITH, S.C., JR., JACOBS, A.K., ADAMS, C.D., ANDERSON, J.L., BULLER, C.E., CREAGER, M.A., ETTINGER, S.M., FAXON, D.P., HALPERIN, J.L., HIRATZKA, L.F., HUNT, S.A., KRUMHOLZ, H.M., KUSHNER, F.G., LYTLE, B.W., NISHIMURA, R.A., ORNATO, J.P., RIEGEL, B., TARKINGTON, L.G. and YANCY, C.W. (2008). "ACC/AHA/HRS 2008 guidelines for device-based therapy of cardiac rhythm abnormalities: A report of the American College of Cardiology/American Heart Association Task Force on Practice Guidelines (Writing Committee to Revise the ACC/AHA/NASPE 2002 Guideline Update for Implantation of Cardiac Pacemakers and Antiarrhythmia Devices) developed in collaboration with the American Association for Thoracic Surgery and Society of Thoracic Surgeons," J. Am. Coll. Cardiol. **51**(21), e1–e62.

EU (1996). European Union. *Criteria for Acceptability of Radiological (Including Radiotherapy) and Nuclear Medicine Installations*, Radiation Protection No 91, http://ec.europa.eu/energy/nuclear/radiation_protection/doc/publication/091_en.pdf (accessed February 1, 2011) (European Union, Luxembourg).

FAIZER, R., DEROSE, G., LAWLOR, D.K., HARRIS, K.A. and FORBES, T.L. (2007). "Objective scoring systems of medical risk: A clinical tool for selecting patients for open or endovascular abdominal aortic aneurysm repair," J. Vasc. Surg. **45**(6), 1102–1108.

FAULKNER, K. and MARSHALL, N.W. (1993). "Personal monitoring of pregnant staff in diagnostic radiology," J. Radiol. Prot. **13**(4), 259–265.

FDA (1994). U.S. Food and Drug Administration. *Avoidance of Serious X-Ray-Induced Skin Injuries to Patients During Fluoroscopically-Guided Procedures*, http://www.fda.gov/downloads/Radiation-EmittingProducts/RadiationEmittingProductsandProcedures/MedicalImaging/MedicalX-Rays/ucm116677.pdf (accessed February 1, 2011) (U.S. Food and Drug Administration, Rockville, Maryland).

FDA (1995a). U.S. Food and Drug Administration. *Recording Information in the Patient's Medical Record That Identifies the Potential for Serious X-Ray-Induced Skin Injuries*, http://www.fda.gov/Radiation-Emitting-

Products/RadiationEmittingProductsandProcedures/MedicalImaging/MedicalX-Rays/ucm117030.htm (accessed February 1, 2011) (U.S. Food and Drug Administration, Rockville, Maryland).

FDA (1995b). U.S. Food and Drug Administration. *Radiation-Induced Skin Injuries from Fluoroscopy*, http://www.fda.gov/Radiation-EmittingProducts/RadiationEmittingProductsandProcedures/MedicalImaging/MedicalX-Rays/ucm116682.htm (accessed February 1, 2011) (U.S. Food and Drug Administration, Rockville, Maryland).

FDA (1999). U.S. Food and Drug Administration. *Resource Manual for Compliance Test Parameters of Diagnostic X-Ray Systems*, http://www.fda.gov/Radiation-EmittingProducts/RadiationEmittingProductsandProcedures/MedicalImaging/MedicalX-Rays/ucm115361.htm (accessed February 1, 2011) (U.S. Food and Drug Administration, Rockville, Maryland).

FDA (2009a). U.S. Food and Drug Administration. "Performance standards for ionizing radiation emitting products. Fluoroscopic equipment," 21 CFR Part 1020.32 (revised April 1), http://www.accessdata.fda.gov/scripts/cdrh/cfdocs/cfcfr/CFRSearch.cfm (accessed February 1, 2011) (U.S. Government Printing Office, Washington).

FDA (2009b). U.S. Food and Drug Administration. "Performance standards for ionizing radiation emitting products. Computed tomography (CT) equipment," 21 CFR Part 1020.33 (revised April 1), http://www.accessdata.fda.gov/scripts/cdrh/cfdocs/cfcfr/CFRSearch.cfm (accessed February 1, 2011) (U.S. Government Printing Office, Washington).

FDA (2009c). U.S. Food and Drug Administration. "Performance standards for ionizing radiation emitting products. Diagnostic x-ray systems and their major components. Leakage radiation from the diagnostic source assembly," 21 CFR Part 1020.30(k) (revised April 1), http://www.accessdata.fda.gov/scripts/cdrh/cfdocs/cfcfr/CFRSearch.cfm (accessed February 1, 2011) (U.S. Government Printing Office, Washington).

FDA (2009d). U.S. Food and Drug Administration. "Performance standards for ionizing radiation emitting products. Diagnostic x-ray systems and their major components. Beam quality," 21 CFR Part 1020.30(m) (revised April 1), http://www.accessdata.fda.gov/scripts/cdrh/cfdocs/cfcfr/CFRSearch.cfm (accessed February 1, 2011) (U.S. Government Printing Office, Washington).

FDA (2009e). U.S. Food and Drug Administration. "Performance standards for ionizing radiation emitting products. Fluoroscopic equipment, Air kerma rates," 21 CFR Part 1020.32(d) (revised April 1), http://www.accessdata.fda.gov/scripts/cdrh/cfdocs/cfcfr/CFRSearch.cfm (accessed February 1, 2011) (U.S. Government Printing Office, Washington).

FDA (2010). U.S. Food and Drug Administration. *White Paper: Initiative to Reduce Unnecessary Radiation Exposure from Medical Imaging*, http://www.fda.gov/Radiation-EmittingProducts/RadiationSafety/RadiationDoseReduction/ucm199994.htm (accessed February 1, 2011) (U.S. Food and Drug Administration, Rockville, Maryland).

FINNERTY, M. and BRENNAN, P.C. (2005). "Protective aprons in imaging departments: Manufacturer stated lead equivalence values require validation," Eur. Radiol. **15**(7), 1477–1484.

FITZGERALD, T.J., JODOIN, M.B., TILLMAN, G., ARONOWITZ, J., PIETERS, R., BALDUCCI, S., MEYER, J., CICCHETTI, M.G., KADISH, S., MCCAULEY, S., SAWICKA, J., URIE, M., LO, Y.C., MAYO, C., ULIN, K., DING, L., BRITTON, M., HUANG, J. and AROUS, E. (2008). "Radiation therapy toxicity to the skin," Dermatol. Clin. **26**(1), 161–172.

FLETCHER, D.W., MILLER, D.L., BALTER, S. and TAYLOR, M.A. (2002). "Comparison of four techniques to estimate radiation dose to skin during angiographic and interventional radiology procedures," J. Vasc. Intervent. Radiol. **13**(4), 391–397.

FRASER-HILL, M.A., RENFREW, D.L. and HILSENRATH, P.E. (1992). "Percutaneous needle biopsy of musculoskeletal lesions. 2. Cost-effectiveness," Am. J. Roentgenol. **158**(4), 813–818.

FRAZIER, T.H., RICHARDSON, J.B., FABRE, V.C. and CALLEN, J.P. (2007). "Fluoroscopy-induced chronic radiation skin injury: A disease perhaps often overlooked," Arch. Dermatol. **143**(5), 637–640.

FRYBACK, D.G. and THORNBURY, J.R. (1991). The efficacy of diagnostic imaging," Med. Decision Making **11**(2), 88–94.

GARRETT, K.M., FULLER, C.E., SANTANA, V.M., SHOCHAT, S.J. and HOFFER, F.A. (2005). "Percutaneous biopsy of pediatric solid tumors," Cancer **104**(3), 644–652.

GAZELLE, G.S. and HAAGA, J.R. (1991). "Biopsy needle characteristics," Cardiovasc. Intervent. Radiol. **14**(1), 13–66.

GEISER, W.R., HUDA, W. and GKANATIOS, N.A. (1997). "Effect of patient support pads on image quality and dose in fluoroscopy," Med. Phys. **24**(3), 377–382.

GELEIJNS, J. and WONDERGEM, J. (2005). "X-ray imaging and the skin: Radiation biology, patient dosimetry and observed effects," Radiat. Prot. Dosim. **114**(1–3), 121–125.

GIANFELICE, D., LEPANTO, L., PERREAULT, P., CHARTRAND-LEFEBVRE, C. and MILETTE, P.C. (2000). "Value of CT fluoroscopy for percutaneous biopsy procedures," J. Vasc. Interv. Radiol. **11**(7), 879–884.

GKANATSIOS, N.A., HUDA, W. and PETERS, K.R. (2002). "Adult patient doses in interventional neuroradiology," Med. Phys. **29**(5), 717–723.

GLADE, G.J., VAHL, A.C., WISSELINK, W., LINSEN, M.A.M. and BALM, R. (2005). "Mid-term survival and costs of treatment of patients with descending thoracic aortic aneurysms; endovascular vs. open repair: A case-control study," Eur. J. Vasc. Endovasc. Surg. **29**(1), 28–34.

GLAZE, S., LEBLANC, A.D. and BUSHONG, S.C. (1984). "Defects in new protective aprons," Radiology **152**(1), 217–218.

GOLDSTEIN, J.A., BALTER, S., COWLEY, M., HODGSON, J. and KLEIN, L.W. (2004). "Occupational hazards of interventional cardiologists: Prevalence of orthopedic health problems in contemporary practice," Catheter. Cardiovasc. Interv. **63**(4), 407–411.

GOODNEY, P.P., SCHERMERHORN, M.L. and POWELL, R.J. (2006). "Current status of carotid artery stenting," J. Vasc. Surg. **43**(2), 406–411.

GRANEL, F., BARBAUD, A., GILLET-TERVER, M.N., REICHERT, S., WEBER, M., DANCHIN, N. and SCMUTZ, J.L. (1998). "Chronic radiodermatitis after interventional cardiac catheterization. Four cases," Ann. Dermatol. Venereol. **125**(6–7), 405–407.

GREENHALGH, R.M., BROWN, L.C., EPSTEIN, D., KWONG, G.P.S., POWELL, J.T., SCULPHER, M.J. and THOMPSON, S.G. (2005). "Endovascular aneurysm repair versus open repair in patients with abdominal aortic aneurysm (EVAR Trial 1): Randomized controlled trial," Lancet **365**(9478), 2179–2186.

GROGAN, E.L., STILES, R.A., FRANCE, D.J., SPEROFF, T., MORRIS, J.A., JR., NIXON, B., GAFFNEY, F.A., SEDDON, R. and PINSON, C.W. (2004). "The impact of aviation-based teamwork training on the attitudes of health-care professionals," J. Am. Coll. Surg. **199**(6), 843–848.

GU, J., BEDNARZ, B., XU, X.G. and JIANG, S.B. (2008). "Assessment of patient organ doses and effective doses using the VIP-man adult male phantom for selected cone-beam CT imaging procedures during image guided radiation therapy," Radiat. Prot. Dosim. **131**(4), 431–443.

GURM, H.S., YADAV, J.S., FAYAD, P., KATZEN, B.T., MISHKEL, G.J., BAJWA, T.K., ANSEL, G., STRICKMAN, N.E., WANG, H., COHEN, S.A., MASSARO, J.M. and CUTLIP, D.E. (2008). "Long-term results of carotid stenting versus endarterectomy in high-risk patients," N. Engl. J. Med. **358**(15), 1572–1579.

HAISSAGUERRE, M., JAIS, P., SHAH, D.C., TAKAHASHI, A., HOCINI, M., QUINIOU, G., GARRIGUE, S., LE MOUROUX, A., LE METAYER, P. and CLEMENTY, J. (1998). "Spontaneous initiation of atrial fibrillation by ecoptic beats originating in the pulmonary veins," N. Eng. J. Med. **339**(10), 659–666.

HALL, E.J. (2002). "Lessons we have learned from our children: Cancer risks from diagnostic radiology," Pediatr. Radiol. **32**(10), 700–706.

HALL, E.J. and GIACCIA, A.J. (2006). *Radiobiology for the Radiologist*, 6th ed. (Lippincott Williams and Wilkins, Philadelphia).

HALL, P., GRANATH, F., LUNDELL, M., OLSSON, K. and HOLM, L.E. (1999). "Lenticular opacities in individuals exposed to ionizing radiation in infancy," Radiat. Res. **152**(2), 190–195.

HALL, E.J., BRENNER, D.J., WORGUL, B. and SMILENOV, L. (2005). "Genetic susceptibility to radiation," Adv. Space Res. **35**(2), 249–253.

HARROD-KIM, P. and WALDMAN, D.L. (2005). "Abnormal portal venous flow at sonography predicts reduced survival after transjugular intrahepatic portosystemic shunt creation," J. Vasc. Interv. Radiol. **16**(11), 1459–1464.

HARSTALL, R., HEINI, P.F., MINI, R.L. and ORLER, R. (2005). "Radiation exposure to the surgeon during fluoroscopically assisted percutaneous vertebroplasty," Spine **30**(16), 1893–1898.

HART, D., HILLIER, M.C. and WALL, B.F. (2009). "National reference doses for common radiographic, fluoroscopic and dental x-ray examinations in the UK," Brit. J. Radiol. **82**(973), 1–12.

HAYASHI, N., SAKAI, T., KITAGAWA, M., INAGAKI, R., YAMAMOTO, T., FUKUSHIMA, T. and ISHII, Y. (1998). "Radiation exposure to interventional radiologists during manual-injection digital subtraction angiography," Cardiovasc. Intervent. Radiol. **21**(3), 240–243.

HEHENKAMP, W.J., VOLKERS, N.A., BIRNIE, E., REEKERS, J.A. and ANKUM, W.M. (2008). "Symptomatic uterine fibroids: Treatment with uterine artery embolization or hysterectomy–results from the randomized clinical embolisation versus hysterectomy (EMMY) trial," Radiology **246**(3), 823–832.

HELLAWELL, G.O., MUTCH, S.J., THEVENDRAN, G., WELLS, E. and MORGAN, R.J. (2005). "Radiation exposure and the urologist: What are the risks?," J. Urol. **174**(3), 948–952.

HENDEL, R.C., PATEL, M.R., KRAMER, C.M. and POON, M. (2006). "ACCF/ACR/SCCT/SCMR/ASNC/NASCI/SCAI/SIR 2006 appropriateness criteria for cardiac computed tomography and cardiac magnetic resonance imaging," J. Am. Coll. Cardiol. **48**(7), 1475–1497.

HENDERSON, R.A., POCOCK, S.J., SHARP, S.J., NANCHAHAL, K., SCULPHER, M.J., BUXTON, M.J. and HAMPTON, J.R. (1998). "Long-term results of RITA-1 trial: Clinical and cost comparisons of coronary angioplasty and coronary-artery bypass grafting. Randomised intervention treatment of angina," Lancet **352**(9138), 1419–1425.

HERAN, M.K.S., MARSHALLECK, F., TEMPLE, M., GRASSI, C.J., CONNOLLY, B., TOWBIN, R.B., BASKIN, K.M., DUBOIS, J., HOGAN, M.J., KUNDU, S., MILLER, D.L., ROEBUCK, D.J., ROSE, S.C., SACKS, D., SIDHU, M., WALLACE, M.J., ZUCKERMAN, D.A. and CARDELLA, J.F. (2010). "Joint quality improvement guidelines for pediatric arterial access and arteriography: From the Societies of Interventional Radiology and Pediatric Radiology," J. Vasc. Interv. Radiol. **21**(1), 32–43.

HEROLD, D.M., HANLON, A.L. and HANKS, G.E. (1999). "Diabetes mellitus: A predictor for late radiation morbidity," Intern. J. Radiat. Oncol. Biol. Phys. **43**(3), 475–479.

HIRD, A.E., WILSON, J., SYMONS, S., SINCLAIR, E., DAVIS, M. and CHOW, E. (2008). "Radiation recall dermatitis: Case report and review of the literature," Current Oncol. **15**(1), 53–62.

HIRSHFELD, J.W., JR., BALTER, S., BRINKER, J.A., KERN, M.J., KLEIN, L.W., LINDSAY, B.D., TOMMASO, C.L., TRACY, C.M., WAGNER, L.K., CREAGER, M.A., ELNICKI, M., LORELL, B.H., RODGERS, G.P., TRACY, C.M. and WEITZ, H.H. (2004). "ACCF/AHA/HRS/SCAI clinical competence statement on physician knowledge to optimize patient safety and image quality in fluoroscopically guided invasive cardiovascular procedures: A report of the American College of Cardiology Foundation/American Heart Association/American College of Physicians Task Force on Clinical Competence and Training," J. Am. Coll. Cardiol. **44**(11), 2259–2282.

HOLMES, D.R., JR., LEON, M.B., MOSES, J.W., POPMA, J.J., CUTLIP, D., FIZGERALD, P.J., BROWN, C., FISCHELL, T., WONG, S.C., MIDEI, M., SNEAD, D. and KUNTZ, R.E. (2004). "Analysis of 1-year clinical outcomes in the SIRIUS trial: A randomized trial of a sirolimus-eluting stent versus a standard stent in patients at high risk for coronary restenosis," Circulation **109**(5), 634–640.

HOPEWELL, J.W. (1980). "The importance of vascular damage in the development of late radiation effects in normal tissues," pages 449 to 459 in *Proceedings of Radiation Biology in Cancer Research*, Meyn, R.E. and Withers, H.R., Eds. (Raven Press, New York).

HOPEWELL, J.W. (1983). "Experimental studies of early and late responses in normal tissues," pages 157 to 166 in *The Biological Basis of Radiotherapy*, Steel, G.G., Adams, G.E. and Peckham, M.J., Eds. (Elsevier, New York).

HOPEWELL, J.W. (1986). "Mechanisms of the action of radiation on skin and underlying tissues," Br. J. Radiol. **19**(Suppl.), 39–47.

HOPEWELL, J.W. (1990). "The skin: Its structure and response to ionizing radiation," Int. J. Radiat. Biol. **57**(4), 751–773.

HOPEWELL, J.W. (1991). "Biological effects of irradiation on skin and recommended dose limits," Radiat. Prot. Dosim. **39**(1/3), 11–24.

HOPEWELL, J.W. (1997). "The volume effect in radiotherapy–its biological significance," Br. J. Radiol. **70**(spec. no), S32–S40.

HOPEWELL, J.W. and YOUNG, C.M.A. (1982). "Effect of field size on the reaction of pig skin to single doses of x rays (authors' reply)," Br. J. Radiol. **55**(660), 936–937.

HOPPER, K.D. (1995). "Percutaneous, radiographically guided biopsy: A history," Radiology **196**(2), 329–333.

HSIEH, W.A., LIN, I.F., CHANG, W.P., CHEN, W.L., HSU, Y.H. and CHEN, M.S. (2010). "Lens opacities in young individuals long after exposure to protracted low-dose-rate gamma radiation in ^{60}Co-contaminated buildings in Taiwan," Radiat. Res. **173**(2), 197–204.

HUBBERT, T.E., VUCICH, J.J. and ARMSTRONG, M.R. (1993). "Lightweight aprons for protection against scattered radiation during fluoroscopy," Am. J. Roentgenol. **161**(5), 1079–1081.

HUBER, S.J. and WYNIA, M.K. (2004). "When pestilence prevails...physician responsibilities in epidemics," Am. J. Bioeth. **4**(1), W5–W11.

HYMES, S.R., STROM, E.A. and FIFE, C. (2006). "Radiation dermatitis: Clinical presentation, pathophysiology, and treatment 2006," J. Am. Acad. Dermatol. **54**(1), 28–46.

IAEA (1996). International Atomic Energy Agency. *International Basic Safety Standards for Protection Against Ionizing Radiation and the Safety of Radiation Sources*, Safety Series No. 115, http://www-pub.iaea.org/MTCD/publications/PDF/Pub996_EN.pdf (accessed February 1, 2011) (International Atomic Energy Agency, Vienna).

IAEA (2009). International Atomic Energy Agency. *Establishing Guidance Levels in X-Ray Guided Medical Interventional Procedures: A*

Pilot Study, Safety Report Series No. 59 (International Atomic Energy Agency, Vienna).

ICRP (1957). International Commission on Radiological Protection. "Exposure of man to ionizing radiation arising from medical procedures; an enquiry into methods of evaluation: A report of the International Commission on Radiological Protection and International Commission on Radiological Units and Measurements," Phys. Med. Biol. **2**(2), 107–151.

ICRP (1977). International Commission on Radiological Protection. *Recommendations of the International Commission on Radiological Protection*, ICRP Publication 26, Ann. ICRP **1**(3) (Elsevier, New York).

ICRP (1991a). International Commission on Radiological Protection. *1990 Recommendations of the International Commission on Radiological Protection*, ICRP Publication 60, Ann. ICRP **21**(1-3) (Elsevier, New York).

ICRP (1991b). International Commission on Radiological Protection. *Radiological Protection in Biomedical Research*, ICRP Publication 62, Ann. ICRP **22**(3) (Elsevier, New York).

ICRP (1991c). International Commission on Radiological Protection. *The Biological Basis for Dose Limitation in the Skin*, ICRP Publication 59, Ann. ICRP **22**(2) (Elsevier, New York).

ICRP (1997). International Commission on Radiological Protection. *General Principles for the Radiation Protection of Workers*, ICRP Publication 75, Ann. ICRP **27**(1) (Elsevier, New York).

ICRP (2000a). International Commission on Radiological Protection. *Avoidance of Radiation Injuries from Medical Interventional Procedures*, ICRP Publication 85, Ann. ICRP **30**(2) (Elsevier, New York).

ICRP (2000b). International Commission on Radiological Protection. *Pregnancy and Medical Radiation*, ICRP Publication 84, Ann. ICRP **30**(1) (Elsevier, New York).

ICRP (2001). International Commission on Radiological Protection. "Diagnostic reference levels in medical imaging: Review and additional advice," pages 33 to 52 in *Radiation and Your Patient - A Guide for Medical Practitioners*, ICRP Supporting Guidance 2, Ann. ICRP **31**(4) (Elsevier, New York).

ICRP (2007a). International Commission on Radiological Protection. *The 2007 Recommendations of the International Commission on Radiological Protection*, ICRP Publication 103, Ann. ICRP **37**(2–4) (Elsevier, New York).

ICRP (2007b). International Commission on Radiological Protection. *Radiological Protection in Medicine*, ICRP Publication 105, Ann. ICRP **37**(6) (Elsevier, New York).

ICRP (2009). International Commission on Radiological Protection. *Adult Reference Computational Phantoms*, ICRP Publication 110, Ann. ICRP **39**(2) (Elsevier, New York).

ICRU (1993). International Commission on Radiation Units and Measurements. *Quantities and Units in Radiation Protection Dosimetry,* ICRU Report 51 (Oxford University Press, Oxford, United Kingdom).

ICRU (2005). International Commission on Radiation Units and Measurements. *Patient Dosimetry for X Rays Used in Medical Imaging*, ICRU Report 74, J. ICRU **5**(2).

IEC (1994). International Electrotechnical Commission. *Protective Devices Against Diagnostic Medical X-Radiation–Part 1: Determination of Attenuation Properties of Materials*, IEC 61331-1 ed1.0 (International Electrotechnical Commission, Geneva).

IEC (1998). International Electrotechnical Commission. *Protective Devices Against Diagnostic Medical X-Radiation–Part 3: Protective Clothing and Protective Devices for Gonads*, IEC 61331-3 ed1.0 (International Electrotechnical Commission, Geneva).

IEC (2000). International Electrotechnical Commission. *Medical Electrical Equipment–Part 2-43: Particular Requirements for the Safety of X-Ray Equipment for Interventional Procedures*, IEC 60601-2-43 ed1.0 (International Electrotechnical Commission, Geneva).

IEC (2004). International Electrotechnical Commission. *Medical Electrical Equipment–Glossary of Defined Terms*, IEC 60788 ed2.0 (International Electrotechnical Commission, Geneva).

IEC (2007). International Electrotechnical Commission. *Radiation Dose Documentation–Part 1: Equipment for Radiography and Radioscopy*, PAS 61910-1 ed1.0 (International Electrotechnical Commission, Geneva).

IEC (2008). International Electrotechnical Commission. *Medical Electrical Equipment–Part 1-3: General Requirements for Basic Safety and Essential Performance–Collateral Standard: Radiation Protection in Diagnostic X-Ray Equipment*, IEC 60601-1-3 ed2.0 (International Electrotechnical Commission, Geneva).

IEC (2010). International Electrotechnical Commission. *Medical Electrical Equipment–Part 2-43: Particular Requirements for the Basic Safety and Essential Performance of X-Ray Equipment for Interventional Procedures*, IEC 60601-2-43 ed2.0 (International Electrotechnical Commission, Geneva).

IHE (2008). Integrating the Healthcare Enterprise. *Radiation Exposure Monitoring (REM) Integration Profile. Trial Implementation Draft*, http://www.ihe.net/Technical_Framework/upload/IHE-RAD_TF_Suppl_Radiation-Exposure-Monitoring_TI_2008-07-03.pdf (accessed February 1, 2011) (Healthcare Information and Management Systems Society, Chicago).

IMANISHI, Y., FUKUI, A., NIIMI, H., ITOH, D., NOZAKI, K., NAKJI, S., ISHIZUKA, K., TABATA, H., FURUYA, Y., UZURA, M., TAKABAMA, H., HASHIZUME, S., ARIMA, S. and NAKAJIMA, Y. (2005). "Radiation-induced temporary hair loss as a radiation damage only occurring in patients who had the combination of MDCT and DSA," Eur. Radiol. **15**(1), 41–46.

IMV (2009a). IMV Medical Information Division. *Benchmark Report, Cardiac Cath, 2008* (IMV Medical Information Division, Des Plaines, Illinois).

IMV (2009b). IMV Medical Information Division. *Benchmark Report, Interventional Angiography, 2008/2009* (IMV Medical Information Division, Des Plaines, Illinois).

IRIE, T., KAJITANI, M. and ITAI, Y.C.T. (2001a). "CT fluoroscopy-guided intervention: Marked reduction of scattered radiation dose to the physician's hand by use of a lead plate and an improved I-I device," J. Vasc. Interv. Radiol. **12**(12), 1417–1421.

IRIE, T., KAJITANI, M., MATSUEDA, K., ARAI, Y., INABA, Y., KUJIRAOKA, Y. and ITAI, Y. (2001b). "Biopsy of lung nodules with use of I-I device under intermittent CT fluoroscopy guidance: Preliminary clinical study," J. Vasc. Interv. Radiol. **12**(2), 215–219.

JACOBSON, G.F., SHABER, R.E., ARMSTRONG, M.A. and HUNG, Y-Y. (2007). "Changes in rates of hysterectomy and uterine conserving procedures for treatment of uterine leiomyoma," Am. J. Obstet. Gynecol. **196**(6), 601.e1–601.e6.

JARVINEN, H., BULS, N., CLERINX, P., JANSEN, J., MILJANIC, S., NIKODEMOVA, D., RANOGAJEC-KOMOR, M. and D'ERRICO, F. (2008). "Overview of double dosimetry procedures for the determination of the effective dose to the interventional radiology staff," Radiat. Protect. Dosim. **129**(1–3), 333–339.

JEAN-BAPTISTE, E., HASSEN-KHODJA, R., BOUILLANNE, P.J., HAUDEBOURG, P., DECLEMY, S. and BATT, M. (2007). "Endovascular repair of infrarenal abdominal aortic aneurysms in high-risk-surgical patients," Eur. J. Vasc. Endovasc. Surg. **34**(2), 145–151.

JENSEN, M.E., EVANS, A.J., MATHIS, J.M., KALLMES, D.F., CLOFT, H.J. and DION, J.E. (1997). "Percutaneous polymethylmethacrylate vertebroplasty in the treatment of osteoporotic vertebral body compression fractures: Technical aspects," Am. J. Neuroradiol. **18**(10), 1897–1904.

JOHNSON, L.W., MOORE, R.J. and BALTER, S. (1992). "Review of radiation safety in the cardiac catheterization laboratory," Cathet. Cardiovasc. Diagn. **25**(3), 186–194.

JOHNSTON, S.C., DUDLEY, R.A., GRESS, D.R. and ONO, L. (1999). "Surgical and endovascular treatment of unruptured cerebral aneurysms at university hospitals," Neurology **52**(9), 1799–1805.

JOHNSTON, S.C., WILSON, C.B., HALBACH, V.V., HIGASHIDA, R.T., DOWD, C.F., MCDERMOTT, M.W., APPLEBURY, C.B., FARLEY, T.L. and GRESS, D.R. (2000). "Endovascular and surgical treatment of unruptured cerebral aneurysms: Comparison of risks," Ann. Neurol. **48**(1), 11–19.

JOHNSTON, W.W. (1984). "Percutaneous fine needle aspiration biopsy of the lung. A study of 1,015 patients," Acta Cytol. **28**(3), 218–224.

JOLLES, B. (1972). "Colorimetric study of radiation induced inflammatory changes in skin," pages 28 to 34 in *Methods in Microcirculation Studies*, Ryan, T.J., Jolles, B. and Holti, G., Eds. (British Microcirculation Society, London).

JOLLES, B. and MITCHELL, R.G. (1947). "Optimal skin tolerance dose levels," Br. J. Radiol. **20**(238), 405–409.

JONES, K.L. (1997). *Smith's Recognizable Patterns of Human Malformation*, 5th ed. (W.B. Saunders Company, Philadelphia).

JOYET, G. and HOHL, K. (1955). "Biological skin reaction in deep therapy as a function of size of the field; law for radiotherapy," Fortschr. Geb. Rontgenstr. **82**(3), 387–400 [in German].

KALLMES, D.F., O, E., ROY, S.S., PICCOLO, R.G., MARX, W.F., LEE, J.K. and JENSEN, M.E. (2003). "Radiation dose to the operator during vertebroplasty: Prospective comparison of the use of 1-cc syringes versus an injection device," Am. J. Neuroradiol. **24**(6), 1257–1260.

KATO, R., KATADA, K., ANNO, H., SUZUKI, S., IDA, Y. and KOGA, S. (1996). "Radiation dosimetry at CT fluoroscopy: Physician's hand dose and development of needle holders," Radiology **201**(2), 576–578.

KATZEN, B.T., DAKE, M.D., MACLEAN, A.A. and WANG, D.S. (2005). "Endovascular repair of abdominal and thoracic aortic aneurysms," Circulation **112**(11), 1663–1675.

KEELING, A.N., NAUGHTON, P.A., OCONNELL, A. and LEE, M.J. (2008). "Does percutaneous transluminal angioplasty improve quality of life?," J. Vasc. Intervent. Radiol. **19**(2), 169–176.

KIM, K.P. and MILLER, D.L. (2009). "Minimising radiation exposure to physicians performing fluoroscopically guided cardiac catheterization procedures: A review," Radiat. Prot. Dosim. **133**(4), 227–233.

KIM, L.G., SCOTT, R.A.P., ASHTON, H.A. and THOMPSON, S.G. (2007). "A sustained mortality benefit from screening for abdominal aortic aneurysm," Ann. Int. Med. **146**(10), 699–706.

KIM, K.P., MILLER, D.L., BALTER, S., KLEINERMAN, R.A., LINET, M.S., KWON, D. and SIMON, S.L. (2008). "Occupational radiation doses to operators performing cardiac catheterization procedures," Health Phys. **94**(3), 211–227.

KIM, C., VASAIWALA, S., HAQUE, F., PRATAP, K. and VIDOVICH, M.I. (2010). "Radiation safety among cardiology fellows," Am. J. Cardiol. **106**(1), 125–128.

KING, J.N., CHAMPLIN, A.M., KELSEY, C.A. and TRIPP, D.A. (2002). "Using a sterile disposable protective surgical drape for reduction of radiation exposure to interventionalists," Am. J. Roentgenol. **178**(1), 153–157.

KING, S.B., III, SMITH, S.C., JR., HIRSHFELD, J.W., JR., JACOBS, A.K., MORRISON, D.A., WILLIAMS, D.O., FELDMAN, T.E., KERN, M.J., O'NEILL, W.W., SCHAFF, H.V., WHITLOW, P.L., ADAMS, C.D., ANDERSON, J.L., BULLER, C.E., CREAGER, M.A., ETTINGER, S.M., HALPERIN, J.L., HUNT, S.A., KRUMHOLZ, H.M., KUSHNER, F.G., LYTLE, B.W., NISHIMURA, R., PAGE, R.L., RIEGEL, B., TARKINGTON, L.G. and YANCY, C.W. (2008). "2007 focused update of the ACC/AHA/SCAI 2005 guideline update for percutaneous coronary intervention: A report of the American College of Cardiology/American Heart Association Task Force on Practice Guidelines," J. Am. Coll. Cardiol. **51**(2), 172–209.

KLEIMAN, N.J. (2007). "Radiation cataract," pages 81 to 95 in *Radiation Protection 145, New Insights in Radiation Risk and Basic Safety Standards* (European Commission, Luxembourg).

KLEIMAN, N.J. and WORGUL, B.V. (1994). "Lens," pages 1 to 39 in *Duane's Foundations of Clinical Ophthalmology,* Vol. 1, Tasman, W. and Jaeger, E.A., Eds. (J.P. Lippincott Co., Philadelphia).

KLEIMAN, N.J., CABRERA, M., DURAN, G., RAMIREZ, R., DURAN, A. and VANO, E. (2009). "Occupational risk of radiation cataract in interventional cardiology," http://www.arvo.org/eweb/DynamicPage.aspx?site=arvo2&webcode=amabstractsearch (accessed February 1, 2011), Invest. Ophthalmol. Vis. Sci. **50**, E-Abstract 511..

KLEIN, B.E., KLEIN, R., LINTON, K.L. and FRANKE, T. (1993). "Diagnostic x-ray exposure and lens opacities: The Beaver Dam Eye Study," Am. J. Public Health **83**(4), 588–590.

KLEIN, L.W., MILLER, D.L., BALTER, S., LASKEY, W., HAINES, D., NORBASH, A., MAURO, M.A. and GOLDSTEIN, J.A. (2009). "Occupational health hazards in the interventional laboratory: Time for a safer environment," Radiology **250**(2), 538–544.

KLIEWER, M.A., SHEAFOR, D.H., PAULSON, E.K., HELSPER, R.S., HERTZBERG, B.S. and NELSON, R.C. (1999). "Percutaneous liver biopsy: A cost-benefit analysis comparing sonographic and CT guidance," Am. J. Roentgenol. **173**(5), 1199–1202.

KLOSE, K.C, MERTENS, R., ALZEN, G., LOER, F. and BOCKING, A. (1990). "CT-guided percutaneous large-bore biopsies in benign and malignant pediatric lesions," Cardiovasc. Intervent. Radiol. **14**(1), 78–83.

KNIGHT, F., GALVIN, R., DAVOREN, R. and MASON, K.P. (2006). "The evolution of universal protocol in interventional radiology," J. Radiol. Nurs. **25**(4), 106–115.

KOCHER, D.C., APOSTOAEI, A.I. and HOFFMAN, F.O. (2005). "Radiation effectiveness factors for use in calculating probability of causation of radiogenic cancers," Health Phys. **89**(1), 3–32.

KOENIG, T.R., WOLFF, D., METTLER, F.A., JR. and WAGNER, L.K. (2001a). "Skin injuries from fluoroscopically guided procedures: Part 1, characteristics of radiation injury," Am. J. Roentgenol. **177**(1), 3–11.

KOENIG, T.R., METTLER, F.A., JR. and WAGNER, L.K. (2001b). "Skin injuries from fluoroscopically guided procedures: Part 2, Review of 73 cases and recommendations for minimizing dose delivered to the patient," Am. J. Roentgenol. **177**(1**),** 13–20.

KOMEMUSHI, A., TANIGAWA, N., KARIVA, S., KOJIMA, H., SHOMURA, Y. and SAWADA, S. (2005). "Radiation exposure to operators during vertebroplasty," J. Vasc. Interv. Radiol. **16**(10), 1327–1332.

KRIZEK, T.J. (1979). "Difficult wounds: Radiation wounds," Clin. Plast. Surg. **6**(4), 541–543.

KRUGER, R. and FACISZEWSKI, T. (2003). "Radiation dose reduction to medical staff during vertebroplasty: A review of techniques and methods to mitigate occupational dose," Spine. **28**(14), 1608–1613.

KRUGER, R.A. and RIEDERER, S.J. (1984). *Basic Concepts of Digital Subtraction Angiography* (Hall Medical Publishers, Boston).

KUMAZAWA, S., NELSON, D.R. and RICHARDSON, A.C.B. (1984). *Occupational Exposure to Ionizing Radiation in the United States: A Comprehensive Review for the Year 1980 and a Summary of Trends for the Years 1960–1985*, EPA 520/1-84-005 (National Technical Information Service, Springfield, Virginia).

KUUKASJARVI, P., RASANEN, P., MALMIVAARA, A., ARONEN, P. and SINTONEN, H. (2007). "Economic evaluation of drug-eluting stents: A systematic literature review and model-based cost-utility analysis," Int. J. Technol. Assess. Health Care **23**(4), 473–479.

LAMBERT, K. and MCKEON, T. (2001). "Inspection of lead aprons: Criteria for rejection," Health Phys. **80**(Suppl. 5), S67–S69.

LAND, C., GILBERT, E., SMITH, J.M., HOFFMAN, F.O., APOSTOAEI, I., THOMAS, B. and KOCHER, D.C. (2003). *Report of the NCI-CDC Working Group to Revise the 1985 NIH Radioepidemiological Tables*, NIH Publication No. 03-5387, http://dceg.cancer.gov/files/NIH_No_03-5387.pdf (accessed February 1, 2011) (National Cancer Institute, Bethesda, Maryland).

LAYTON, K.F., KALLMES, D.F., CLOFT, H.J., SCHUELER, B.A. and STURCHIO, G.M. (2006). "Radiation exposure to the primary operator during endovascular surgical neuroradiology procedures," Am. J. Neuroradiol. **27**(4), 742–743.

LEE, C., LODWICK, D., HURTADO, J., PAFUNDI, D., WILLIAMS, J.L. and BOLCH, W.E. (2010). "The UF family of reference hybrid phantoms for computational radiation dosimetry," Phys. Med. Biol. **55**(2), 339–363.

LEGIEHN, G.M. and HERAN, M.K.S. (2008). "Venous malformations: Classification, development, diagnosis, and interventional radiologic management," Radiol. Clin. North Am. **46**(3), 545–597.

LEINFELDER, P.J. and KERR, H.D. (1936). "Roentgen-ray cataract: An experimental, clinical, and microscopic study," Am. J. Ophthalmol. **19**, 739–756.

LETT, J.T., LEE, A.C. and COX, A.B. (1991). "Late cataractogenesis in rhesus monkeys irradiated with protons and radiogenic cataract in other species," Radiat. Res. **126**(2), 147–156.

LEUNG, K.C. and MARTIN, C.J. (1996). "Effective doses for coronary angiography," Br. J. Radiol. **69**(821), 426–431.

LEWIS, E.C., CONNOLLY, B., TEMPLE, M., JOHN, P., CHAIT, P.G., VAUGHAN, J. and AMARAL, J.G. (2008). "Growth outcomes and complications after radiologic gastrostomy in 120 children," Pediatr. Radiol. **38**(9), 963–970.

LIPSITZ, E.C., VEITH, F.J., OHKI, T., HELLER, S., WAIN, R.A., SUGGS, W.D., LEE, J.C., KWEI, S., GOLDSTEIN, K., RABIN, J., CHANG, D. and MEHTA, M. (2000). "Does the endovascular repair of aortoiliac aneurysms pose a radiation safety hazard to vascular surgeons?," J. Vasc. Surg. **32**(4), 704–710.

LIVINGSTONE, R.S. and MAMMEN, T.G. (2005). "Evaluation of radiation dose to patients during abdominal embolizations," Indian J. Med. Sci. **59**(12), 528–534.

LOGRONO, R., KURTYCZ, D.F.I., SPROAT, I.A., SHALKHAM, J.E., STEWART, J.A. and INHORN, S.L. (1998). "Multidisciplinary approach to deep-seated lesions requiring radiologically-guided fine-needle aspiration," Diagn. Cytopathol. **18**(5), 338–342.

LUNING, M., SCHRODER, K., WOLFF, H., KRANZ, D. and HOPPE, E. (1991). "Percutaneous biopsy of the liver," Cardiovasc. Intervent. Radiol. **14**(1), 40–42.

MACKEE, G.M. and CIPOLLARO, A.C. (1947). *X-Rays and Radium in the Treatment of Diseases of the Skin* (Lea and Febiger, Philadelphia).

MACKENZIE, A., EMERTON, D., LAWINSKI, C., COLE, H., HONEY, I. and BLAKE, P. (2006). *Cardiovascular Imaging Systems: A Comparative Report*, Report 06044, ed. 5 (KCARE, King's College Hospital, London).

MAHESH, M. (2001). "Fluoroscopy: Patient radiation exposure issues," Radiographics **21**(4), 1033–1045.

MAKARY, M.A., AL-ATTARl, A., HOLZMUELLER, C.G., SEXTON, J.B., SYIN, D., GILSON, M.M., SULKOWSKI, M.S. and PRONOVOST, P.J. (2007). "Needlestick injuries among surgeons in training," N. Engl. J. Med. **356**(26), 2693–2699.

MANTOVANI, L., D'ERCOLE, L., LISCIANDRO, F., QUARETTI, P., AZZARETTI, A., RODOLICO, G., SALUZZO, C.M., SPINAZZOLA, A., DI MARIA, F., OTTOLENGHI, A., THYRION, F.Z. and ANDREUCCI, L. (2006). "Radiochromic films for improved evaluation of patient dose in liver interventions," J. Vasc. Interv. Radiol. **17**(5), 855–862.

MARINE, J.E. (2007). "Catheter ablation therapy for supraventricular arrhythmias," JAMA **298**(23), 2768–2778.

MARSHALL, N.W. and FAULKNER, K. (1992). "The dependence of the scattered radiation dose to personnel on technique factors in diagnostic radiology," Br. J. Radiol. **65**(769), 44–49.

MARSHALL, N.W., FAULKNER, K. and CLARKE, P. (1992). "An investigation into the effect of protective devices on the dose to radiosensitive organs in the head and neck," Br. J. Radiol. **65**(777), 799–802.

MARSHALL, N.W., NOBLE, J. and FAULKNER, K. (1995). "Patient and staff dosimetry in neuroradiological procedures," Br. J. Radiol. **68**(809), 495–501.

MARTIN, C.J. (2007). "Effective dose: How should it be applied to medical exposures?," Brit. J. Radiol. **80**(956), 639-647.

MARTIN, C.J. (2009). "A review of radiology staff doses and dose monitoring requirements," Radiat. Prot. Dosim. **136**(3), 140–157.

MARX, M.V., NIKLASON, L. and MAUGER, E.A. (1992). "Occupational radiation exposure to interventional radiologists: A prospective study," J. Vasc. Intervl. Radiol. **3**(4), 597–606.

MAVRIKOU, I., KOTTOU, S., TSAPAKI, V. and NEOFOTISTOU, V. (2008). "High patient doses in interventional cardiology due to physicians' negligence: How can they be prevented?," Radiat. Prot. Dosim. **129**(1–3), 67–70.

MCCAFFREY, J.P., SHEN, H., DOWNTON, B. and MAINEGRA-HING, E. (2007). "Radiation attenuation by lead and nonlead materials used in radiation shielding garments," Med. Phys. **34**(2), 530–537.

MCCOLLOUGH, C.H., CHRISTNER, J.A. and KOFLER, J.M. (2010). "How effective is effective dose as a predictor of radiation risk?," Am. J. Roentgenol. **194**(4), 890–896.

MCFADDEN, S.L., MOONEY, R.B. and SHEPHERD, P.H. (2002). "X-ray dose and associated risks from radiofrequency catheter ablation procedures," Br. J. Radiol. **75**(891), 253–265.

MCLAREN, C.A. and ROEBUCK, D.J. (2003). "Interventional radiology for renovascular hypertension in children," Tech. Vasc. Interv. Radiol. **6**(4), 150–157.

MCPARLAND, B.J. (1998a). "A study of patient radiation doses in interventional radiological procedures," Br. J. Radiol. **71**(842), 175–185.

MCPARLAND, B.J. (1998b). "Entrance skin dose estimates derived from dose-area-product measurements in interventional radiological procedures," Br. J. Radiol. **71**(852), 1288–1295.

MERRIAM, G.R., JR. and FOCHT, E.F. (1962). "A clinical and experimental study of the effect of single and divided doses of radiation on cataract production," Trans. Am. Ophthalmol. Soc. **60**, 35–52.

MERRIAM, G.R., JR. and WORGUL, B.V. (1983). "Experimental radiation cataract – its clinical relevance," Bull. N.Y. Acad. Med. **59**(4), 372–392.

MICHEL, R. and ZORN, M.J. (2002). "Implementation of an x-ray radiation protective equipment inspection program," Health Phys. **82**(Suppl. 2), S51–S53.

MIKSYS, N., GORDON, C.L., THOMAS, K. and CONNOLLY, B.L. (2010). "Estimating effective dose to pediatric patients undergoing interventional radiology procedures using anthropomorphic phantoms and MOSFET dosimeters," Am. J. Roentgenol. **194**(5), 1315–1322.

MILLER, D.L (2008). "Overview of contemporary interventional fluoroscopy procedures," Health Phys. **95**(5), 638–644.

MILLER, D.L., BALTER, S., NOONAN, P.T. and GEORGIA, J.D. (2002). "Minimizing radiation-induced skin injury in interventional radiology procedures," Radiology **225**(2), 329–336.

MILLER, D.L., BALTER, S., COLE, P.E., LU, H.T., SCHUELER, B.A., GEISINGER, M., BERENSTEIN, A., ALBERT, R., GEORGIA, J.D., NOONAN, P.T., CARDELLA, J.F., ST. GEORGE, J., RUSSELL, E.J., MALISCH, T.W., VOGELZANG, R.L., MILLER, G.L., III and ANDERSON, J. (2003a). "Radiation doses in interventional radiology procedures: The RAD-IR study. Part I: Overall measures of dose," J. Vasc. Interv. Radiol. **14**(6), 711–727.

MILLER, D.L., BALTER, S., COLE, P.E., LU, H.T., BERENSTEIN, A., ALBERT, R., SCHULER, B.A., GEORGIA, J.D., NOONAN, P.T., RUSSELL, E., MALISCH, T., VOGELZANG, R.L., GEISINGER, M., ST. GEORGE, J., CARDELLA, J.F., MILLER, G. and ANDERSON, J.

(2003b). "Radiation doses in interventional radiology procedures: The RAD-IR study. Part II. Skin dose," J. Vasc. Interv. Radiol. **14**(8), 977–990.

MILLER, D.L., BALTER, S., WAGNER, L.K., CARDELLA, J.F., CLARK, T.W.I., NEITHAMER, C.D., JR., SCHWARTZBERG, M.S., SWAN, T.L., TOWBIN, R.B., RHOLL, K.S. and SACKS, D. (2004). "Quality improvement guidelines for recording patient radiation dose in the medical record," J. Vasc. Interv. Radiol. **15**(5), 423–429.

MILLER, D.L., KWON, D. and BONAVIA, G.H. (2009). "Reference levels for patient radiation doses in interventional radiology: Proposed initial values for U.S. practice," Radiology **253**(3), 753–764.

MILLER, D.L., BALTER, S., SCHUELER, B.A., WAGNER, L.K., STRAUSS, K.J. and VANO, E. (2010a). "Clinical radiation management for fluoroscopically guided interventional procedures," Radiology **257**(2), 321–332.

MILLER, D.L., VANO E., BARTAL, G., BALTER, S., DIXON, R., PADOVANI, R., SCHUELER, B., CARDELLA, J.F. and DE BAERE, T. (2010b). "Occupational radiation protection in interventional radiology: A joint guideline of the Cardiovascular and Interventional Radiology Society of Europe and the Society of Interventional Radiology," Cardiovasc. Intervent. Radiol. **33**(2), 230–239.

MINAMOTO, A., TANIGUCHI, H., YOSHITANI, N., MUKAI, S., YOKOYAMA, T., KUMAGAMI, T., TSUDA, Y., MISHIMA, H.K., AMEMIYA, T., NAKASHIMA, E., NERIISHI, K., HIDA, A., FUJIWARA, S., SUZUKI, G. and AKAHOSHI, M. (2004). "Cataract in atomic bomb survivors," Int. J. Radiat. Biol. **80**(5), 339–345.

MOLYNEUX, A., KERR, R., STRATTON, I., SANDERCOCK, P., CLARKE, M., SHRIMPTON. J. and HOLMAN, R. (2002). "International Subarachnoid Aneurysm Trial (ISAT) of neurosurgical clipping versus endovascular coiling in 2143 patients with ruptured intracranial aneurysms: A randomised trial," Lancet **360**(9342), 1267–1274.

MONACO, J.L., BOWEN, K., TADROS, P.N. and WITT, P.D. (2003). "Iatrogenic deep musculocutaneous radiation injury following percutaneous coronary intervention," J. Invasive Cardiol. **15**(8), 451–453.

MOORE, W.E., FERGUSON, G. and ROHRMANN, C. (1980). "Physical factors determining the utility of radiation safety glasses," Med. Phys. **7**(1), 8–12.

MOORE, B., VANSONNENBERG, E., CASOLA, G. and NOVELLINE, R.A. (1992). "The relationship between back pain and lead apron use in radiologists," Am. J. Roentgenol. **158**(1), 191–193.

MORRELL, R.E. and ROGERS, A.T. (2006). "A mathematical model for patient skin dose assessment in cardiac catheterization procedures," Br. J. Radiol. **79**(945), 756–761.

MORRIS, G.M. and HOPEWELL, J.W. (1986). "Changes in the cell kinetics of pig epidermis after repeated daily doses of x rays," Br. J. Radiol. **19**(Suppl.), 34–38.

MORRIS, G.M. and HOPEWELL, J.W. (1988). "Changes in the cell kinetics of pig epidermis after single doses of x rays," Br. J. Radiol. **61**(723), 205–211.

MOSS, W.T. (1959). *Therapeutic Radiology; Rationale, Technique, Results* (C.V. Mosby, St. Louis, Missouri).

MOUSTAFA, H.F. and HOPEWELL, J.W. (1979). "Blood flow clearance changes in pig skin after single doses of x rays," Br. J. Radiol. **52**(614), 138–144.

MURASE, E., SIEGELMAN, E.S., OUTWATER, E.K., PEREZ-JAFFE, L.A. and TURECK, R.W. (1999). "Uterine leiomyomas: Histopathologic features, MR imaging findings, differential diagnosis, and treatment," Radiographics **19**(5), 1179–1197.

MUSSON, D.M. and HELMREICH, R.L. (2004). "Team training and resource management in health care: Current issues and future directions," Har. Health Policy Rev. **5**(1), 25–35.

NAKASHIMA, E., NERIISHI, K. and MINAMOTO, A. (2006). "A reanalysis of atomic-bomb cataract data, 2000-2002: A threshold analysis," Health Phys. **90**(2), 154–160.

NA/NRC (1990). National Academies/National Research Council. *Health Effects of Exposure to Low Levels of Ionizing Radiation*, BEIR V (National Academies Press, Washington).

NA/NRC (2006). National Academies/National Research Council. *Health Risks from Exposure to Low Levels of Ionizing Radiation*, BEIR VII, Phase 2 (National Academies Press, Washington).

NAWFEL, R.D., JUDY, P.F., SILVERMAN, S.G., HOOTON, S., TUNCALI, K. and ADAMS, D.F. (2000). "Patient and personnel exposure during CT fluoroscopy-guided interventional procedures," Radiology **216**(1), 180–184.

NCI (2005). National Cancer Institute. *Interventional Fluoroscopy: Reducing Radiation Risks for Patients and Staff*, NIH Publication No. 05-5286, http://www.cancer.gov/images/Documents/45bae965-697a-4de8-9dae-b77222e0e79d/InterventionalFluor.pdf (accessed February 1, 2011) (National Cancer Institute, Bethesda, Maryland).

NCI (2006). National Cancer Institute. *Common Terminology Criteria for Adverse Events v3.0 (CTCAE)*, http://ctep.cancer.gov/protocoldevelopment/electronic_applications/docs/ctcaev3.pdf (accessed February 1, 2011) (National Cancer Institute, Bethesda, Maryland).

NCRP (1976). National Council on Radiation Protection and Measurements. *Structural Shielding Design and Evaluation for Medical Use of X-Rays and Gamma Rays of Energies up to 10 MeV*, NCRP Report No. 49 (National Council on Radiation Protection and Measurements, Bethesda, Maryland).

NCRP (1987). National Council on Radiation Protection and Measurements. *Ionizing Radiation Exposure of the Population of the United States*, NCRP Report No. 93 (National Council on Radiation Protection and Measurements, Bethesda, Maryland).

NCRP (1989a). National Council on Radiation Protection and Measurements. *Exposure of the U.S. Population from Occupational Radiation*,

NCRP Report No. 101 (National Council on Radiation Protection and Measurements, Bethesda, Maryland).

NCRP (1989b). National Council on Radiation Protection and Measurements. *Radiation Protection for Medical and Allied Health Personnel*, NCRP Report No. 105 (National Council on Radiation Protection and Measurements, Bethesda, Maryland).

NCRP (1993). National Council on Radiation Protection and Measurements. *Limitation of Exposure to Ionizing Radiation*, NCRP Report No. 116 (National Council on Radiation Protection and Measurements, Bethesda, Maryland).

NCRP (1995a). National Council on Radiation Protection and Measurements. *An Introduction to Efficacy in Diagnostic Radiology and Nuclear Medicine (Justification of Medical Radiation Exposure)*, NCRP Commentary No. 13 (National Council on Radiation Protection and Measurements, Bethesda, Maryland).

NCRP (1995b). National Council on Radiation Protection and Measurements. *Use of Personal Monitors to Estimate Effective Dose Equivalent and Effective Dose to Workers for External Exposure to Low-LET Radiation*, NCRP Report No. 122 (National Council on Radiation Protection and Measurements, Bethesda, Maryland).

NCRP (1997). National Council on Radiation Protection and Measurements. *Uncertainties in Fatal Cancer Risk Estimates Used in Radiation Protection*, NCRP Report No. 126 (National Council on Radiation Protection and Measurements, Bethesda, Maryland).

NCRP (1998). National Council on Radiation Protection and Measurements. *Operational Radiation Safety Program*, NCRP Report No. 127 (National Council on Radiation Protection and Measurements, Bethesda, Maryland).

NCRP (1999). National Council on Radiation Protection and Measurements. *Biological Effects and Exposure Limits for "Hot Particles,"* NCRP Report No. 130 (National Council on Radiation Protection and Measurements, Bethesda, Maryland).

NCRP (2000a). National Council on Radiation Protection and Measurements. *Radiation Protection for Procedures Performed Outside the Radiology Department*, NCRP Report No. 133 (National Council on Radiation Protection and Measurements, Bethesda, Maryland).

NCRP (2000b). National Council on Radiation Protection and Measurements. *Radiation Protection Guidance for Activities in Low-Earth Orbit*, NCRP Report No. 132 (National Council on Radiation Protection and Measurements, Bethesda, Maryland).

NCRP (2004). National Council on Radiation Protection and Measurements. *Structural Shielding Design for Medical X-Ray Imaging Facilities*, NCRP Report No. 147 (National Council on Radiation Protection and Measurements, Bethesda, Maryland).

NCRP (2005). National Council on Radiation Protection and Measurements. *Extrapolation of Radiation-Induced Cancer Risks from Nonhuman Experimental Systems to Humans*, NCRP Report No. 150

(National Council on Radiation Protection and Measurements, Bethesda, Maryland).

NCRP (2009). National Council on Radiation Protection and Measurements. *Ionizing Radiation Exposure of the Population of the United States*, NCRP Report No. 160 (National Council on Radiation Protection and Measurements, Bethesda, Maryland).

NEEMAN, Z., DROMI, S.A., SARIN, S. and WOOD, B.J. (2006). "CT fluoroscopy shielding: Decreases in scattered radiation for the patient and operator," J. Vasc. Interv. Radiol. **17**(12), 1999–2004.

NEMA (2008). National Electrical Manufacturers Association. *Primary User Controls for Interventional Angiography X-Ray Equipment*, NEMA Standards Publication XR 24-2008 (National Electrical Manufacturers Association, Rosslyn, Virginia).

NERIISHI, K., NAKASHIMA, E., MINAMOTO, A., FUJIWARA, S., AKAHOSHI, M., MISHIMA, H.K., KITAOKA, T. and SHORE, R.E. (2007). "Postoperative cataract cases among atomic bomb survivors: Radiation dose response and threshold," Radiat. Res. **168**(4), 404–408.

NHS (1993). National Health Service. "What do we mean by appropriate health care?" Qual. Health Care **2**, 117–123.

NICKOLOFF, E.L., KHANDJI, A. and DUTTA, A. (2000). "Radiation doses during CT fluoroscopy," Health Phys. **79**(6), 675–681.

NICKOLOFF, E.L., LU, Z.F., DUTTA, A., SO, J., BALTER, S. and MOSES, J. (2007). "Influence of flat-panel fluoroscopic equipment variables on cardiac radiation doses," Cardiovasc. Intervent. Radiol. **30**(2), 169–176.

NIKOLIC, B., SPIES, J.B., LUNDSTEN, M.J. and ABBARA, S. (2000). "Patient radiation dose associated with uterine artery embolization," Radiology **214**(1), 121–125.

NIOSH (2004). National Institute for Occupational Safety and Health. *Worker Health Chartbook 2004*, NIOSH Publication No. 2004–146, http://www.cdc.gov/niosh/docs/2004-146 (accessed February 1, 2011) (National Institute for Occupational Safety and Health, Cincinnati, Ohio).

NIST (2008). National Institute of Standards and Technology. *About the National Voluntary Laboratory Accreditation Program (NVLAP)*, http://ts.nist.gov/standards/accreditation/index.cfm (accessed February 1, 2011) (National Institute of Standards and Technology, Gaithersburg, Maryland).

NORBASH, A.M., BUSICK, D. and MARKS, M.P. (1996). "Techniques for reducing interventional neuroradiologic skin dose: Tube position rotation and supplemental beam filtration," Am. J. Neuroradiol. **17**(1), 41–49.

NRC (1998). U.S. Nuclear Regulatory Commission. "Planned special exposures," 10 CFR Part 20.1206 (amended July 23), http://www.nrc.gov/reading-rm/doc-collections/cfr/part020/part020-1206.html (accessed February 1, 2011) (U.S. Nuclear Regulatory Commission, Washington).

NRC (1999). U.S. Nuclear Regulatory Commission. *Regulatory Guide 8.13 – Instruction Concerning Prenatal Radiation Exposure*, rev. 3, http://www.nrc.gov/reading-rm/doc-collections/reg-guides/occupational-health/rg/

8-13 (accessed February 1, 2011) (U.S. Nuclear Regulatory Commission, Washington).

O'BRIEN, B. and VAN DER PUTTEN, W. (2008). "Quantification of risk-benefit in interventional radiology," Radiat. Prot. Dosim. **129**(1–3), 59–62.

O'DEA, T.J., GEISE, R.A. and RITENOUR, E.R. (1999). "The potential for radiation-induced skin damage in interventional neuroradiological procedures: A review of 522 cases using automated dosimetry," Med. Phys. **26**(9), 2027–2033.

ORIOL, M.D. (2006). "Crew resource management. Applications in healthcare organizations," J. Nurs. Admin. **36**(9), 402–406.

ORTH, R.C., WALLACE, M.J. and KUO, M.D. (2008). "C-arm cone-beam CT: General principles and technical considerations for use in interventional radiology," J. Vasc. Interv. Radiol. **19**(6), 814–821.

PADOVANI, R. and QUAI, E. (2005). "Patient dosimetry approaches in interventional cardiology and literature dose data review," Radiat. Prot. Dosim. **117**(1–3), 217–221.

PARK, T.H., EICHLING, J.O., SCHECTMAN, K.B., BROMBERG, B.I., SMITH, J.M. and LINDSAY, B.D. (1996). "Risk of radiation induced skin injuries from arrhythmia ablation procedures," Pacing Clin. Electrophysiol. **19**(9), 1363–1369.

PATEL, M.R., DEHMER, G.J., HIRSHFELD, J.W., SMITH, P.K. and SPERTUS, J.A. (2009). "ACCF/SCAI/STS/AATS/AHA/ASNC 2009 appropriateness criteria for coronary revascularization," J. Am. Coll. Cardiol. **53**(6), 530–553.

PATERSON, R. (1948). *The Treatment of Malignant Disease by Radium X-Rays: Being a Practice of Radiotherapy* (Arnold, London).

PAULSON, E.K., SHEAFOR, D.H., ENTERLINE, D.S., MCADAMS, H.P. and YOSHIZUMI, T.T. (2001). "CT fluoroscopy-guided interventional procedures: Techniques and radiation dose to radiologists," Radiology **220**(1), 161–167.

PERISINAKIS, K., THEOCHAROPOULOS, N., DAMILAKIS, J., KATONIS, P., PAPADOKOSTAKIS, G., HADJIPAVLOU, A. and GOURTSOYIANNIS, N. (2004). "Estimation of patient dose and associated radiogenic risks from fluoroscopically guided pedicle screw insertion," Spine **29**(14), 1555–1560.

PERISINAKIS, K., THEOCHAROPOULOS, N., DAMILAKIS, J., MANIOS, E., VARDAS, P. and GOURTSOYIANNIS, N. (2005). "Fluoroscopically guided implantation of modern cardiac resynchronization devices: Radiation burden to the patient and associated risks," J. Am. Coll. Cardiol. **46**(12), 2335–2339.

PETERZOL, A., QUAI, E., PADOVANI, R., BERNARDI, G., KOTRE, C.J. and DOWLING, A. (2005). "Reference levels in PTCA as a function of procedure complexity," Radiat. Prot. Dosim. **117**(1–3), 54–58.

PEYNIRCIOGLU, B., CANYIGIT, M., ERGUN, O., PAMUK, G.A. and CIL, B.E. (2007). "Radiologically placed venous ports in children," J. Vasc. Interv. Radiol. **18**(4), 1389–1394.

PLEIS, J.R. and LETHBRIDGE-CEJKU, M. (2007). "Summary health statistics for U.S. adults: National Health Interview Survey, 2006," http://www.cdc.gov/nchs/data/series/sr_10/sr10_235. pdf (accessed February 1, 2011), Vital Health Stat. **10**(235).

POLUDNIOWSKI, G., LANDRY, G., DEBLOIS, F., EVANS, P.M. and VERHAEGEN, F. (2009). "SpekCalc: A program to calculate photon spectra from tungsten anode x-ray tubes," Phys. Med. Biol. **54**(19), N433–N438.

PRON, G., MOCARSKI, E., BENNETT, J., VILOS, G., COMMON, A., ZAIDI, M., SNIDERMAN, K., ASCH, M., KOZAK, R., SIMONS, M., TRAN, C. and KACHURA, J. (2003). "Tolerance, hospital stay, and recovery after uterine artery embolization for fibroids: The Ontario Uterine Fibroid Embolization Trial," J. Vasc. Interv. Radiol. **14**(10), 1243–1250.

PRUSS-USTUN, A., RAPITI, E. and HUTIN, Y. (2005). "Estimation of the global burden of disease attributable to contaminated sharps injuries among health-care workers," Am. J. Ind. Med. **48**(6), 482–490.

PUTNIK, K., STADLER, P., SCHAEFER, C. and KOELBL, O. (2006). "Enhanced radiation sensitivity and radiation recall dermatitis (RRD) after hypericin therapy – case report and review of literature," Radiat. Oncol. **1**(1), 32–36.

RAMPADO, O. and ROPOLO, R. (2005). "Entrances skin dose distribution maps for interventional neuroradiological procedures: A preliminary study," Radiat. Prot. Dosim. **117**(1–3), 256–259.

RAMSDALE, M.L., WALKER, W.J. and HORTON, P.W. (1990). "Extremity doses during interventional radiology," Clin. Radiol. **41**(1), 34–36.

RANDALL, M.G. and HORN, B. (1977). "A systems approach to acceptance testing of diagnostic x-ray equipment," Proc. SPIE **127**, 158–166.

RAUCH, P.L. (1982). "Performance characteristics of diagnostic x-ray generators," pages 126 to 156 in *Acceptance Testing of Radiological Imaging Equipment*, AAPM Symposium Proceedings No. 1, Lin, P.J.P., Kriz, R.J., Rauch, P.L., Strauss, K.J. and Rossi, R.P., Eds. (American Institute of Physics, New York).

RAUCH, P.L. and BLOCK, R.W. (1976). "Acceptance testing experience with fourteen new installations," Proc. SPIE **96**, 12–18.

RAUCH, P.L. and BLOCK, R.W. (1977). "Performance evaluation of the falling load technique," Proc. SPIE **127**, 145–151.

RAUCH, P.L. and STRAUSS, K.S. (1998). "X-ray generator, tube, collimator, positioner, and table," pages 61-82 in *RSNA Categorical Course in Diagnostic Radiology Physics: Cardiac Catheterization Imaging* (Radiological Society of North America, Oak Brook, Illinois).

ROACH, J.P., PARTRICK, D.A., BRUNY, J.L., ALLSHOUSE, M.J., KARRER, F.M. and ZIEGLER, M.M. (2007). "Complicated appendicitis in children: A clear role for drainage and delayed appendectomy," Am. J. Surg. **194**(6), 769–773.

ROEBUCK, D. (2009). "Paediatric interventional radiology," Pediatr. Radiol. **39**(Suppl. 3), 491–495.

ROECK, W.W. (1994). "X-ray room preparation: Layout and design considerations," pages 965 to 992 in *Specification, Acceptance Testing and Quality Control of Diagnostic X-Ray Imaging Equipment*, AAPM Monograph No. 20, Seibert, J.A., Barnes, G.T. and Gould, R.G., Eds. (American Institute of Physics, New York).

ROSE, J.S. (1983). *Invasive Radiology: Risks and Patient Care* (Year Book Medical Publishers, Inc., Chicago).

ROSENTHAL, L.S., MAHESH, M., BECK, T.J., SAUL, J.P., MILLER, J.M., KAY, N., KLEIN, L.S., HUANG, S., GILLETE, P., PRYSTOWSKY, E., CARLSON, M., BERGER, R.D., LAWRENCE, H., YONG, P. and CALKINS, H. (1998). "Predictors of fluoroscopy time and estimated radiation exposure during radiofrequency catheter ablation procedures," Am. J. Cardiol. **182**(4), 451–458.

ROSS, A.M., SEGAL, J., BORENSTEIN, D., JENKINS, E. and CHO, S. (1997). "Prevalence of spinal disc disease among interventional cardiologists," Am. J. Cardiol. **79**(1), 68–70.

ROSSI, R.P. (1982). "Acceptance testing of radiographic x-ray generators," pages 110 to 125 in *Acceptance Testing of Radiological Imaging Equipment*, AAPM Symposium Proceedings No. 1, Lin, P.J.P., Kriz, R.J., Rauch, P.L., Strauss, K.J. and Rossi, R.P., Eds. (American Institute of Physics, New York).

ROSSI, R.P. (1994). "X-ray generator and automatic exposure control device acceptance testing," pages 267 to 301 in *Specification, Acceptance Testing and Quality Control of Diagnostic X-Ray Imaging Equipment*, AAPM Monograph No. 20, Seibert, J.A., Barnes, G.T. and Gould, R.G., Eds. (American Institute of Physics, New York).

ROSSI, R.P. and HENDEE, W.R. (1980). "Facility planning and preparation," pages 88 to 97 in *The Selection and Performance Evaluation of Radiologic Equipment*, Hendee, W.R., Ed. (Williams and Wilkins, Baltimore).

ROSSI, R.P., LIN, P-J.P., RAUSCH, P.L. and STRAUSS, K.J. (1985). *Performance Specifications and Acceptance Testing for X-Ray Generators and Automatic Exposure Control Devices*, AAPM Report No. 14 (American Institute of Physics, New York).

RUDERMAN, C., TRACY, C.S., BENSIMON, C.M., BERNSTEIN, M., HAWRYLUCK, L., SHAUL, R.Z. and UPSHUR, R.E.G. (2006). "On pandemics and the duty to care: Whose duty? Who cares?," BMC Med. Ethics. **7**(1), 5–10.

RUIZ-CRUCES, R., PEREZ-MARTINEZ, M., MARTIN-PALANCA, A., FLORES, A., CRISTOFOL, J., MARTINEZ-MORILLO, M. and DIEZ DE LOS RIOS, A. (1997). "Patient dose in radiologically guided interventional vascular procedures: Conventional versus digital systems," Radiology **205**(2), 385–393.

RUIZ CRUCES, R., GARCIA-GRANADOS, J., DIAZ ROMERO, F.J. and HERNANDEZ ARMAS, J. (1998). "Estimation of effective dose in some digital angiographic and interventional procedures," Br. J. Radiol. **71**(841), 42–47.

RUTHERFORD, R.B. (2006). "Randomized EVAR trials and advent of level I evidence: A paradigm shift in management of large abdominal aortic aneurysms?," Sem. Vasc. Surg. **19**(2), 69–74.

SCHLATTL, H., ZANKL, M., EDER, H. and HOESCHEN, C. (2007). "Shielding properties of lead-free protective clothing and their impact on radiation doses," Med. Phys. **34**(11), 4270–4280.

SCHUELER, B.A., VRIEZE, T.J., BJARNASON, H. and STANSON, A.W. (2006). "An investigation of operator exposure in interventional radiology," Radiographics **26**(5), 1533–1541.

SCHUELER, B.A., STURCHIO, G., LANDSWORTH, R., HINDAL, M. and MAGNUSON, D. (2009). "Does new lightweight leaded eyewear provide adequate radiation protection for fluoroscopists?," (abstract) Med. Phys. **36**(6), 2747–2748.

SEBIRE, N.J. and ROEBUCK, D.J. (2006). "Pathological diagnosis of paediatric tumours from image-guided needle core biopsies: A systematic review," Pediatr. Radiol. **36**(5), 426–431.

SEIBERT, J.A. (2004). "Tradeoffs between image quality and dose," Pediatr. Radiol. **34**(Suppl. 3), 183–195.

SEIBERT, J.A. (2006). "Flat-panel detectors: How much better are they?," Pediatr. Radiol. **36**(Suppl. 2), 173–181.

SEPKOWITZ, K.A. and EISENBERG, L. (2005). "Occupational deaths among healthcare workers," Emerg. Infect. Dis. **11**(7), 1003–1008.

SERRUYS, P.W., MORICE, M.C., KAPPETEIN, A.P., COLOMBO, A., HOLMES, D.R., MACK, M.J., STAHLE, E., FELDMAN, T.E., VAN DEN BRAND, M., BASS, E.J., VAN DYCK, N., LEADLEY, K., DAWKINS, K.D. and MOHR, F.W. (2009). "Percutaneous coronary intervention versus coronary artery bypass grafting for severe coronary artery disease," N. Engl. J. Med. **360**(10), 961–972.

SERVOMAA, A. and KARPPINEN, J. (2001). "The dose-area product and assessment of the occupational dose in interventional radiology," Radiat. Prot. Dosim. **96**(1–3), 235–236.

SHOPE, T.B. (1996). "Radiation-induced skin injuries from fluoroscopy," Radiographics **16**(5), 1195–1199.

SHORE, R.E. and WORGUL, B.V. (1999). "Overview of the epidemiology of radiation cataracts," pages 183 to 189 in *Ocular Radiation Risk Assessment in Populations Exposed to Environmental Radiation Contamination*, Junk, A.K., Kundiev, Y., Vitte, P. and Worgul, B.V., Eds. (Kluwer Academic Publishers, The Netherlands).

SIDHU, M.K., GOSKE, M.J., COLEY, B.J., CONNOLLY, B., RACADIO, J., YOSHIZUMI, T.T., UTLEY, T. and STRAUSS, K.J. (2009). "Image Gently, Step Lightly: Increasing radiation dose awareness in pediatric interventions through an international social marketing campaign," J. Vasc. Interv. Radiol. **20**(9), 1115–1119.

SIEBER, V.K. and HOPEWELL, J.W. (1990). "Radiation-induced temporary partial epilation in the pig: A biological indicator of radiation dose and dose distribution to the skin," Radiat. Protect. Dosim. **30**(2), 117–120.

SIEBER, V.K., WELLS, J., REZVANI, M. and HOPEWELL, J.W. (1986). "Radiation induced damage to the cells of pig hairs: A biological indicator of radiation dose and dose distribution in skin," Radiat. Prot. Dosim. **16**(4), 301–306.

SIEBER, V.K., SUGDEN, E.M., ALCOCK, C.J. and BELTON, R.R. (1992). "Reduction in the diameter of human hairs following irradiation," Br. J. Radiol. **65**(770), 148–151.

SIGURDSON, A.J., DOODY, M.M., RAO, R.S., FREEDMAN, D.M., ALEXANDER, B.H., HAUPTMANN, M., MOHAN, A.K., YOSHINAGA, S., HILL, D.A., TARONE, R., MABUCHI, K., RON, E. and LINET, M.S. (2003). "Cancer incidence in the US radiologic technologists health study, 1983-1998," Cancer **97**(12), 3080–3089.

SIISKONEN, T., TAPIOVAARA, M., KOSUNEN, A., LEHTINEN, M. and VARTIAINEN, E. (2007). "Monte Carlo simulations of occupational radiation doses in interventional radiology," Brit. J. Radiol. **80**(954), 460–468.

SILVERMAN, S.G., MUELLER, P.R., PINKNEY, L.P., KOENKER, R.M. and SELTZER, S.E. (1993). "Predictive value of image-guided adrenal biopsy: Analysis of results of 101 biopsies," Radiology **187**(3), 715–718.

SILVERMAN, S.G., DEUSON, T.E., KANE, N., ADAMS, D.F., SELTZER, S.E., PHILLIPS, M.D., KHORASANI, R., ZINNER, M.J. and HOLMAN, B.L. (1998). "Percutaneous abdominal biopsy: Cost-identification analysis," Radiology **206**(2), 429–435.

SISTROM, C.L. (2008). "In support of the ACR appropriateness criteria," J. Am. Coll. Radiol. **5**(5), 630–635.

SMANS, K., STRUELENS, L., HOORNAERT, M.T., BLEESER, F., BULS, N., BERUS, D., CLERINX, P., MALCHAIR, F., VANHAVERE, F. and BOSMANS, H. (2008). "A study of the correlation between dose area product and effective dose in vascular radiology," Radiat. Prot. Dosim. **130**(3), 300–308.

SMITH, T.P., PRESSON, T.L., HENEGHAN, M.A. and RYAN, J.M. (2003). "Transjugular biopsy of the liver in pediatric and adult patients using an 18-gauge automated core biopsy needle: A retrospective review of 410 consecutive procedures," Am. J. Roentgenol. **180**(1), 167–172.

SMITH, S.C., JR., FELDMAN, T.E., HIRSHFELD, J.W., JR., JACOBS, A.K., KERN, M.J., KING, S.B., III, MORRISON, D.A., O'NEILL, W.W., SCHAFF, H.V., WHITLOW, P.L. and WILLIAMS, D.O. (2006). "ACC/AHA/SCAI 2005 guideline update for percutaneous coronary intervention," J. Am. Coll. Cardiol. **47**(1), e1–e121.

SOVIK, E., KLOW, N.E., HELLESNES, J. and LYKKE, J. (1996). "Radiation-induced skin injury after percutaneous transluminal coronary angioplasty," Acta. Radiol. **37**(3), 305–306.

SPELIC, D.C., KACZMAREK, R.V., HILOHI, M.C. and MOYAL, A.E. (2010). "Nationwide surveys of chest, abdomen, lumbosacral spine radiography, and upper gastrointestinal fluoroscopy: A summary of findings," Health Phys. **98**(3), 498–514.

SPIES, J.B., COOPER, J.M., WORTHINGTON-KIRSCH, R., LIPMAN, J.C., MILLS, B.B. and BENENATI, J.F. (2004). "Outcome of uterine embolization and hysterectomy for leiomyomas: Results of a multicenter study," Am. J. Obstet. Gynecol. **191**(1), 22–31.

SRINIVAS, Y. and WILSON, D.L. (2002). “Image quality evaluation of flat panel and image intensifier digital magnification in x-ray fluoroscopy,” Med. Phys. **29**(7), 1611–1621.

STARCHMAN, D.E., JOHNSON, R.G., HYKES, D.L. and HOWLAND, W.J. (1976). “The role of kVp accuracy of diagnostic x-ray units and other performance parameters in quality assurance,” Proc. SPIE **96**, 31–36.

STAVAS, J.M., SMITH, T.P., DELONG, D.M., MILLER, M.J., SUHOCKI, P.V. and NEWMAN, G.E. (2006). “Radiation hand exposure during restoration of flow to the thrombosed dialysis access graft,” J. Vasc. Interv. Radiol. **17**(10), 1611–1617.

STECKER, M.S., BALTER, S., TOWBIN, R.B., MILLER, D.L., VANO, E., BARTAL, G., ANGLE, J.F., CHAO, C.P., COHEN, A.M., DIXON, R.G., GROSS, K., HARTNELL, G.G., SCHUELER, B., STATLER, J.D., DE BAERE, T. and CARDELLA, J.F. (2009). “Guidelines for patient radiation dose management,” J. Vasc. Interv. Radiol. **20**(7 Suppl.), S263–S273.

STERNBERGH, W.C., III and MONEY, S.R. (2000). “Hospital cost of endovascular versus open repair of abdominal aortic aneurysms: A multicenter study,” J. Vasc. Surg. **31**(2), 237–244.

STEVEN, D., ROSTOCK, T., SERVATIUS, H., HOFFMANN, B., DREWITZ, I., MULLERLEILE, K., MEINERTZ, T. and WILLEMS, S. (2008). “Robotic versus conventional ablation for common-type atrial flutter: A prospective randomized trial to evaluate the effectiveness of remote catheter navigation,” Heart Rhythm **5**(11), 1556–1560.

STIFTER, E., SACU, S., BENESCH, T. and WEGHAUPT, H. (2005). “Impairment of visual acuity and reading performance and the relationship with cataract type and density,” Invest. Ophthalmol. Vis. Sci. **46**(6), 2071–2075.

STIFTER, E., SACU, S., THALER, A. and WEGHAUPT, H. (2006). “Contrast acuity in cataracts of different morphology and association to self-reported visual function,” Invest. Opthalmol. Vis. Sci. **47**(12), 5412–5422.

STISOVA, V. (2004). “Effective dose to patient during cardiac interventional procedures (Prague workplaces),” Radiat. Prot. Dosim. **111**(3), 271–274.

STONE, M.S., ROBSON, K.J. and LEBOIT, P.E. (1998). “Subacute radiation dermatitis from fluoroscopy during coronary artery stenting: Evidence for cytotoxic lymphocyte mediated apoptosis,” J. Am. Acad. Dermatol. **38**(Suppl. 2), 333–336.

STORM, E.S., MILLER, D.L., HOOVER, L.J., GEORGIA, J.D. and BIVENS, T. (2006). “Radiation doses from venous access procedures,” Radiology **238**(3), 1044–1050.

STRATAKIS, J., DAMILAKIS, J., HATZIDAKIS, A., THEOCHAROPOULOS, N. and GOURTSOYIANNIS, N. (2006). “Occupational radiation exposure from fluoroscopically guided percutaneous transhepatic biliary procedures,” J. Vasc. Interv. Radiol. **17**(5), 863–871.

STRAUSS, K.J. (1982). "X-ray room barrier acceptance testing," pages 255 to 274 in *Acceptance Testing of Radiological Imaging Equipment*, AAPM Symposium Proceedings No. 1, Lin, P.J.P., Kriz, R.J., Rauch, P.L., Strauss, K.J. and Rossi, R.P., Eds. (American Institute of Physics, New York).

STRAUSS, K.J. (1996). "Radiographic equipment and components: Technology overview and quality improvement," pages 21 to 48 in *Syllabus: A Categorical Course in Physics-Technology Update and Quality Improvement of Diagnostic X-Ray Imaging Equipment*, Gould, R.G. and Boone, J.M., Eds. (Radiological Society of North America, Oak Brook, Illinois).

STRAUSS, K.J. (2002). "Interventional equipment acquisition process: Cradle to grave," pages 797 to 848 in *Intravascular Brachtherapy Fluoroscopically Guided Interventions*, Belter, S., Chan, R.C. and Shoppe, T.B., Eds. (Medical Physics Publishing, Madison, Wisconsin).

STRAUSS, K.J. (2006a). "Pediatric interventional radiography equipment: Safety considerations," Pediatr. Radiol. **36**(Suppl. 2), 126–135.

STRAUSS, K.J. (2006b). "Interventional suite and equipment management: Cradle to grave," Pediatr. Radiol. **36**(Suppl. 2), 221–236.

STRAUSS, K.J. and KASTE, S.C. (2006). "The ALARA (as low as reasonably achievable) concept in pediatric interventional and fluoroscopic imaging: Striving to keep radiation doses as low as possible during fluoroscopy of pediatric patients – A white paper executive summary," Radiology **240**(3), 621–622.

STRAUSS, K.J., DAMIANO, M.M. and RULLI, R.S. (1996). "Electrical and mechanical hazards," pages 143 to 164 in *Complications in Diagnostic Imaging and Interventional Radiology*, Ansell, G., Batman, M.A., Kaufman, J.A. and Wilkins, R.A., Eds. (Blackwell Science, Cambridge, Massachusetts).

STROUPE, K.T., MORRISON, D.A., HLATKY, M.A., BARNETT, P.G., CAO, L., LYTTLE, C., HYNES, D.M. and HENDERSON, W.G. (2006). "Cost-effectiveness of coronary artery bypass grafts versus percutaneous coronary intervention for revascularization of high-risk patients," Circulation **114**(12), 1251–1257.

STRUELENS, L., VANHAVERE, F., BOSMANS, H., VAN LOON, R. and MOL, H. (2005). "Skin dose measurements on patients for diagnostic and interventional neuroradiology: A multicentre study," Radiat. Prot. Dosim. **114**(1–3), 143–146.

SUZUKI, S., FURUI, S., KOHTAKE, H., YOKOYAMA, N., KOZUMA, K., YAMAMOTO, Y. and ISSHIKI, T. (2006). "Radiation exposure to patient's skin during percutaneous coronary intervention for various lesions, including chronic total occlusion," Circ. J. **70**(1), 44–48.

SUZUKI, S., FURUI, S., ISSHIKI, T., KOZUMA, K., KOYAMA, Y., YAMAMOTO, H., OCHIAI, M., ASAKURA, Y. and IKARI, Y. (2008). "Patients' skin dose during percutaneous coronary intervention for chronic total occlusion," Catheter Cardiovasc. Interv. **71**(2), 160–164.

SY, K., DIPCHAND, A., ATENAFU, E., CHAIT, P., BANNISTER, L., TEMPLE, M., JOHN, P., CONNOLLY, B. and AMARAL, J.G. (2008).

"Safety and effectiveness of radiologic percutaneous gastrostomy and gastro jejunostomy in children with cardiac disease," Am. J. Roentgenol. **191**(4), 1169–1174.

SYNOWITZ, M. and KIWIT, J. (2006). "Surgeon's radiation exposure during percutaneous vertebroplasty," J. Neurosurg. Spine **4**(2), 106–109.

THEODORAKOU, C. and HORROCKS, J.A. (2003). "A study on radiation doses and irradiated areas in cerebral embolization," Br. J. Radiol. **76**(908), 546–552.

THIERRY-CHEF, I., SIMON, S.L. and MILLER, D.L. (2006). "Radiation dose and cancer risk among pediatric patients undergoing interventional neuroradiology procedures," Pediatr. Radiol. **36**(Suppl. 2), 159–162.

THOMADSEN, B.R., PALIWAL, B.R., PETEREIT, D.G. and RANALLO, F.N. (2000). "Radiation injury from x-ray exposure during brachytherapy localization," Med. Phys. **27**(7), 1681–1684.

TJC (2006). The Joint Commission. *Radiation Overdose as a Reviewable Sentinel Event* (The Joint Commission, Oakbrook Terrace, Illinois).

TJC (2007). The Joint Commission. *Sentinel Event Policy and Procedures*, (The Joint Commission, Oakbrook Terrace, Illinois).

TJC (2010). The Joint Commission. *The Universal Protocol for Preventing Wrong Site, Wrong Procedure, and Wrong Person Surgery. Guidance for Health Care Professionals* (The Joint Commission, Oakbrook Terrace, Illinois).

TRIANNI, A., CHIZZOLA, G., TOH, H., QUAI, E., CRAGNOLINI, E., BERNARDI, G., PROCLEMER, A. and PADOVANI, R. (2005). "Patient skin dosimetry in haemodynamic and electrophysiology interventional cardiology," Radiat. Prot. Dosim. **117**(1–3), 241–246.

TROUT, E.D. (1977). "Isodose curves in a phantom due to diagnostic quality x-radiation," Health Phys. **33**(5), 359–367.

TSALAFOUTAS, I.A., GONI, H., MANIATIS, P.N., PAPPAS, P., BOUZAS, N. and TZORTZIS, G. (2006). "Patient doses from noncardiac diagnostic and therapeutic interventional procedures," J. Vasc. Interv. Radiol. **17**(9), 1489–1498.

TSALAFOUTAS, I.A., TSAKAPI, V., KALIAKMANIS, A., PNEUMATICOS, S., TSORONIS, F., KOULENTIANOS, E.D. and PAPACHRISTOU, G. (2008). "Estimation of radiation doses to patients and surgeons from various fluoroscopically guided orthopaedic surgeries," Radiat. Prot. Dosim. **128**(1), 112–119.

TSAPAKI, V., TRIANTOPOULOU, C., MANIATIS, P., KOTTOU, S., TSALAFOUTAS, J. and PAPAILOU, J. (2008). "Patient skin dose assessment during CT-guided interventional procedures," Radiat. Prot. Dosim. **129**(1–3), 29–31.

TURESSON, I. and NOTTER, G. (1986). "Dose-response and dose-latency relationships for human skin after various fractionation schedules," Br. J. Cancer **7**(Suppl.), 67–72.

UNSCEAR (1986). United Nations Scientific Committee on the Effects of Atomic Radiation. *UNSCEAR 1986 Report. Genetic and Somatic Effects of Ionizing Radiation, UNSCEAR 1986 Report to the General*

Assembly, with Annexes, http://www.unscear.org/unscear/en/publications/1986.html (accessed February 1, 2011) (United Nations Publications, New York).

UNSCEAR (2000a). United Nations Scientific Committee on the Effects of Atomic Radiation. *UNSCEAR 2000 Report. Sources and Effects of Ionizing Radiation, Volume I: Sources. UNSCEAR 2000 Report to the General Assembly, with Scientific Annexes,* http://www.unscear.org/unscear/en/publications/2000_1.html (accessed February 1, 2011) (United Nations Publications, New York).

UNSCEAR (2000b). United Nations Scientific Committee on the Effects of Atomic Radiation. *UNSCEAR 2000 Report. Sources and Effects of Ionizing Radiation, Volume II: Effects. UNSCEAR 2000 Report to the General Assembly, with Scientific Annexes,* http://www.unscear.org/unscear/en/publications/2000_2.html (accessed February 1, 2011) (United Nations Publications, New York).

UNSCEAR (2001). United Nations Scientific Committee on the Effects of Atomic Radiation. *UNSCEAR 2001 Report. Hereditary Effects of Radiation. UNSCEAR 2001 Report to the General Assembly, with Scientific Annex,* http://www.unscear.org/unscear/en/publications/2001.html (accessed February 1, 2011) (United Nations Publications, New York).

VAN DEN AARDWEG, G.J.M.J., HOPEWELL, J.W. and SIMMONDS, R.H. (1988). "Repair and recovery in the epithelial and vascular connective tissues of pig skin after irradiation," Radiother. Oncol. **11**(1), 73–82.

VAN HEYNINGEN, R. (1975). "What happens to the human lens in cataract?," Sci. Amer. **233**(6), 70–81.

VANO, E. and GONZALEZ, L. (2001). "Approaches to establishing reference levels in interventional radiology," Radiat. Prot. Dosim. **94**(1–2), 109–112.

VANO, E., GUIBELALDE, E., FERNANDEZ, J.M., GONZALEZ, L. and TEN, J.L. (1997). "Patient dosimetry in interventional radiology using slow films," Br. J. Radiol. **70**(830), 195–200.

VANO, E., ARRANZ, L., SASTRE, J.M., MORO, C., LEDO, A., GARATE, M.T. and MINGUEZ, I. (1998a). "Dosimetric and radiation protection considerations based on some cases of patient skin injuries in interventional cardiology," Br. J. Radiol. **71**(845), 510–516.

VANO, E., GONZALEZ, L., BENEYTEZ, F. and MORENO, F. (1998b). "Lens injuries induced by occupational exposure in non-optimized interventional radiology laboratories," Br. J. Radiol. **71**(847), 728–733.

VANO, E., GONZALEZ, L., GUIBELALDE, E., FERNANDEZ, J.M. and TEN, J.I. (1998c). "Radiation exposure to medical staff in interventional and cardiac radiology," Br. J. Radiol. **71**(849), 954–960.

VANO, E., GONZALEZ, L., TEN, J.I., FERNANDEZ, J.M., GUIBELALDE, E. and MACAYA, C. (2001). "Skin dose and dose-area product values for interventional cardiology procedures," Br. J. Radiol. **74**(877), 48–55.

VANO, E., PRIETO, C., FERNANDEZ, J.M., GONZALEZ, L., SABATE, M. and GALVAN, C. (2003). "Skin dose and dose-area product values in patients undergoing intracoronary brachytherapy," Br. J. Radiol. **76**(901), 32–38.

VANO, E., GONZALEZ, L., FERNANDEZ, J.M., ALFONSO, F. and MACAYA, C. (2006). "Occupational radiation doses in interventional cardiology: A 15-year follow-up," Br. J. Radiol. **79**(941), 383–188.

VANO, E., JARVINEN, H., KOSUNEN, A., BLY, R., MALONE, J., DOWLING, A., LARKIN, A., PADOVANI, R., BOSMANS, H., DRAGUSIN, O., JASCHKE, W., TORBICA, P., BACK, C., SCHREINER, A., BOKOU, C., KOTTOU, S., TSAPAKI, V., JANKOWSKI, J., PAPIERZ, S., DOMIENIK, J., WERDUCH, A., NIKODEMOVA, D., SALAT, D., KEPLER, K., BOR, M.D., VASSILEVA, J., BORISOVA, R., PELLET, S. and CORBETT, R.H. (2008a). "Patient dose in interventional radiology: A European survey," Radiat. Prot. Dosim. **129**(1–3), 39–45.

VANO, E., GONZALEZ, L., FERNANDEZ, J.M. and HASKAL, Z.J. (2008b). "Eye lens exposure to radiation in interventional suites: Caution is warranted," Radiology **248**(3), 945–953.

VANO, E., SANCHEZ, R., FERNANDEZ, J.M., GALLEGO, J.J., VERDU, J.F., GONZALEZ DE GARAY, M., AZPIAZU, A., SEGARRA, A., HERNANDEZ, M.T., CANIS, M., DIAZ, F., MORENO, F. and PALMERO, J. (2009). "Patient dose reference levels for interventional radiology. A national approach," Cardiovasc. Intervent. Radiol. **32**(1), 19–24.

VANO, E., KLEIMAN, N.J., DURAN, A., REHANI, M.M., ECHEVERRI, D. and CABRERA, M. (2010). "Radiation cataract risk in interventional cardiology personnel," Radiat. Res. **174**(4), 490–495.

VEHMAS, T. (1997). "Radiation exposure during standard and complex interventional procedures," Br. J. Radiol. **70**(831), 296–298.

VERDUN, F.R., AROUA, A., TRUEB, P.H.R., VOCK, P. and VALLEY, J.F. (2005). "Diagnostic and interventional radiology: A strategy to introduce reference dose level taking into account the national practice," Radiat. Prot. Dosim. **114**(1–3), 188–191.

VLIETSTRA, R.E., WAGNER, L.K., KOENIG, T. and METTLER, F. (2004). "Radiation burns as a severe complication of fluoroscopically guided cardiological interventions," J. Interven. Cardiol. **17**(3), 1–12.

VOLKERS, N.A., HEHENKAMP, W.J.K., SMIT, P., ANKUM, W.M., REEKERS, J.A. and BIRNIE, E. (2008). "Economic evaluation of uterine artery embolization versus hysterectomy in the treatment of symptomatic uterine fibroids: Results from the randomized EMMY trial," J. Vasc. Interv. Radiol. **19**(7), 1007–1017.

VON BOETTICHER, H., LACHMUND, J. and HOFFMANN, W. (2009). "Cardiac catheterization: Impact of face and neck shielding on new estimates of effective dose," Health Phys. **97**(6), 622–627.

VON ESSEN, C.F. (1969). "Radiation tolerance of the skin," Acta Radiol. Ther. Phys. Biol. **8**(4), 311–330.

WAGNER, L.K. (2007). "Radiation injury is a potentially serious complication to fluoroscopically-guided complex interventions," Biomed. Imag. Intervent. J. **3**(2), e22.

WAGNER, L.K. and ARCHER, B.R. (2004). *Minimizing Risks from Fluoroscopic X-Rays: Bioeffects, Instrumentation, and Examination*, 4th ed. (Partners in Radiation Management, The Woodlands, Texas).

WAGNER, L.K. and HAYMAN, L.A. (1982). "Pregnancy and women radiologists," Radiology **145**(2), 559–562.

WAGNER, L.K. and MULHERN, O.R. (1996). "Radiation-attenuating surgical gloves: Effects of scatter and secondary electron production," Radiology **200**(1), 45–48.

WAGNER, L.K., ARCHER, B.R. and COHEN, A.M. (2000). "Management of patient skin dose in fluoroscopically guided interventional procedures," J. Vasc. Interv. Radiol. **11**(1), 25–33.

WALL, B.F. and SHRIMPTON, P.C. (1998). "The historical development of reference doses in diagnostic radiology," Radiat. Prot. Dosim. **80**(1–3), 15–20.

WALL, B.F., KENDALL, G.M., EDWARDS, A.A., BOUFFLER, S., MUIRHEAD, C.R. and MEARA, J.R. (2006). "What are the risks from medical x-rays and other low dose radiation?," Br. J. Radiol. **79**(940), 285–294.

WALLACE, M.J., KUO, M.D., GLAIBERMAN, C., BINKERT, C.A., ORTH, R.C. and SOULEZ, G. (2008). "Three dimensional C-arm cone-beam CT: Applications in the interventional suite," J. Vasc. Interv. Radiol. **19**(6), 799–813.

WELCH, T.J., SHEEDY, P.F., II, JOHNSON, C.D., JOHNSON, C.M. and STEPHENS, D.H. (1989). "CT-guided biopsy: Prospective analysis of 1,000 procedures," Radiology **171**(2), 493–496.

WHITBY, M. and MARTIN, C.J. (2005). "A study of the distribution of dose across the hands of interventional radiologists and cardiologists," Br. J. Radiol. **78**(927), 219–229.

WHITE, C.C. (1999). "Health care professionals and treatment of HIV-positive patients. Is there an affirmative duty to treat under common law, the Rehabilitation Act, or the Americans with Disabilities Act?," J. Leg. Med. **20**(1), 67–113.

WHO (2000). World Health Organization. *Efficacy and Radiation Safety in Interventional Radiology* (World Health Organization, Geneva).

WIEBERS, D.O., WHISNANT, J.P., HUSTON, J., III, MEISSNER, I., BROWN, R.D., JR., PIEPGRAS, D.G., FORBES, G.S., THIELEN, K., NICHOLS, D., O'FALLON, W.M., PEACOCK, J., JAEGER, L., KASSELL, N.F., KONGABLE-BECKMAN, G.L. and TORNER, J.C. (2003). "Unruptured intracranial aneurysms: Natural history, clinical outcome, and risks of surgical and endovascular treatment," Lancet **362**(9378), 103–110.

WILDE, G. and SJOSTRAND, J. (1997). "A clinical study of radiation cataract formation in adult life following gamma irradiation of the lens in early childhood," Br. J. Ophthalmol. **81**(4), 261–266.

WILLIAMS, J.R. (1996). "Scatter dose estimation based on dose-area product and the specification of radiation barriers," Br. J. Radiol. **69**(827), 1032–1037.

WILLIAMS, J. (1997). "The interdependence of staff and patient doses in interventional radiology," Br. J. Radiol. **70**(833), 498–503.

WITHERS, H.R. (1967). "The dose-survival relationship for irradiation of epithelial cells of mouse skin," Br. J. Radiol. **40**(471), 187–194.

WONG, L. and REHM, J. (2004). "Images in clinical medicine. Radiation injury from a fluoroscopic procedure," N. Engl. J. Med. **350**(25), e23.

WORGUL, B.V., MERRIAM, G.R., JR. and MEDVEDOVSKY, C. (1989). "Cortical cataract development – an expression of primary damage to the lens epithelium," Lens Eye Tox. Res. **6**(4), 559–571.

WORGUL, B.V., KUNDIEV, Y., LIKHTAREV, I., SERGIENKO, N., WEGENER, A. and MEDVEDOVSKY, C.P. (1996). "Use of subjective and non-subjective methodologies to evaluate lens radiation damage in exposed populations – an overview," Radiat. Environ. Biophys. **35**(3), 137–144.

WORGUL, B.V., HASKAL, Z.J. and JUNK, A.K. (2004). "Interventional radiology carries occupational risk for cataracts," RSNA News **14**(6), 5–6.

WORGUL, B.V., KLEIMAN, N.J. and DAVID, J.D. (2005a). "A positive and a negative bystander effect influences cataract outcome in the irradiated lens," http://www.arvo.org/eweb/DynamicPage.aspx?site=arvo2&webcode=amabstractsearch (accessed February 1, 2011), Invest. Ophthalmol. Vis. Sci. **46**, E-Abstract 832.

WORGUL, B.V., SMILENOV, L., BRENNER, D.J., VAZQUEZ, M. and HALL, E.J. (2005b). "Mice heterozygous for the ATM gene are more sensitive to both x-ray and heavy ion exposure than are wild types," Adv. Space Res. **35**(2), 254–259.

WORGUL, B.V., KUNDIYEV, Y.I., SERGIYENKO, N.M., CHUMAK, V.V., VITTE, P.M., MEDVEDOVSKY, C., BAKHANOVA, E.V., JUNK, A.K., KYRYCHENKO, O.Y., MUSIJACHENKO, N.V., SHYLO, S.A., VITTE, O.P., XU, S., XUE, X. and SHORE, R.E. (2007). "Cataracts among Chernobyl clean-up workers: Implications regarding permissible eye exposures," Radiat. Res. **167**(2), 233–243.

WU, I.C. and ORBACH, D.B. (2009). "Neurointerventional management of high-flow vascular malformations of the head and neck," Neuroimaging Clin. N. Am. **19**(2), 219–240.

WYNIA, M.K. and GOSTIN, L.O. (2004). "Ethical challenges in preparing for bioterrorism: Barriers within the health care system," Am. J. Public Health **94**(7), 1096–1102.

WYSE, D.G., WALDO, A.L., DIMARCO, J.P., DOMANSKI, M.J., ROSENBERG, Y., SCHRON, E.B., KELLEN, J.C., GREENE, H.L., MICKEL, M.C., DALQUIST, J.E. and CORLEY, S.D. (2002). "A comparison of rate control and rhythm control in patients with atrial fibrillation," N. Engl. J. Med. **347**(23), 1825–1833.

YAFFE, M.J., MAWDSLEY, G.E., LILLEY, M., SERVANT, R. and REH, G. (1991). "Composite materials for x-ray protection," Health Phys. **60**(5), 661–664.

YOSHINAGA, S., MABUCHI, K., SIGURDSON, A.J., DOODY, M.M. and RON, E. (2004). "Cancer risks among radiologists and radiologic technologists: Review of epidemiologic studies," Radiology **233**(2), 313–321.

ZIPES, D.P., CAMM, A.J., BORGGREFE, M., BUXTON, A.E., CHAITMAN, B., FROMER, M., GREGORATOS, G., KLEIN, G., MOSS, A.J., MYERBURG, R.J., PRIORI, S.G., QUINONES, M.A., RODEN, D.M., SILKA, M.J. and TRACY, C. (2006). "ACC/AHA/ESC 2006 guidelines for management of patients with ventricular arrhythmias and the prevention of sudden cardiac death: A report of the American College of Cardiology/American Heart Association Task Force and the European Society of Cardiology Committee for Practice Guidelines," J. Am. Coll. Cardiol. **48**(5), e247–e346.

ZUGER, A. and MILES, S.H. (1987). "Physicians, AIDS, and occupational risk. Historic traditions and ethical obligations," JAMA **258**(14), 1924–1928.

ZWEERS, D., GELEIJNS, J., AARTS, N.J.M., HARDAM, L.J., LAMERIS, J.S., SCHULTZ, F.W. and SCHULTZE KOOL, L.J. (1998). "Patient and staff radiation dose in fluoroscopy-guided TIPS procedures and dose reduction, using dedicated fluoroscopy exposure settings," Br. J. Radiol. **71**(846), 672–676.

The NCRP

The National Council on Radiation Protection and Measurements is a nonprofit corporation chartered by Congress in 1964 to:

1. Collect, analyze, develop and disseminate in the public interest information and recommendations about (a) protection against radiation and (b) radiation measurements, quantities and units, particularly those concerned with radiation protection.
2. Provide a means by which organizations concerned with the scientific and related aspects of radiation protection and of radiation quantities, units and measurements may cooperate for effective utilization of their combined resources, and to stimulate the work of such organizations.
3. Develop basic concepts about radiation quantities, units and measurements, about the application of these concepts, and about radiation protection.
4. Cooperate with the International Commission on Radiological Protection, the International Commission on Radiation Units and Measurements, and other national and international organizations, governmental and private, concerned with radiation quantities, units and measurements and with radiation protection.

The Council is the successor to the unincorporated association of scientists known as the National Committee on Radiation Protection and Measurements and was formed to carry on the work begun by the Committee in 1929.

The participants in the Council's work are the Council members and members of scientific and administrative committees. Council members are selected solely on the basis of their scientific expertise and serve as individuals, not as representatives of any particular organization. The scientific committees, composed of experts having detailed knowledge and competence in the particular area of the committee's interest, draft proposed recommendations. These are then submitted to the full membership of the Council for careful review and approval before being published.

The following comprise the current officers and membership of the Council:

Officers

President	Thomas S. Tenforde
Senior Vice President	Kenneth R. Kase
Secretary and Treasurer	David A. Schauer

Members

John F. Ahearne
E. Stephen Amis, Jr.
Sally A. Amundson
Kimberly E. Applegate
Benjamin R. Archer
Stephen Balter
Steven M. Becker
Joel S. Bedford
Mythreyi Bhargavan
Eleanor A. Blakely
William F. Blakely
Wesley E. Bolch
Thomas B. Borak
Andre Bouville
Leslie A. Braby
James A. Brink
Brooke R. Buddemeier
Jerrold T. Bushberg
John F. Cardella
Charles E. Chambers
Polly Y. Chang
S.Y. Chen
Lawrence L. Chi
Mary E. Clark
Michael L. Corradini
Allen G. Croff
Paul M. DeLuca
Christine A. Donahue
Stephen A. Feig
Alan J. Fischman
Patricia A. Fleming
John R. Frazier
Donald P. Frush
Ronald E. Goans
Robert L. Goldberg
Milton J. Guiberteau
Raymond A. Guilmette
Roger W. Harms
Martin Hauer-Jensen
Kathryn D. Held
Roger W. Howell
Hank C. Jenkins-Smith
Timothy J. Jorgensen
Kenneth R. Kase
Ann R. Kennedy
William E. Kennedy, Jr.
David C. Kocher
Ritsuko Komaki
Amy Kronenberg
Susan M. Langhorst
John J. Lanza
Edwin M. Leidholdt, Jr.
James C. Lin
Martha S. Linet
Jill A. Lipoti
Paul A. Locke
Jay H. Lubin
C. Douglas Maynard
Debra McBaugh
Ruth E. McBurney
Charles W. Miller
Donald L. Miller
William H. Miller
William F. Morgan
Stephen V. Musolino
David S. Myers
Bruce A. Napier
Gregory A. Nelson
Andrea K. Ng
Carl J. Paperiello
Terry C. Pellmar
R. Julian Preston
Kathryn H. Pryor
Jerome C. Puskin
Abram Recht
Adela Salame-Alfie
Beth A. Schueler
J. Anthony Seibert
Stephen M. Seltzer
Edward A. Sickles
Steven L. Simon
Christopher G. Soares
Michael G. Stabin
Daniel J. Strom
Tammy P. Taylor
Thomas S. Tenforde
Julie K. Timins
Richard E. Toohey
Elizabeth L. Travis
Fong Y. Tsai
Louis K. Wagner
Chris G. Whipple
Robert C. Whitcomb, Jr.
Stuart C. White
Gayle E. Woloschak
Shiao Y. Woo
Andrew J. Wyrobek
X. George Xu
R. Craig Yoder
Marco A. Zaider

Distinguished Emeritus Members

Warren K. Sinclair, *President Emeritus;* Charles B. Meinhold, *President Emeritus*
S. James Adelstein, *Honorary Vice President*
W. Roger Ney, *Executive Director Emeritus*
William M. Beckner, *Executive Director Emeritus*

Seymour Abrahamson
Lynn R. Anspaugh
John A. Auxier
William J. Bair
Harold L. Beck
Bruce B. Boecker
John D. Boice, Jr.
Robert L. Brent
Antone L. Brooks
Randall S. Caswell
J. Donald Cossairt
James F. Crow
Gerald D. Dodd
Sarah S. Donaldson
William P. Dornsife
Keith F. Eckerman
Thomas S. Ely
R.J. Michael Fry
Thomas F. Gesell
Ethel S. Gilbert
Joel E. Gray
Robert O. Gorson
Arthur W. Guy
Eric J. Hall
Naomi H. Harley
William R. Hendee
F. Owen Hoffman
Donald G. Jacobs
Bernd Kahn
Charles E. Land
John B. Little
Roger O. McClellan
Barbara J. McNeil
Fred A. Mettler, Jr.
Kenneth L. Miller
Dade W. Moeller
A. Alan Moghissi
Wesley L. Nyborg
John W. Poston, Sr.
Andrew K. Poznanski
Genevieve S. Roessler
Marvin Rosenstein
Lawrence N. Rothenberg
Henry D. Royal
Michael T. Ryan
William J. Schull
Roy E. Shore
Paul Slovic
John E. Till
Lawrence W. Townsend
Robert L. Ullrich
Arthur C. Upton
Richard J. Vetter
F. Ward Whicker
Susan D. Wiltshire
Marvin C. Ziskin

Lauriston S. Taylor Lecturers

Charles E. Land (2010) *Radiation Protection and Public Policy in an Uncertain World*
John D. Boice, Jr. (2009) *Radiation Epidemiology: The Golden Age and Remaining Challenges*
Dade W. Moeller (2008) *Radiation Standards, Dose/Risk Assessments, Public Interactions, and Yucca Mountain: Thinking Outside the Box*
Patricia W. Durbin (2007) *The Quest for Therapeutic Actinide Chelators*
Robert L. Brent (2006) *Fifty Years of Scientific Research: The Importance of Scholarship and the Influence of Politics and Controversy*
John B. Little (2005) *Nontargeted Effects of Radiation: Implications for Low-Dose Exposures*
Abel J. Gonzalez (2004) *Radiation Protection in the Aftermath of a Terrorist Attack Involving Exposure to Ionizing Radiation*
Charles B. Meinhold (2003) *The Evolution of Radiation Protection: From Erythema to Genetic Risks to Risks of Cancer to ?*
R. Julian Preston (2002) *Developing Mechanistic Data for Incorporation into Cancer Risk Assessment: Old Problems and New Approaches*
Wesley L. Nyborg (2001) *Assuring the Safety of Medical Diagnostic Ultrasound*
S. James Adelstein (2000) *Administered Radioactivity: Unde Venimus Quoque Imus*
Naomi H. Harley (1999) *Back to Background*
Eric J. Hall (1998) *From Chimney Sweeps to Astronauts: Cancer Risks in the Workplace*
William J. Bair (1997) *Radionuclides in the Body: Meeting the Challenge!*
Seymour Abrahamson (1996) *70 Years of Radiation Genetics: Fruit Flies, Mice and Humans*
Albrecht Kellerer (1995) *Certainty and Uncertainty in Radiation Protection*
R.J. Michael Fry (1994) *Mice, Myths and Men*
Warren K. Sinclair (1993) *Science, Radiation Protection and the NCRP*
Edward W. Webster (1992) *Dose and Risk in Diagnostic Radiology: How Big? How Little?*
Victor P. Bond (1991) *When is a Dose Not a Dose?*
J. Newell Stannard (1990) *Radiation Protection and the Internal Emitter Saga*
Arthur C. Upton (1989) *Radiobiology and Radiation Protection: The Past Century and Prospects for the Future*
Bo Lindell (1988) *How Safe is Safe Enough?*
Seymour Jablon (1987) *How to be Quantitative about Radiation Risk Estimates*
Herman P. Schwan (1986) *Biological Effects of Non-ionizing Radiations: Cellular Properties and Interactions*
John H. Harley (1985) *Truth (and Beauty) in Radiation Measurement*
Harald H. Rossi (1984) *Limitation and Assessment in Radiation Protection*
Merril Eisenbud (1983) *The Human Environment—Past, Present and Future*
Eugene L. Saenger (1982) *Ethics, Trade-Offs and Medical Radiation*
James F. Crow (1981) *How Well Can We Assess Genetic Risk? Not Very*
Harold O. Wyckoff (1980) *From "Quantity of Radiation" and "Dose" to "Exposure" and "Absorbed Dose"—An Historical Review*
Hymer L. Friedell (1979) *Radiation Protection—Concepts and Trade Offs*
Sir Edward Pochin (1978) *Why be Quantitative about Radiation Risk Estimates?*

Herbert M. Parker (1977) *The Squares of the Natural Numbers in Radiation Protection*

Currently, the following committees are actively engaged in formulating recommendations:

Program Area Committee 1: Basic Criteria, Epidemiology, Radiobiology, and Risk

SC 1-13 Impact of Individual Susceptibility and Previous Radiation Exposure on Radiation Risk for Astronauts

SC 1-15 Radiation Safety in NASA Lunar Missions'

SC 1-16 Uncertainties in the Estimation of Radiation Risks and Probability of Disease Causation

SC 1-17 Second Cancers and Cardiopulmonary Effects After Radiotherapy

SC 1-18 Use of Ionizing Radiation Screen Systems for Detection of Radioactive Materials that Could Represent a Threat to Homeland Security

SC 1-19 Health Protection Issues Associated with Use of Active Detection Technology Security Systems for Detection of Radioactive Threat Materials

SC 1-20 Biological Effectiveness of Photons as a Function of Energy

Program Area Committee 2: Operational Radiation Safety

SC 2-5 Investigation of Radiological Incidents

Program Area Committee 3: Nuclear and Radiological Security and Safety

Program Area Committee 4: Radiation Protection in Medicine

SC 4-2 Population Monitoring and Decontamination Following a Nuclear/Radiological Incident

SC 4-3 Diagnostic Reference Levels in Medical Imaging: Recommendations for Application in the United States

SC 4-4 Risks of Ionizing Radiation to the Developing Embryo, Fetus and Nursing Infant

Program Area Committee 5: Environmental Radiation and Radioactive Waste Issues

SC 5-1 Approach to Optimizing Decision Making for Late-Phase Recovery from Nuclear or Radiological Terrorism Incidents

SC 64-22 Design of Effective Effluent and Environmental Monitoring Programs

Program Area Committee 6: Radiation Measurements and Dosimetry

In recognition of its responsibility to facilitate and stimulate cooperation among organizations concerned with the scientific and related aspects of radiation protection and measurement, the Council has created a category of NCRP Collaborating Organizations. Organizations or groups of organizations that are national or international in scope and are concerned with scientific problems involving radiation quantities, units, measurements and effects, or radiation protection may be admitted to collaborating status by the Council. Collaborating Organizations provide a means by which NCRP can gain input into its activities from a wider segment of society. At the same time, the relationships with the Collaborating Organizations facilitate wider dissemination of information about the Council's activities, interests and concerns. Collaborating Organizations have the opportunity to comment on draft reports (at the time that these are submitted to the members of the Council). This is intended to capitalize on the fact that Collaborating Organizations are in an excellent position to both contribute to the identification of what needs to be treated in NCRP reports and to identify problems that might result from proposed recommendations. The present Collaborating Organizations with which NCRP maintains liaison are as follows:

American Academy of Dermatology
American Academy of Environmental Engineers
American Academy of Health Physics
American Academy of Orthopaedic Surgeons
American Association of Physicists in Medicine
American Bracytherapy Society
American College of Cardiology
American College of Medical Physics
American College of Nuclear Physicians
American College of Occupational and Environmental Medicine
American College of Radiology
American Conference of Governmental Industrial Hygienists
American Dental Association
American Industrial Hygiene Association
American Institute of Ultrasound in Medicine
American Medical Association
American Nuclear Society
American Pharmaceutical Association
American Podiatric Medical Association
American Public Health Association
American Radium Society
American Roentgen Ray Society
American Society for Radiation Oncology
American Society of Emergency Radiology
American Society of Health-System Pharmacists
American Society of Nuclear Cardiology
American Society of Radiologic Technologists
Association of Educators in Imaging and Radiological Sciences
Association of University Radiologists
Bioelectromagnetics Society
Campus Radiation Safety Officers
College of American Pathologists
Conference of Radiation Control Program Directors, Inc.
Council on Radionuclides and Radiopharmaceuticals
Defense Threat Reduction Agency
Electric Power Research Institute
Federal Aviation Administration
Federal Communications Commission
Federal Emergency Management Agency
Genetics Society of America
Health Physics Society
Institute of Electrical and Electronics Engineers, Inc.

Institute of Nuclear Power Operations
International Brotherhood of Electrical Workers
National Aeronautics and Space Administration
National Association of Environmental Professionals
National Center for Environmental Health/Agency for Toxic Substances
National Electrical Manufacturers Association
National Institute for Occupational Safety and Health
National Institute of Standards and Technology
Nuclear Energy Institute
Office of Science and Technology Policy
Paper, Allied-Industrial, Chemical and Energy Workers International Union
Product Stewardship Institute
Radiation Research Society
Radiological Society of North America
Society for Cardiovascular Angiography and Interventions
Society for Pediatric Radiology
Society for Risk Analysis
Society of Cardiovascular Computed Tomography
Society of Chairmen of Academic Radiology Departments
Society of Interventional Radiology
Society of Nuclear Medicine
Society of Radiologists in Ultrasound
Society of Skeletal Radiology
U.S. Air Force
U.S. Army
U.S. Coast Guard
U.S. Department of Energy
U.S. Department of Housing and Urban Development
U.S. Department of Labor
U.S. Department of Transportation
U.S. Environmental Protection Agency
U.S. Navy
U.S. Nuclear Regulatory Commission
U.S. Public Health Service
Utility Workers Union of America

NCRP has found its relationships with these organizations to be extremely valuable to continued progress in its program.

Another aspect of the cooperative efforts of NCRP relates to the Special Liaison relationships established with various governmental organizations that have an interest in radiation protection and measurements. This liaison relationship provides: (1) an opportunity for participating organizations to designate an individual to provide liaison between the organization and NCRP; (2) that the individual designated will receive copies of draft NCRP reports (at the time that these are submitted to the members of the Council) with an invitation to comment, but not vote; and (3) that new NCRP efforts might be discussed with liaison individuals as appropriate, so that they might have an opportunity to make suggestions on new studies and related matters. The following organizations participate in the Special Liaison Program:

Australian Radiation Laboratory
Bundesamt fur Strahlenschutz (Germany)
Canadian Association of Medical Radiation Technologists
Canadian Nuclear Safety Commission
Central Laboratory for Radiological Protection (Poland)
China Institute for Radiation Protection
Commissariat a l'Energie Atomique (France)
Commonwealth Scientific Instrumentation Research Organization (Australia)
European Commission
Health Council of the Netherlands
Health Protection Agency
International Commission on Non-ionizing Radiation Protection
International Commission on Radiation Units and Measurements
International Commission on Radiological Protection
International Radiation Protection Association
Japanese Nuclear Safety Commission
Japan Radiation Council
Korea Institute of Nuclear Safety
Russian Scientific Commission on Radiation Protection
South African Forum for Radiation Protection
World Association of Nuclear Operators
World Health Organization, Radiation and Environmental Health

NCRP values highly the participation of these organizations in the Special Liaison Program.

The Council also benefits significantly from the relationships established pursuant to the Corporate Sponsor's Program. The program facilitates the interchange of information and ideas and corporate sponsors provide valuable fiscal support for the Council's program. This developing program currently includes the following Corporate Sponsors:

3M
GE Healthcare
Global Dosimetry Solutions, Inc.
Landauer, Inc.
Nuclear Energy Institute

The Council's activities have been made possible by the voluntary contribution of time and effort by its members and participants and the generous support of the following organizations:

3M Health Physics Services
Agfa Corporation
Alfred P. Sloan Foundation
Alliance of American Insurers
American Academy of Dermatology
American Academy of Health Physics
American Academy of Oral and Maxillofacial Radiology
American Association of Physicists in Medicine
American Cancer Society
American College of Medical Physics

American College of Nuclear Physicians
American College of Occupational and Environmental Medicine
American College of Radiology
American College of Radiology Foundation
American Dental Association
American Healthcare Radiology Administrators
American Industrial Hygiene Association
American Insurance Services Group
American Medical Association
American Nuclear Society
American Osteopathic College of Radiology
American Podiatric Medical Association
American Public Health Association
American Radium Society
American Roentgen Ray Society
American Society for Radiation Oncology
American Society for Therapeutic Radiology and Oncology
American Society of Radiologic Technologists
American Veterinary Medical Association
American Veterinary Radiology Society
Association of Educators in Radiological Sciences, Inc.
Association of University Radiologists
Battelle Memorial Institute
Canberra Industries, Inc.
Chem Nuclear Systems
Center for Devices and Radiological Health
College of American Pathologists
Committee on Interagency Radiation Research and Policy Coordination
Commonwealth Edison
Commonwealth of Pennsylvania
Consolidated Edison
Consumers Power Company
Council on Radionuclides and Radiopharmaceuticals
Defense Nuclear Agency
Defense Threat Reduction Agency
Duke Energy Corporation
Eastman Kodak Company
Edison Electric Institute
Edward Mallinckrodt, Jr. Foundation
EG&G Idaho, Inc.
Electric Power Research Institute
Electromagnetic Energy Association
Federal Emergency Management Agency
Florida Institute of Phosphate Research
Florida Power Corporation
Fuji Medical Systems, U.S.A., Inc.
Genetics Society of America
Global Dosimetry Solutions
Health Effects Research Foundation (Japan)
Health Physics Society

ICN Biomedicals, Inc.
Institute of Nuclear Power Operations
James Picker Foundation
Martin Marietta Corporation
Motorola Foundation
National Aeronautics and Space Administration
National Association of Photographic Manufacturers
National Cancer Institute
National Electrical Manufacturers Association
National Institute of Standards and Technology
New York Power Authority
Philips Medical Systems
Picker International
Public Service Electric and Gas Company
Radiation Research Society
Radiological Society of North America
Richard Lounsbery Foundation
Sandia National Laboratory
Siemens Medical Systems, Inc.
Society of Nuclear Medicine
Society of Pediatric Radiology
Southern California Edison Company
U.S. Department of Energy
U.S. Department of Labor
U.S. Environmental Protection Agency
U.S. Navy
U.S. Nuclear Regulatory Commission
Victoreen, Inc.
Westinghouse Electric Corporation

Initial funds for publication of NCRP reports were provided by a grant from the James Picker Foundation.

NCRP seeks to promulgate information and recommendations based on leading scientific judgment on matters of radiation protection and measurement and to foster cooperation among organizations concerned with these matters. These efforts are intended to serve the public interest and the Council welcomes comments and suggestions on its reports or activities.

NCRP Publications

NCRP publications can be obtained online in both hard- and soft-copy (downloadable PDF) formats at http://NCRPpublications.org. Professional societies can arrange for discounts for their members by contacting NCRP. Additional information on NCRP publications may be obtained from the NCRP website (http://NCRPonline.org) or by telephone (800-229-2652, ext. 25) and fax (301-907-8768). The mailing address is:

NCRP Publications
7910 Woodmont Avenue
Suite 400
Bethesda, MD 20814-3095

Abstracts of NCRP reports published since 1980, abstracts of all NCRP commentaries, and the text of all NCRP statements are available at the NCRP website. Currently available publications are listed below.

NCRP Reports

No.	Title
8	*Control and Removal of Radioactive Contamination in Laboratories* (1951)
22	*Maximum Permissible Body Burdens and Maximum Permissible Concentrations of Radionuclides in Air and in Water for Occupational Exposure* (1959) [includes Addendum 1 issued in August 1963]
25	*Measurement of Absorbed Dose of Neutrons, and of Mixtures of Neutrons and Gamma Rays* (1961)
27	*Stopping Powers for Use with Cavity Chambers* (1961)
30	*Safe Handling of Radioactive Materials* (1964)
32	*Radiation Protection in Educational Institutions* (1966)
35	*Dental X-Ray Protection* (1970)
36	*Radiation Protection in Veterinary Medicine* (1970)
37	*Precautions in the Management of Patients Who Have Received Therapeutic Amounts of Radionuclides* (1970)
38	*Protection Against Neutron Radiation* (1971)
40	*Protection Against Radiation from Brachytherapy Sources* (1972)
41	*Specification of Gamma-Ray Brachytherapy Sources* (1974)
42	*Radiological Factors Affecting Decision-Making in a Nuclear Attack* (1974)
44	*Krypton-85 in the Atmosphere—Accumulation, Biological Significance, and Control Technology* (1975)
46	*Alpha-Emitting Particles in Lungs* (1975)

83 *The Experimental Basis for Absorbed-Dose Calculations in Medical Uses of Radionuclides* (1985)
84 *General Concepts for the Dosimetry of Internally Deposited Radionuclides* (1985)
86 *Biological Effects and Exposure Criteria for Radiofrequency Electromagnetic Fields* (1986)
87 *Use of Bioassay Procedures for Assessment of Internal Radionuclide Deposition* (1987)
88 *Radiation Alarms and Access Control Systems* (1986)
89 *Genetic Effects from Internally Deposited Radionuclides* (1987)
90 *Neptunium: Radiation Protection Guidelines* (1988)
92 *Public Radiation Exposure from Nuclear Power Generation in the United States* (1987)
93 *Ionizing Radiation Exposure of the Population of the United States* (1987)
94 *Exposure of the Population in the United States and Canada from Natural Background Radiation* (1987)
95 *Radiation Exposure of the U.S. Population from Consumer Products and Miscellaneous Sources* (1987)
96 *Comparative Carcinogenicity of Ionizing Radiation and Chemicals* (1989)
97 *Measurement of Radon and Radon Daughters in Air* (1988)
99 *Quality Assurance for Diagnostic Imaging* (1988)
100 *Exposure of the U.S. Population from Diagnostic Medical Radiation* (1989)
101 *Exposure of the U.S. Population from Occupational Radiation* (1989)
102 *Medical X-Ray, Electron Beam and Gamma-Ray Protection for Energies Up to 50 MeV (Equipment Design, Performance and Use)* (1989)
103 *Control of Radon in Houses* (1989)
104 *The Relative Biological Effectiveness of Radiations of Different Quality* (1990)
105 *Radiation Protection for Medical and Allied Health Personnel* (1989)
106 *Limit for Exposure to "Hot Particles" on the Skin* (1989)
107 *Implementation of the Principle of As Low As Reasonably Achievable (ALARA) for Medical and Dental Personnel* (1990)
108 *Conceptual Basis for Calculations of Absorbed-Dose Distributions* (1991)
109 *Effects of Ionizing Radiation on Aquatic Organisms* (1991)
110 *Some Aspects of Strontium Radiobiology* (1991)
111 *Developing Radiation Emergency Plans for Academic, Medical or Industrial Facilities* (1991)
112 *Calibration of Survey Instruments Used in Radiation Protection for the Assessment of Ionizing Radiation Fields and Radioactive Surface Contamination* (1991)
113 *Exposure Criteria for Medical Diagnostic Ultrasound: I. Criteria Based on Thermal Mechanisms* (1992)
114 *Maintaining Radiation Protection Records* (1992)
115 *Risk Estimates for Radiation Protection* (1993)
116 *Limitation of Exposure to Ionizing Radiation* (1993)

146 *Approaches to Risk Management in Remediation of Radioactively Contaminated Sites* (2004)
147 *Structural Shielding Design for Medical X-Ray Imaging Facilities* (2004)
148 *Radiation Protection in Veterinary Medicine* (2004)
149 *A Guide to Mammography and Other Breast Imaging Procedures* (2004)
150 *Extrapolation of Radiation-Induced Cancer Risks from Nonhuman Experimental Systems to Humans* (2005)
151 *Structural Shielding Design and Evaluation for Megavoltage X- and Gamma-Ray Radiotherapy Facilities* (2005)
152 *Performance Assessment of Near-Surface Facilities for Disposal of Low-Level Radioactive Waste* (2005)
153 *Information Needed to Make Radiation Protection Recommendations for Space Missions Beyond Low-Earth Orbit* (2006)
154 *Cesium-137 in the Environment: Radioecology and Approaches to Assessment and Management* (2006)
155 *Management of Radionuclide Therapy Patients* (2006)
156 *Development of a Biokinetic Model for Radionuclide-Contaminated Wounds and Procedures for Their Assessment, Dosimetry and Treatment* (2006)
157 *Radiation Protection in Educational Institutions* (2007)
158 *Uncertainties in the Measurement and Dosimetry of External Radiation* (2007)
159 *Risk to the Thyroid from Ionizing Radiation* (2008)
160 *Ionizing Radiation Exposure of the Population of the United States* (2009)
161 *Management of Persons Contaminated with Radionuclides* (2008)
162 *Self Assessment of Radiation-Safety Programs* (2009)
163 *Radiation Dose Reconstruction: Principles and Practices* (2009)
164 *Uncertainties in Internal Radiation Dose Assessment* (2009)
165 *Responding to a Radiological or Nuclear Terrorism Incident: A Guide for Decision Makers* (2010)
168 *Radiation Dose Management for Fluoroscopically-Guided Interventional Medical Procedures* (2010)

Binders for NCRP reports are available. Two sizes make it possible to collect into small binders the "old series" of reports (NCRP Reports Nos. 8–30) and into large binders the more recent publications (NCRP Reports Nos. 32–165, 168). Each binder will accommodate from five to seven reports. The binders carry the identification "NCRP Reports" and come with label holders which permit the user to attach labels showing the reports contained in each binder.

The following bound sets of NCRP reports are also available:

Volume I. NCRP Reports Nos. 8, 22
Volume II. NCRP Reports Nos. 23, 25, 27, 30
Volume III. NCRP Reports Nos. 32, 35, 36, 37
Volume IV. NCRP Reports Nos. 38, 40, 41
Volume V. NCRP Reports Nos. 42, 44, 46
Volume VI. NCRP Reports Nos. 47, 49, 50, 51

Volume VII. NCRP Reports Nos. 52, 53, 54, 55, 57
Volume VIII. NCRP Report No. 58
Volume IX. NCRP Reports Nos. 59, 60, 61, 62, 63
Volume X. NCRP Reports Nos. 64, 65, 66, 67
Volume XI. NCRP Reports Nos. 68, 69, 70, 71, 72
Volume XII. NCRP Reports Nos. 73, 74, 75, 76
Volume XIII. NCRP Reports Nos. 77, 78, 79, 80
Volume XIV. NCRP Reports Nos. 81, 82, 83, 84, 85
Volume XV. NCRP Reports Nos. 86, 87, 88, 89
Volume XVI. NCRP Reports Nos. 90, 91, 92, 93
Volume XVII. NCRP Reports Nos. 94, 95, 96, 97
Volume XVIII. NCRP Reports Nos. 98, 99, 100
Volume XIX. NCRP Reports Nos. 101, 102, 103, 104
Volume XX. NCRP Reports Nos. 105, 106, 107, 108
Volume XXI. NCRP Reports Nos. 109, 110, 111
Volume XXII. NCRP Reports Nos. 112, 113, 114
Volume XXIII. NCRP Reports Nos. 115, 116, 117, 118
Volume XXIV. NCRP Reports Nos. 119, 120, 121, 122
Volume XXV. NCRP Report No. 123I and 123II
Volume XXVI. NCRP Reports Nos. 124, 125, 126, 127
Volume XXVII. NCRP Reports Nos. 128, 129, 130
Volume XXVIII. NCRP Reports Nos. 131, 132, 133
Volume XXIX. NCRP Reports Nos. 134, 135, 136, 137
Volume XXX. NCRP Reports Nos. 138, 139
Volume XXXI. NCRP Report No. 140
Volume XXXII. NCRP Reports Nos. 141, 142, 143
Volume XXXIII. NCRP Report No. 144
Volume XXXIV. NCRP Reports Nos. 145, 146, 147
Volume XXXV. NCRP Reports Nos. 148, 149
Volume XXXVI. NCRP Reports Nos. 150, 151, 152
Volume XXXVII, NCRP Reports Nos. 153, 154, 155
Volume XXXVIII, NCRP Reports Nos. 156, 157, 158
Volume XXXIX, NCRP Reports Nos. 159, 160
Volume XL. NCRP Report No. 161 (Vol I and II)
Volume XLI. NCRP Reports Nos. 162, 163

(Titles of the individual reports contained in each volume are given previously.)

NCRP Commentaries

No.	Title
1	*Krypton-85 in the Atmosphere—With Specific Reference to the Public Health Significance of the Proposed Controlled Release at Three Mile Island* (1980)
4	*Guidelines for the Release of Waste Water from Nuclear Facilities with Special Reference to the Public Health Significance of the Proposed Release of Treated Waste Waters at Three Mile Island* (1987)
5	*Review of the Publication, Living Without Landfills* (1989)
6	*Radon Exposure of the U.S. Population—Status of the Problem (1991)*

Proceedings of the Annual Meeting

No. Title

9 *New Dosimetry at Hiroshima and Nagasaki and Its Implications for Risk Estimates*, Proceedings of the Twenty-Third Annual Meeting held on April 8-9, 1987 (including Taylor Lecture No. 11) (1988)

10 *Radon*, Proceedings of the Twenty-Fourth Annual Meeting held on March 30-31, 1988 (including Taylor Lecture No. 12) (1989)

11 *Radiation Protection Today—The NCRP at Sixty Years*, Proceedings of the Twenty-Fifth Annual Meeting held on April 5-6, 1989 (including Taylor Lecture No. 13) (1990)

12 *Health and Ecological Implications of Radioactively Contaminated Environments*, Proceedings of the Twenty-Sixth Annual Meeting held on April 4-5, 1990 (including Taylor Lecture No. 14) (1991)

13 *Genes, Cancer and Radiation Protection,* Proceedings of the Twenty-Seventh Annual Meeting held on April 3-4, 1991 (including Taylor Lecture No. 15) (1992)

14 *Radiation Protection in Medicine,* Proceedings of the Twenty-Eighth Annual Meeting held on April 1-2, 1992 (including Taylor Lecture No. 16) (1993)

15 *Radiation Science and Societal Decision Making,* Proceedings of the Twenty-Ninth Annual Meeting held on April 7-8, 1993 (including Taylor Lecture No. 17) (1994)

16 *Extremely-Low-Frequency Electromagnetic Fields: Issues in Biological Effects and Public Health*, Proceedings of the Thirtieth Annual Meeting held on April 6-7, 1994 (not published).

17 *Environmental Dose Reconstruction and Risk Implications,* Proceedings of the Thirty-First Annual Meeting held on April 12-13, 1995 (including Taylor Lecture No. 19) (1996)

18 *Implications of New Data on Radiation Cancer Risk*, Proceedings of the Thirty-Second Annual Meeting held on April 3-4, 1996 (including Taylor Lecture No. 20) (1997)

19 *The Effects of Pre- and Postconception Exposure to Radiation*, Proceedings of the Thirty-Third Annual Meeting held on April 2-3, 1997, Teratology **59**, 181–317 (1999)

20 *Cosmic Radiation Exposure of Airline Crews, Passengers and Astronauts*, Proceedings of the Thirty-Fourth Annual Meeting held on April 1-2, 1998, Health Phys. **79**, 466–613 (2000)

21 *Radiation Protection in Medicine: Contemporary Issues*, Proceedings of the Thirty-Fifth Annual Meeting held on April 7-8, 1999 (including Taylor Lecture No. 23) (1999)

22 *Ionizing Radiation Science and Protection in the 21st Century,* Proceedings of the Thirty-Sixth Annual Meeting held on April 5-6, 2000, Health Phys. **80**, 317–402 (2001)

23 *Fallout from Atmospheric Nuclear Tests—Impact on Science and Society,* Proceedings of the Thirty-Seventh Annual Meeting held on April 4-5, 2001, Health Phys. **82**, 573–748 (2002)

24 *Where the New Biology Meets Epidemiology: Impact on Radiation Risk Estimates*, Proceedings of the Thirty-Eighth Annual Meeting held on April 10-11, 2002, Health Phys. **85**, 1–108 (2003)

25 *Radiation Protection at the Beginning of the 21st Century–A Look Forward*, Proceedings of the Thirty-Ninth Annual Meeting held on April 9–10, 2003, Health Phys. **87**, 237–319 (2004)

26 *Advances in Consequence Management for Radiological Terrorism Events*, Proceedings of the Fortieth Annual Meeting held on April 14–15, 2004, Health Phys. **89**, 415–588 (2005)
27 *Managing the Disposition of Low-Activity Radioactive Materials*, Proceedings of the Forty-First Annual Meeting held on March 30–31, 2005, Health Phys. **91**, 413–536 (2006)
28 *Chernobyl at Twenty*, Proceedings of the Forty-Second Annual Meeting held on April 3–4, 2006, Health Phys. **93**, 345–595 (2007)
29 *Advances in Radiation Protection in Medicine*, Proceedings of the Forty-Third Annual Meeting held on April 16-17, 2007, Health Phys. **95**, 461–686 (2008)
30 *Low Dose and Low Dose-Rate Radiation Effects and Models*, Proceedings of the Forty-Fourth Annual Meeting held on April 14–15, 2008, Health Phys. **97**, 373–541 (2009)

Lauriston S. Taylor Lectures

No. Title

1 *The Squares of the Natural Numbers in Radiation Protection by* Herbert M. Parker (1977)
2 *Why be Quantitative about Radiation Risk Estimates?* by Sir Edward Pochin (1978)
3 *Radiation Protection—Concepts and Trade Offs* by Hymer L. Friedell (1979) [available also in *Perceptions of Risk*, see above]
4 *From "Quantity of Radiation" and "Dose" to "Exposure" and "Absorbed Dose"—An Historical Review* by Harold O. Wyckoff (1980)
5 *How Well Can We Assess Genetic Risk? Not Very* by James F. Crow (1981) [available also in *Critical Issues in Setting Radiation Dose Limits*, see above]
6 *Ethics, Trade-offs and Medical Radiation* by Eugene L. Saenger (1982) [available also in *Radiation Protection and New Medical Diagnostic Approaches*, see above]
7 *The Human Environment—Past, Present and Future* by Merril Eisenbud (1983) [available also in *Environmental Radioactivity*, see above]
8 *Limitation and Assessment in Radiation Protection* by Harald H. Rossi (1984) [available also in *Some Issues Important in Developing Basic Radiation Protection Recommendations*, see above]
9 *Truth (and Beauty) in Radiation Measurement* by John H. Harley (1985) [available also in *Radioactive Waste*, see above]
10 *Biological Effects of Non-ionizing Radiations: Cellular Properties and Interactions* by Herman P. Schwan (1987) [available also in *Nonionizing Electromagnetic Radiations and Ultrasound*, see above]
11 *How to be Quantitative about Radiation Risk Estimates* by Seymour Jablon (1988) [available also in *New Dosimetry at Hiroshima and Nagasaki and its Implications for Risk Estimates*, see above]
12 *How Safe is Safe Enough?* by Bo Lindell (1988) [available also in *Radon*, see above]

13 *Radiobiology and Radiation Protection: The Past Century and Prospects for the Future* by Arthur C. Upton (1989) [available also in *Radiation Protection Today,* see above]

14 *Radiation Protection and the Internal Emitter Saga* by J. Newell Stannard (1990) [available also in *Health and Ecological Implications of Radioactively Contaminated Environments*, see above]

15 *When is a Dose Not a Dose?* by Victor P. Bond (1992) [available also in *Genes, Cancer and Radiation Protection*, see above]

16 *Dose and Risk in Diagnostic Radiology: How Big? How Little?* by Edward W. Webster (1992) [available also in *Radiation Protection in Medicine*, see above]

17 *Science, Radiation Protection and the NCRP* by Warren K. Sinclair (1993) [available also in *Radiation Science and Societal Decision Making,* see above]

18 *Mice, Myths and Men* by R.J. Michael Fry (1995)

19 *Certainty and Uncertainty in Radiation Research* by Albrecht M. Kellerer. Health Phys. **69**, 446–453 (1995)

20 *70 Years of Radiation Genetics: Fruit Flies, Mice and Humans* by Seymour Abrahamson. Health Phys. **71**, 624–633 (1996)

21 *Radionuclides in the Body: Meeting the Challenge* by William J. Bair. Health Phys. **73**, 423–432 (1997)

22 *From Chimney Sweeps to Astronauts: Cancer Risks in the Work Place* by Eric J. Hall. Health Phys. **75**, 357–366 (1998)

23 *Back to Background: Natural Radiation and Radioactivity Exposed* by Naomi H. Harley. Health Phys. **79**, 121–128 (2000)

24 *Administered Radioactivity: Unde Venimus Quoque Imus* by S. James Adelstein. Health Phys. **80**, 317–324 (2001)

25 *Assuring the Safety of Medical Diagnostic Ultrasound* by Wesley L. Nyborg. Health Phys. **82**, 578–587 (2002)

26 *Developing Mechanistic Data for Incorporation into Cancer and Genetic Risk Assessments: Old Problems and New Approaches* by R. Julian Preston. Health Phys. **85**, 4–12 (2003)

27 *The Evolution of Radiation Protection–From Erythema to Genetic Risks to Risks of Cancer to ?* by Charles B. Meinhold, Health Phys. **87**, 240–248 (2004)

28 *Radiation Protection in the Aftermath of a Terrorist Attack Involving Exposure to Ionizing Radiation* by Abel J. Gonzalez, Health Phys. **89**, 418–446 (2005)

29 *Nontargeted Effects of Radiation: Implications for Low Dose Exposures* by John B. Little, Health Phys. **91**, 416–426 (2006)

30 *Fifty Years of Scientific Research: The Importance of Scholarship and the Influence of Politics and Controversy* by Robert L. Brent, Health Phys. **93**, 348–379 (2007)

31 *The Quest for Therapeutic Actinide Chelators* by Patricia W. Durbin, Health Phys. **95**, 465–492 (2008)

32 *Yucca Mountain Radiation Standards, Dose/Risk Assessments, Thinking Outside the Box, Evaluations, and Recommendations* by Dade W. Moeller, Health Phys. **97**, 376–391 (2009)

Symposium Proceedings

No.	Title
1	*The Control of Exposure of the Public to Ionizing Radiation in the Event of Accident or Attack*, Proceedings of a Symposium held April 27-29, 1981 (1982)
2	*Radioactive and Mixed Waste—Risk as a Basis for Waste Classification,* Proceedings of a Symposium held November 9, 1994 (1995)
3	*Acceptability of Risk from Radiation—Application to Human Space Flight,* Proceedings of a Symposium held May 29, 1996 (1997)
4	*21st Century Biodosimetry: Quantifying the Past and Predicting the Future*, Proceedings of a Symposium held February 22, 2001, Radiat. Prot. Dosim. **97**(1), (2001)
5	*National Conference on Dose Reduction in CT, with an Emphasis on Pediatric Patients*, Summary of a Symposium held November 6-7, 2002, Am. J. Roentgenol. **181**(2), 321–339 (2003)

NCRP Statements

No.	Title
1	"Blood Counts, Statement of the National Committee on Radiation Protection," Radiology **63**, 428 (1954)
2	"Statements on Maximum Permissible Dose from Television Receivers and Maximum Permissible Dose to the Skin of the Whole Body," Am. J. Roentgenol., Radium Ther. and Nucl. Med. **84**, 152 (1960) and Radiology **75**, 122 (1960)
3	*X-Ray Protection Standards for Home Television Receivers, Interim Statement of the National Council on Radiation Protection and Measurements* (1968)
4	*Specification of Units of Natural Uranium and Natural Thorium, Statement of the National Council on Radiation Protection and Measurements* (1973)
5	*NCRP Statement on Dose Limit for Neutrons* (1980)
6	*Control of Air Emissions of Radionuclides* (1984)
7	*The Probability That a Particular Malignancy May Have Been Caused by a Specified Irradiation* (1992)
8	*The Application of ALARA for Occupational Exposures* (1999)
9	*Extension of the Skin Dose Limit for Hot Particles to Other External Sources of Skin Irradiation* (2001)
10	*Recent Applications of the NCRP Public Dose Limit Recommendation for Ionizing Radiation* (2004)

Other Documents

The following documents were published outside of the NCRP report, commentary and statement series:

Somatic Radiation Dose for the General Population, Report of the Ad Hoc Committee of the National Council on Radiation Protection and

Measurements, 6 May 1959, Science **131** (3399), February 19, 482–486 (1960)

Dose Effect Modifying Factors in Radiation Protection, Report of Subcommittee M-4 (Relative Biological Effectiveness) of the National Council on Radiation Protection and Measurements, Report BNL 50073 (T-471) (1967) Brookhaven National Laboratory (National Technical Information Service, Springfield, Virginia)

Residential Radon Exposure and Lung Cancer Risk: Commentary on Cohen's County-Based Study, Health Phys. **87**(6), 656–658 (2004)